Gregory L. Moss

Purdue University

Lab Manual: A Design Approach

to accompany

DIGITAL SYSTEMS:

PRINCIPLES AND APPLICATIONS

Eleventh Edition

By Ronald J. Tocci, Neal S. Widmer, & Gregory L. Moss

Prentice Hall

Boston Columbus Indianapolis New York San Francisco Upper Saddle River

Amsterdam Cape Town Dubai London Madrid Milan Munich Paris Montreal Toronto

Delhi Mexico City Sao Paulo Sydney Hong Kong Seoul Singapore Taipei Tokyo

Editorial Director: Vern Anthony
Acquisitions Editor: Wyatt Morris
Editorial Assistant: Chris Reed
Director of Marketing: David Gesell
Senior Marketing Manager: Alicia Wozniak
Marketing Assistant: Les Roberts
Senior Managing Editor: JoEllen Gohr
Project Manager: Rex Davidson
Senior Operations Supervisor: Pat Tonneman
Operations Specialist: Laura Weaver
Art Director: Diane Ersnberger
Cover Designer: Integra
Printer/Binder: Demand Production Center
Cover Printer: Demand Production Center

Quartus® II screens are reprinted courtesy of Altera Corporation.

Altera is a trademark and service mark of Altera Corporation in the United States and other countries. Altera products are the intellectual property of Altera Corporation and are protected by copyright laws and one or more U.S. and foreign patents and patent applications. This text is set in Times Roman by Gregory L. Moss.

Copyright 2011 Pearson Education, Inc., publishing as Prentice Hall, 1 Lake Street, Upper Saddle River, New Jersey, 07458. All rights reserved. Manufactured in the United States of America. This publication is protected by Copyright, and permission should be obtained from the publisher prior to any prohibited reproduction, storage in a retrieval system, or transmission in any form or by any means, electronic, mechanical, photocopying, recording, or likewise. To obtain permission(s) to use material from this work, please submit a written request to Pearson Education, Inc., Permissions Department, 1 Lake Street, Upper Saddle River, New Jersey 07458.

Many of the designations by manufacturers and seller to distinguish their products are claimed as trademarks. Where those designations appear in this book, and the publisher was aware of a trademark claim, the designations have been printed in initial caps or all caps.

10 9 8 7 6 5 4 3 2 1

Prentice Hall
is an imprint of

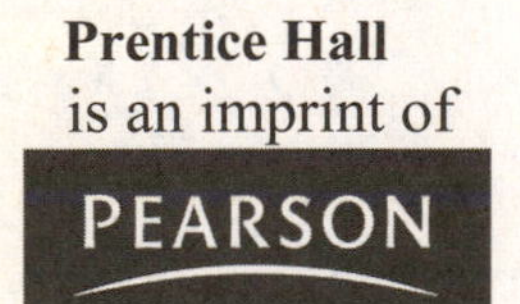

www.pearsonhighered.com

ISBN-13: 978-0-13-215381-2
ISBN-10: 0-13-215381-5

CONTENTS

To my expanding family,
Marita, David, Ryan, Christy, and Jeannie

PREFACE

This laboratory manual is written for students in an introductory digital electronics course that emphasizes logic circuit analysis, applications, and design. This lab manual, which accompanies the eleventh edition of *Digital Systems: Principles and Applications* by Ronald J. Tocci, Neal S. Widmer, and Gregory L. Moss, has been extensively revised but continues the digital logic design approach for beginners that has been presented in previous editions. The lab assignments consist of circuit projects that range from investigating basic logic concepts to synthesizing circuits for new applications. Most digital projects will be implemented using complex programmable logic devices (CPLDs) or field programmable gate arrays (FPGAs). The projects are intended to challenge all students and to provide them with some directed laboratory experience that develops insight into digital principles, applications, and techniques of logic circuit analysis and design.

New to This Edition

The revisions found in this edition continue to follow our primary goal of presenting the fundamentals of digital system analysis and design using the technology that our graduates would encounter in industry today. Major changes have been made to 9 (out of 23) of the topical units contained in this manual and minor changes have been made to several others. Some of the more significant changes in this edition are:

- Students are gradually introduced to the Quartus II design software starting with Unit 1 and progressing through Unit 5 of the lab manual.
- The introductory lab projects now use PLDs instead of standard logic chips, which are rapidly becoming obsolete.
- Unit 7 Testing Flip-flops and Sequential Circuit Applications has 50% new projects.
- The Quartus megafunction adder is used in Unit 9 Arithmetic Circuit Applications along with mostly new projects.
- Counter operation is introduced using megafunctions in a totally new Unit 11 Creating and Testing Counter Functions.
- Counter applications using megafunctions are explored in a totally new Unit 13 Applications Using LPM_Counter Megafunctions.
- Many new examples, especially using LPM megafunctions, and simulations are included throughout the lab manual.
- New FPGA digital development boards are utilized.
- Quartus II tutorials have been updated for the latest version of the design software and include many screen shots to help students become familiar with the software.

Digital Design Software and Hardware Tools

PLD programming is performed with Quartus II, an industry-leading, digital system development package from Altera Corporation. Detailed step-by-step tutorials on the procedures for using Quartus II are available from Pearson's Electronics Textbooks web portal: **http://www.pearsonhighered.com/electronics**.

With Quartus II, hierarchical digital designs can be entered using schematics and hardware description languages (HDLs). This lab manual actually supports two HDL languages, VHDL and AHDL. VHDL is an industry-standard language and has the advantage that it can be used with other development system software, but it is a bit more difficult to master. AHDL, on the other hand, is a proprietary language from Altera that is quite similar to VHDL but has several features that make hardware design easier for beginners. A major objective of this manual is to provide the greatest possible flexibility to fit your course goals. Quartus II also provides design verification through circuit simulation and timing analysis. A compiled digital design can be downloaded at a student's lab station directly to a target CPLD or FPGA device via the USB port on a personal computer. A free Web Edition of Quartus II is available at **http://www.altera.com**.

There are many relatively low cost digital development boards for Altera FPGAs or CPLDs that can be utilized in the digital laboratory. This lab manual recommends using one of the Terasic Technologies (**http://www.terasic.com**) development boards such as the DE0, DE1, or DE2. Photographs of the DE1 and DE2 boards are shown on page 2.

Lab Manual Organization

This lab manual is divided into 23 major topical units, with each unit containing an introductory discussion of the digital topic addressed, several circuit and application examples, and a variety of laboratory projects. Many examples of alternative design solutions are given in several of the lab units. The digital components used in our circuit applications include basic logic gates and flip-flops (Quartus primitives library) the standard parts from the maxplus2 library, LPMs (Library of Parameterized Modules) in the megafunction library, and HDLs. The Quartus II design software is first introduced using schematic capture design entry and is followed by units that discuss writing HDL code for combinational and sequential circuits. Introductions to each of the two HDL languages are given in separate lab manual units for your selection. Later lab units that deal with common digital functions, such as decoders, encoders, multiplexers, counters, and shift registers, each have a variety of lab projects that use a mix of schematic and HDL design entry. Both HDL language solutions to examples are given, and they are clearly labeled with AHDL or VHDL.

The applications-oriented lab projects are designed to provide beginning students with extensive experience in the analysis and design of digital logic circuits. The manual provides an extensive selection of projects for either a two-semester introductory digital course sequence or, by reducing the breadth and depth of topics covered, a single-semester digital course. The sequencing of topics in this manual follows the textbook by Tocci, Widmer, and Moss; however, the order of many lab units or projects can be rearranged to better fit a specific course's schedule. The lab manual is designed for course flexibility by including many different lab projects with varying levels of difficulty. To provide more depth, some lab units may be assigned over more than one laboratory period. There is no expectation that any course would have sufficient time for students to perform each of the laboratory projects provided. Rather, instructors can select and rearrange the topics and projects to fit their particular course objectives. Likewise, two HDLs are presented so that your school can decide which language approach can best suit your course and program goals.

Personal computers and electronic design automation software have changed how digital systems are designed and developed in industry today. Programmable logic devices and logic circuit development software are popular and extremely important digital technologies. These technologies need to be included in the educational experience of future electronics workers. Accordingly, the custom implementation of logic circuits using FPGAs or CPLDs is emphasized in this manual.

A list of laboratory equipment, integrated circuits, and other necessary components is found in the Equipment List on page ix. Data sheets for most standard logic devices can be found at **http://www.ti.com** (Texas Instruments). Data sheets for the Altera CPLD and FPGA devices can be found on the web at **http://www.altera.com**.

I am very grateful to the Altera Corporation, whose continued support has helped to make this laboratory manual possible.

Gregory L. Moss

NOTES TO LAB INSTRUCTORS

Quartus Tutorial Files

Quartus II tutorial files are available at **www.pearsonhighered.com/electronics**. These tutorials give detailed, step-by-step procedures for design entry (using schematic capture and hardware description languages), simulation, and device programming. The following tutorials are provided in Adobe PDF format:

Quartus Tutorial 1 – Schematic
Quartus Tutorial 2 – Simulation
Quartus Tutorial 3 – Hierarchical
Quartus Tutorial 4 – HDL
Quartus Notes

Quartus II Software

The Web Edition of Quartus II is a free version of the commercial software and the latest version can be downloaded at **www.altera.com**. This software package contains all of the tools necessary to design and develop CPLD/FPGA solutions for logic circuit design.

Project Files for Lab Manual Units 1 & 2

Archived project files for Lab Units 1 and 2 are available at **www.pearsonhighered.com/electronics**. There are 3 project versions for each Lab Unit since each of the supported Terasic (**www.terasic.com**) boards (DE0, DE1, and DE2) has a different FPGA device and pin assignments. Lab instructors should select the appropriate version to match the Terasic board used in their laboratory and download the archived project files. Then restore the project, compile, and distribute the complete project folder for students to use with Units 1 and 2. These lab projects can be easily revised to support other development boards.

EQUIPMENT LIST

Recommended Laboratory Equipment and Software

Digital multimeter
Signal generator
Personal computer
Frequency counter
Oscilloscope (4-trace)
Altera Quartus II development software
Terasic DE0, DE1, or DE2 FPGA development board or equivalent
Digital breadboarding system

Digital Integrated Circuits

Quantity	*Part #*	*Logic families*	*Description*
1	7404	ALS, LS, HC, HCT	Hex INVERTERs
1	7414	LS, HC, HCT	Hex Schmitt-trigger INVERTERs
1	74138	ALS, LS, HC, HCT	3-line-to-8-line DECODER/DEMUX
1	74221	LS	Dual MONOSTABLE MULTIVIBRATOR
1	74244	ALS, LS, HC, HCT	Octal 3-state BUFFER
2	2114		Static RAM (1K × 4) [9114]

Linear Integrated Circuits

1	NE555	Timer
1	AD557	8-bit digital-to-analog converter
1	ADC0804	8-bit analog-to-digital converter

Miscellaneous Components

Resistors (1/4 watt):

1.0 kΩ	1.1 kΩ	3.3 kΩ
10 kΩ	27 kΩ	33 kΩ
47 kΩ	62 kΩ	68 kΩ
72 kΩ	82 kΩ	

Capacitors:

10 μf	0.1 μf (×2)	0.01 μf (×2)
0.001 μf	0.0047 μf	150 pf

Potentiometers (10-turn):

10 kΩ

Rectifier diode: 1N4001 (×2)

SPDT switch

Grayhill 88BA2-ND (or equivalent) Keypad (4 × 4 matrix)

UNIT 1

INTRODUCTION TO THE DE0/DE1/DE2 DEVELOPMENT & EDUCATION BOARDS

Objectives

- To describe the function and operation of a typical digital development board.
- To test basic logic gates using logic switches and lights.

Digital Circuit Testing

Digital circuits are designed and tested using development boards that can be programmed for a desired function using software on a PC. Two digital development and education boards manufactured by Terasic Technologies, the DE1 and DE2, are shown in Fig. 1-1. There are many other development boards available. Typical digital development boards will include lamp monitors, logic switches, pulsers, and a clock to test your logic circuits.

Power supply

A power source for the digital devices will be required. The DE1 and DE2 boards have voltage regulators to provide the correct voltage for the chips on the board, but a power source still must be connected to the board. A USB cable from a PC is needed to program the boards and may also be able to provide sufficient power to the board. An external power supply may also be used if you wish to disconnect the PC from the DE1 or DE2. **Note that you will lose the programmed circuit design if you turn off the power to these boards!**

Fig. 1-1 Terasic DE1 (top) and DE2 boards (photographs by G. L. Moss)

Lamp monitors (LEDs)

The lamp monitors on the DE1 and DE2 are LEDs that can be wired to various points in the digital circuit being tested to indicate the voltage level at that circuit node. The lamp monitors will light when a digital "HIGH" voltage is applied to them. See Fig. 1-2.

Logic switches

The logic switches are wired as inputs to the circuit being tested to provide either of the two input logic levels (voltages). A switch in the "down" position will provide a logic "LOW" voltage, while a switch in the "up" position will provide a logic "HIGH" voltage.

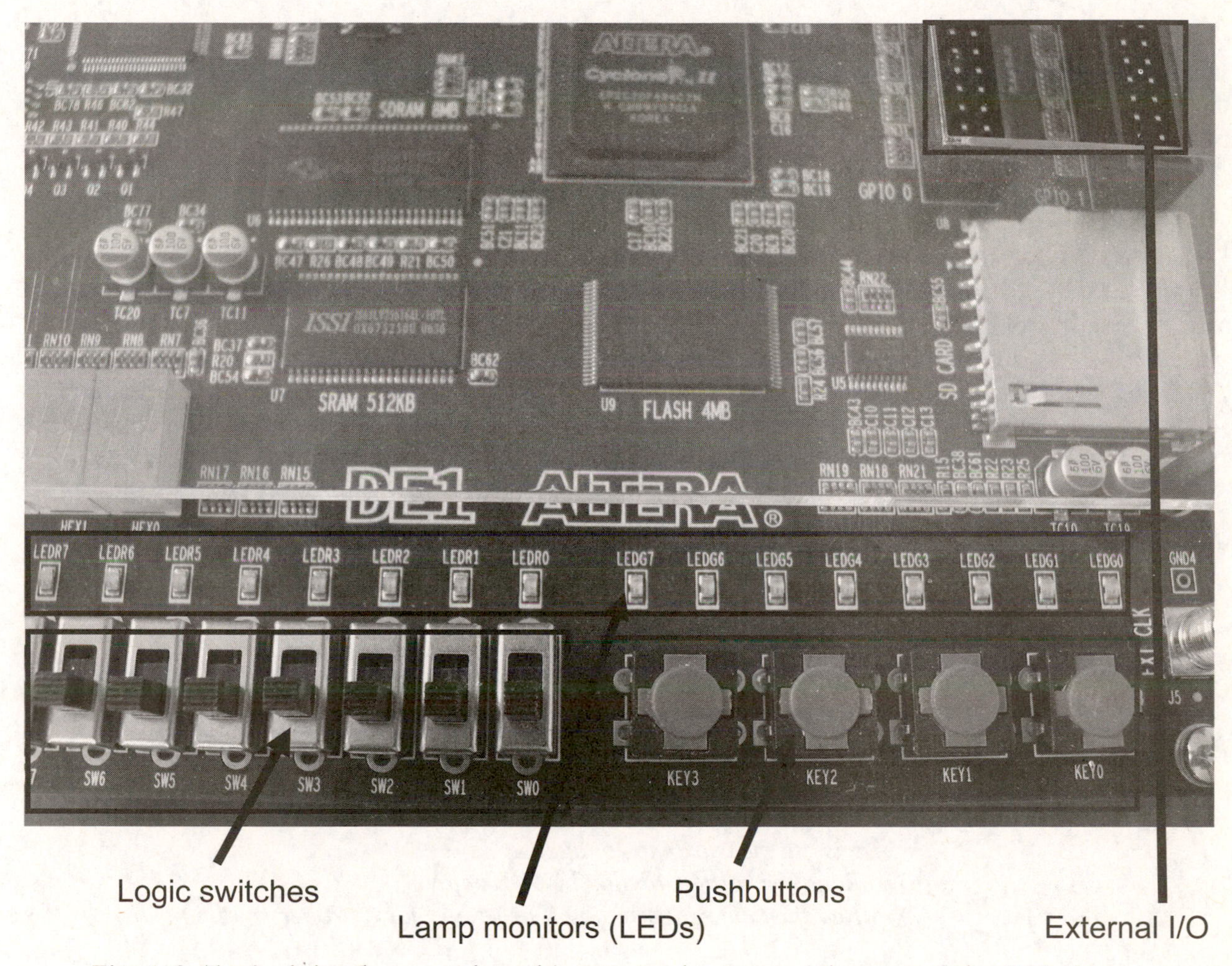

Fig. 1-2 Typical development board inputs and outputs (photograph by G. L. Moss)

Pushbuttons

A pushbutton is used to provide a momentary "bounce free" logic input to the test circuit. The pushbutton will output a "normal" logic level until it is pressed to apply the opposite input condition.

Clock

A clock provides a repetitive pulse waveform of a certain frequency that can be used for the timing control of some digital circuits.

External I/O

A digital development board will usually provide a way to connect external input signals to the circuit being tested and to connect the outputs to other circuits and electronic test equipment.

Measuring Voltage with a Digital Multi-Meter (DMM)

The potential difference or voltage between the positive and negative terminals of a battery or power supply connected to a circuit will cause current to flow through the circuit. The basic unit to measure potential difference is the volt (V). A potential difference or voltage drop occurs across the various devices in a circuit when current flows through them. The magnitude (and polarity with respect to a reference point in the circuit) of a potential difference is measured with an instrument called a voltmeter. You must be extremely careful when making voltage measurements, because the measurement is made on a "live" (powered) circuit. The voltmeter test leads (probes) are placed across (in parallel with) the device or power source whose voltage is to be measured. The procedure to measure DC voltages using a typical digital voltmeter is:

(1) *Connect the test leads to the DMM*
(2) *Set the function switch to measure DC voltages*
(3) *Set the range switch for the maximum voltage anticipated (unless the meter is autoranging)*
(4) *Connect (or touch) the black test lead to the reference point of the circuit (or component)*
(5) *Connect (or touch) the red test lead to the point in the circuit where you wish to measure the voltage*
(6) *Read the voltage on the digital display*
(7) *Readjust the range setting if necessary for a proper reading*

Ask your lab instructor if you have any questions concerning the use of the DMM.

Laboratory Projects

Investigate the features of the digital development board and Quartus II design software by performing the following tasks. Your lab instructor should provide the project files for this lab (see Notes to Lab Instructors on page viii of this lab manual). Consult your lab instructor if you have any problems or questions concerning the laboratory procedures or software operation.

1.1 Setting up the test circuit

(a) **Your lab instructor will give you the name and storage location of the project folder that contains the necessary files for this lab.** The information in this folder will be used to program the test circuit into the development board. Copy this complete project folder to your student hard drive work space as directed by your lab instructor.

(b) Connect a DE0/DE1/DE2 to the PC using the USB cable. Power the DE0/DE1/DE2 and turn-on by pressing the red button at the top-left side of the board. The built-in test program should display a light pattern on the LEDs and a count on the 7–segment display. See your lab instructor if there is a problem.

(c) Start Quartus II on the PC. The main screen will open with a "Getting Started With Quartus II Software" window on top. Click the "Open Existing Project" button. Navigate to the project folder. Select the Lab1 project file and click the Open button. After the project has loaded into Quartus, click the Programmer button on the toolbar (see screen capture below).

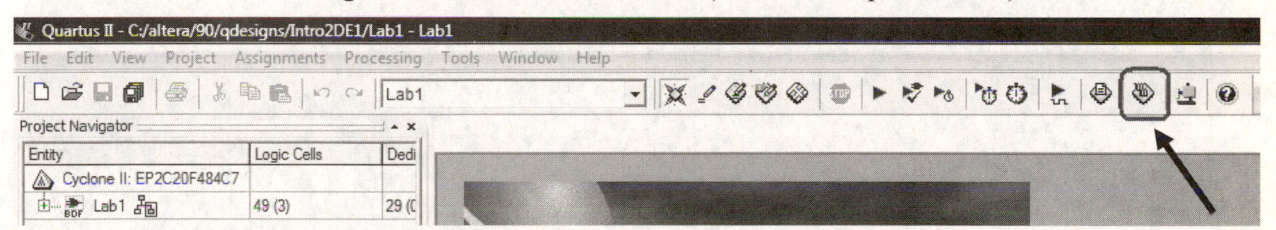

(d) The programmer window should open. If the programmer window does not look like the screen capture below, consult your lab instructor. Click the Start button. The LEDs should dim and the pattern should stop after a couple of seconds. The DE0/DE1/DE2 should be programmed with our circuit and will be wired to selected switches and LED lamps. Continue with lab project 1.2 to test the circuit on the DE1/DE2. **Do not turn off the power!**

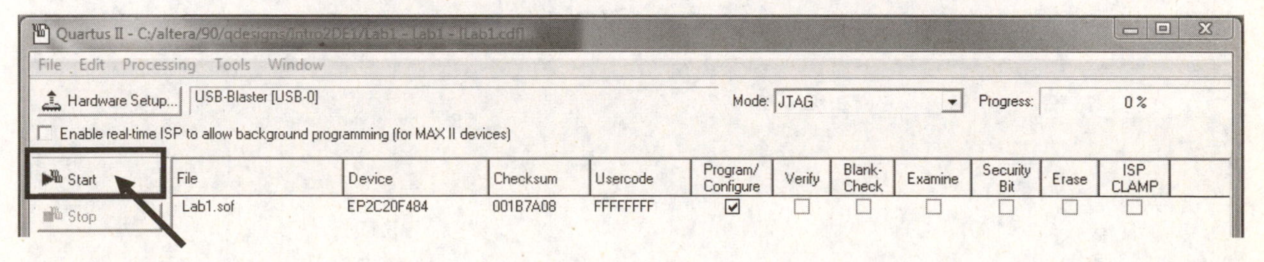

1.2 Identify board

Which digital development board are you using? DE____

1.3 Logic switches

How many logic switches are on your development board? ________

Complete the table below using logic switch SW1, which is wired to LEDG1.

Logic Switch SW1	LEDG1 (on/off)	Logic level (HIGH/LOW)	Voltage at connector pin
Down			
Up			

Carefully measure and record the voltage (with respect to ground) on pin 40 of GPIO 1 (JP2) for each position of logic switch SW1. See photo below. This pin is wired to SW1 by our test circuit.

1.4 Lamp monitors (LEDs)

Determine the number of lamp monitors on your development board.

LED label	Color	Number
LEDR	Red	
LEDG	Green	

1.5 Pushbuttons

How many pushbuttons are on your development board? ________

Complete the table below using pushbutton #1, which is wired directly to to LEDG2. Pushbutton #1 is also the input to a NOT gate whose output is connected to LEDG3.

Pushbutton #1	LEDG2 (on/off)	LEDG3 (on/off)
Normal		
Pressed		

What is the "normal" logic level for a pushbutton on the DE0/DE1/DE2? __________

1.6 Clock

The test circuit signal applied to LEDG0 is called a clock. SW9, SW8, and SW7 have been wired to control this signal. SW9 is the on/off control for the clock.

What SW9 logic level turns on the clock signal produced by our circuit? __________

What is the logic level of the clock signal when LEDG0 is on? __________

Describe the clock operation at lowest frequency. ________________________

____

What inputs are applied to SW8 and SW7 for the slowest clock frequency?

SW8 = _______ SW7 = _______

What happens to the clock frequency for other input combinations of SW8 and SW7?

____

What do you think the clock is doing when both SW8 and SW7 are HIGH?

____

1.7 Simple logic circuits

Four simple logic circuits have also been constructed with our Lab1 test circuit. The circuit outputs are labeled W, X, Y, and Z. These outputs have been wired to the four LEDs listed in the Truth Table below. Two logic switches (SW3 and SW2) are used to provide the A and B inputs to the four logic circuits. There are four input combinations possible with the two switches. Determine the truth table and write the logic expression as a function of inputs A and B for each circuit.

Truth Table:

A (SW3)	B (SW2)	W (LEDG4)	X (LEDG5)	Y (LEDG6)	Z (LEDG7)
0	0				
0	1				
1	0				
1	1				

Logic expressions:

W = ______________________________

X = ______________________________

Y = ______________________________

Z = ______________________________

UNIT 2

TESTING COMBINATIONAL LOGIC CIRCUITS USING DE0/DE1/DE2 BOARDS

Objectives

- To test combinational logic circuits to determine the functional operation of the circuits.
- To identify common logic functions produced by various circuit configurations from the resulting truth table.

Combinational Logic Circuits

Logic Gates

A logic gate is the simplest device used to construct digital circuits. The output voltage or logic level for each type of gate is a function of the applied input(s). Various types of logic gates are available (including inverters, ANDs, ORs, NANDs, and NORs), each with its own unique logic function. Combinational logic circuits are constructed by interconnecting various logic gates to implement a particular circuit function.

Logic Circuit Testing and Verification

Logic circuits can be tested using a digital development board such as the Terasic DE0, DE1, or DE2. The logic circuit design will be entered into Quartus II where a separate logic switch will be assigned to each input and a lamp monitor to the circuit's output. Quartus will process the information before being used to program the development board. Apply various input logic levels with the switches and monitor the output produced by the circuit with the lamp. List the resultant functional operation of the circuit in a truth table. Determine if the circuit is operating properly.

Example 2-1

Determine the logic expression and predict the theoretical operation for circuit Q in Fig. 2-1. Identify the common logic function produced by this circuit.

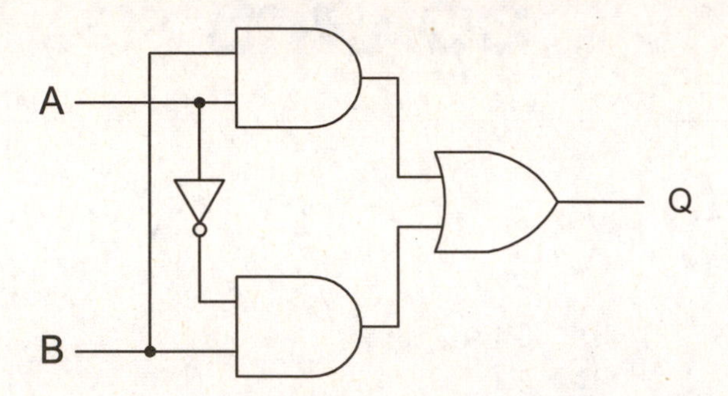

Fig. 2-1 Labeled schematic for Example 2-1

Start by writing the logic expression for each intermediate gate output:

The output of the inverter: $\overline{A}$

The output of the top AND: $A\ B$

The output of the bottom AND: $\overline{A}\ B$

Create a truth table that has a column for each of the intermediate gate output functions that were identified above and then add a final column for the overall circuit output Q (see Table 2-1). Predict the theoretical output of each function for the given input conditions. The circuit output will be the result of ORing the two AND gate outputs.

A	B	$\overline{A}$	A B	$\overline{A}$ B	Q
0	0	1	0	0	0
0	1	1	0	1	1
1	0	0	0	0	0
1	1	0	1	0	1

Table 2-1 Truth table prediction for Example 2-1

The logic expression for Q is given below. By inspecting the truth table, we can see that the circuit's logic function is the same as the input B. Therefore, we can also say that the circuit function can be simplified to: Q = B.

$$Q = A \cdot B + \overline{A} \cdot B = B$$

Laboratory Projects

Your lab instructor will give you the name and storage location of the project folder that contains the necessary files for this lab. (See Notes to Lab Instructors on page viii.) Copy this complete project folder to your student hard drive work space as directed by your lab instructor. Connect your development board to the PC using the USB cable and turn on the power. "Open Existing Project" with Quartus (*press Ctrl+J, if continuing from another project*). Navigate to the project folder and select the Lab2 project file. Open the Quartus Programmer and click Start. Test and demonstrate the given circuits. The following logic switches and LEDs have been assigned in this project.

A	B	T	Z	Y	X	J	K	W	V
SW1	SW0	LEDG3	LEDG2	LEDG1	LEDG0	LEDG7	LEDG6	LEDG5	LEDG4

2.1 Simple circuits

To test these circuits on the DE0/DE1/DE2, set SW9 to a LOW logic level. Write the logic expression for each circuit. Record the test results in a truth table. Determine a simpler equivalent logic expression for each circuit.

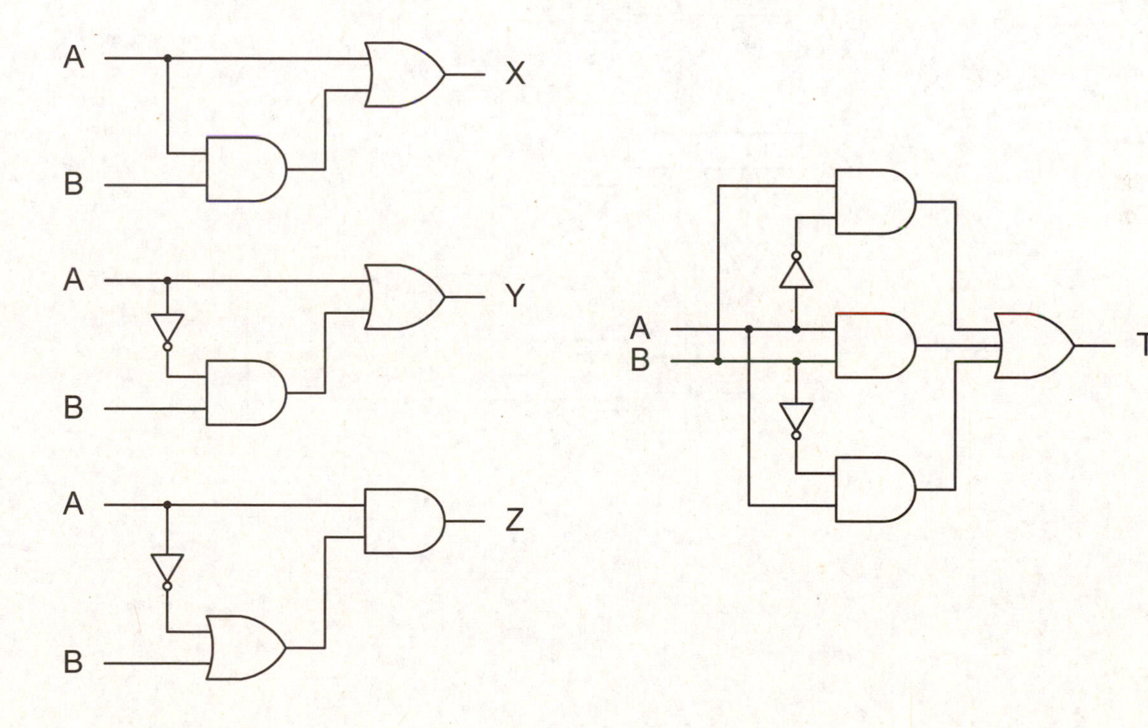

2.2 More circuit functions

To test these circuits on the DE0/DE1/DE2, set SW9 to a HIGH logic level. Write the logic expression for each circuit. Record the test results in a truth table. Label each circuit function with its logic gate name (NAND, NOR, XOR, XNOR). Describe when circuit V produces a LOW output. Describe when circuit W produces a LOW output. Describe when circuit J produces a HIGH output. Describe when circuit K produces a HIGH output. What is the functional relationship between circuits J and K?

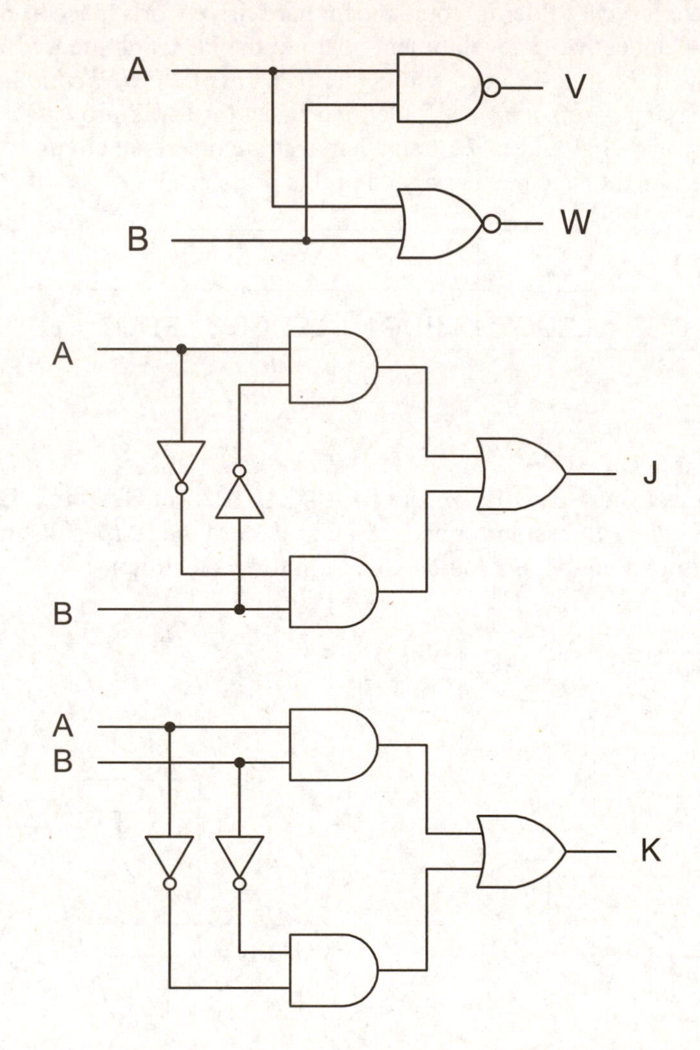

UNIT 3

SCHEMATIC CAPTURE & ANALYSIS OF COMBINATIONAL LOGIC CIRCUITS

Objectives

- To analyze combinational logic circuits and predict their operation.
- To perform design entry of combinational logic circuits using schematic capture.
- To construct and test combinational logic circuits by programming an FPGA or a CPLD.

Combinational Logic Circuit Analysis

The theoretical operation of a combinational logic circuit can be predicted by analyzing the circuit's output for every possible input combination. The circuit analysis for each input combination is performed by determining the resultant output of each gate, working from the input side of the circuit to the output. A common logic circuit configuration consists of AND gates that feed an OR gate. This standard circuit arrangement can be easily described in a Sum-Of-Product (SOP) Boolean expression that also provides a simple shortcut to speed up the circuit analysis process.

Logic circuits may be functionally equivalent. They may perform the same function (i.e., their logic truth tables are identical) but be constructed from different logic gates or interconnected in an entirely different manner. In fact we will find that often a complex logic circuit can be replaced with a much simpler one that performs the identical function.

Programmable Logic Devices

With programmable logic devices (PLDs), logic circuit designers can go from a conceptual design to customized functional parts in a matter of minutes. A PLD is a digital IC that is capable of being programmed to provide a specific logic function. The PLD family of devices consists of a variety of device architectures and configurations. The two most common architectures use either a Look-Up Table (LUT) or an AND/OR logic gate array. These logic structures are repeated many times, often thousands of times, within a PLD and provide the building blocks to create any desired set of logic functions for an application. The building blocks are programmed to achieve the desired logic function using PLD development software. Many different PLD families and part numbers are available, providing a wide variety of choices in the number of inputs and outputs that are available, the number of SOP product terms that can be handled, and the ability to produce registered outputs. The programmable flexibility of PLDs typically allows circuit designers to replace many different standard IC chips, which are now largely obsolete, with a single PLD package. PLDs are available for either one-time-only programming or are erasable and reprogrammable. High-capacity programmable logic devices contain the equivalent of at least hundreds of logic gates and flip-flops, and some families of PLDs contain hundreds of thousands. Depending upon their specific architectural characteristics, these high-capacity PLDs are called complex programmable logic devices (CPLDs) or field programmable gate arrays (FPGAs). FPGAs often employ the LUT architecture and CPLDs frequently use the AND/OR logic array. FPGAs and CPLDs should be handled carefully because they are constructed with MOS (metal oxide semiconductor) transistors and, as such, can be easily damaged by static electricity.

Altera Quartus II Development Software

A logic design is described using various types of computer files that are processed by development system software such as Altera's Quartus II. The Web Edition of Quartus II is a free version of the commercial software and can be downloaded from http://www.altera.com. This software package contains all of the tools necessary to design and develop PLD solutions for logic circuit design. After design entry, the logic design is compiled using the computer software. In this step, the computer determines how to fit your design into a target PLD. The designer will then typically simulate the design with the computer software to verify that it satisfies the specifications. Any functional errors will require that the design be modified and appropriate design files changed. Design changes can be implemented simply by altering the design files. The updated design is re-compiled and then re-simulated for verification. When the designer is satisfied that the design meets the specifications, a PLD is programmed with the design information and the prototype device is then ready to be tested in your application.

This unit will introduce Altera's Quartus II development software using schematic capture for design entry. An example design project with step-by-step procedures is available at www.pearsonhighered.com/electronics in the Tutorials folder. The tutorial is designed to run in one window on your PC while Quartus II is running in another window so that you can easily switch back and forth between the tutorial instructions and the actual software.

Example 3-1 (*Tutorial 1*)

Analyze the circuit in Fig. 3-1 and determine its truth table. Construct the circuit in an FPGA or CPLD using schematic capture.

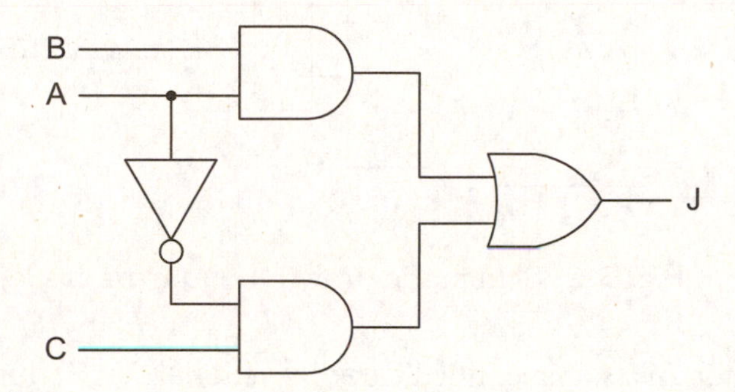

Fig. 3-1 Schematic for circuit analysis example

The logic expression for the given logic circuit is:

$$J = A\,B + \overline{A}\,C$$

Since the logic circuit has 3 input variables, a truth table (see Table 3-1) listing all 8 possible combinations is constructed. Create a separate output column in the truth table for each logic gate in the circuit and label the column to represent the gate function or output node name. The resultant output for every gate in the circuit can then be determined for each of the 8 possible input combinations. This gate output information is added to the truth table results for each of the logic gates.

A	B	C	$\overline{A}$	A B	$\overline{A}$ C	J
0	0	0	1	0	0	0
0	0	1	1	0	1	1
0	1	0	1	0	0	0
0	1	1	1	0	1	1
1	0	0	0	0	0	0
1	0	1	0	0	0	0
1	1	0	0	1	0	1
1	1	1	0	1	0	1

Table 3-1 Truth table analysis for example circuit

The given schematic can be drawn using Quartus II software. Quartus II uses a library of basic logic devices called primitives to create the schematic for this circuit. The resulting schematic is shown in Fig. 3-2. Note that input and output symbols are included in the schematic to provide ports that will be connected to pins on a PLD device. This design file will then be compiled by Quartus II so that an FPGA or CPLD may be programmed with this design. The input and output pins for this design may be connected to logic switches and lamp monitors to test the design.

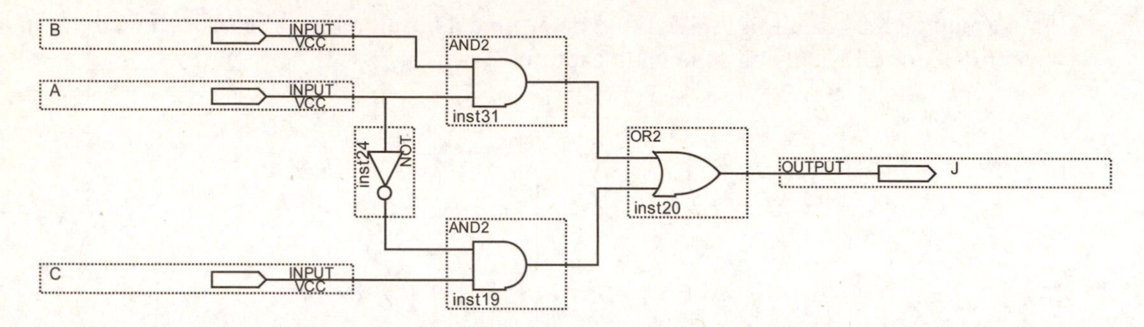

Fig. 3-2 Schematic drawn in Quartus II

A tutorial for this example (titled Quartus Tutorial 1 – Schematic) is available from the publisher's textbook web site www.pearsonhighered.com/electronics. Navigate to the *pdf* tutorial file and follow the step-by-step procedures. **Note that some specific details will be dependent upon the hardware that is used in your laboratory. Your lab instructor will provide this information for you.**

Laboratory Projects

3.1 Example 3-1 (Tutorial)
Program an FPGA or CPLD with the circuit given in Example 3-1. Refer to Quartus Tutorial 1 – Schematics.

3.2 Equivalent circuits
Analyze the following pairs of equivalent logic circuits. Write the logic expression for each circuit and predict the operation for each one with a truth table. Do the truth tables confirm that the pairs of circuits are equivalent? Construct and test each circuit in an FPGA or CPLD using schematic capture. Do the circuit pairs function the same?

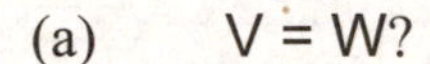

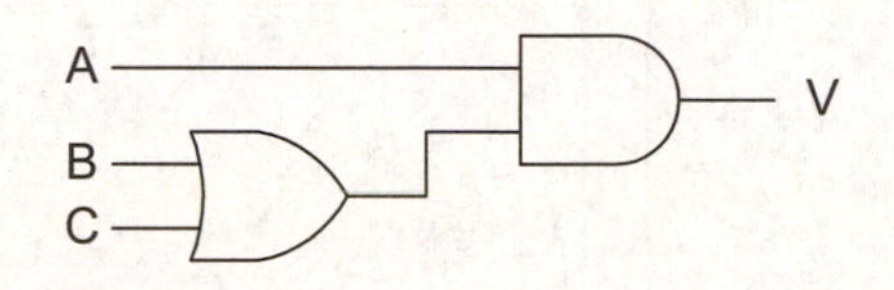

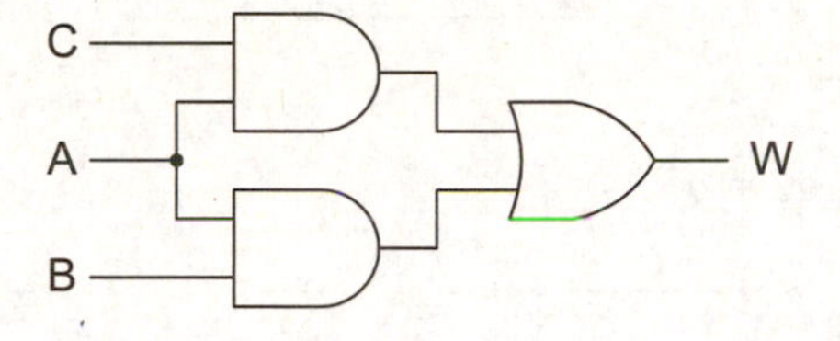

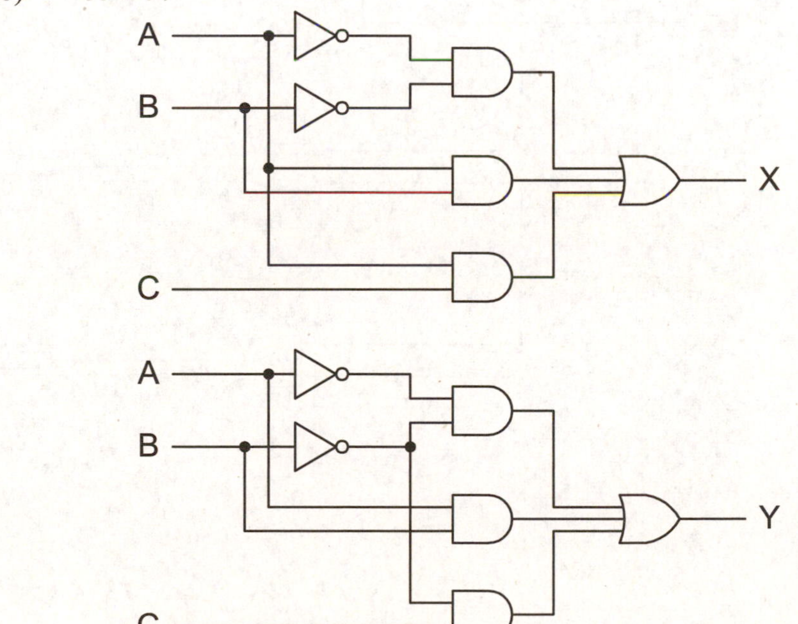

3.3 DeMorgan's theorem

Each of the two logic circuits below has multiple outputs. Prove that the three outputs for each individual circuit are equivalent by programming the circuit in an FPGA or CPLD and testing it with switches and lights.

(a) p1 = p2 = p3?

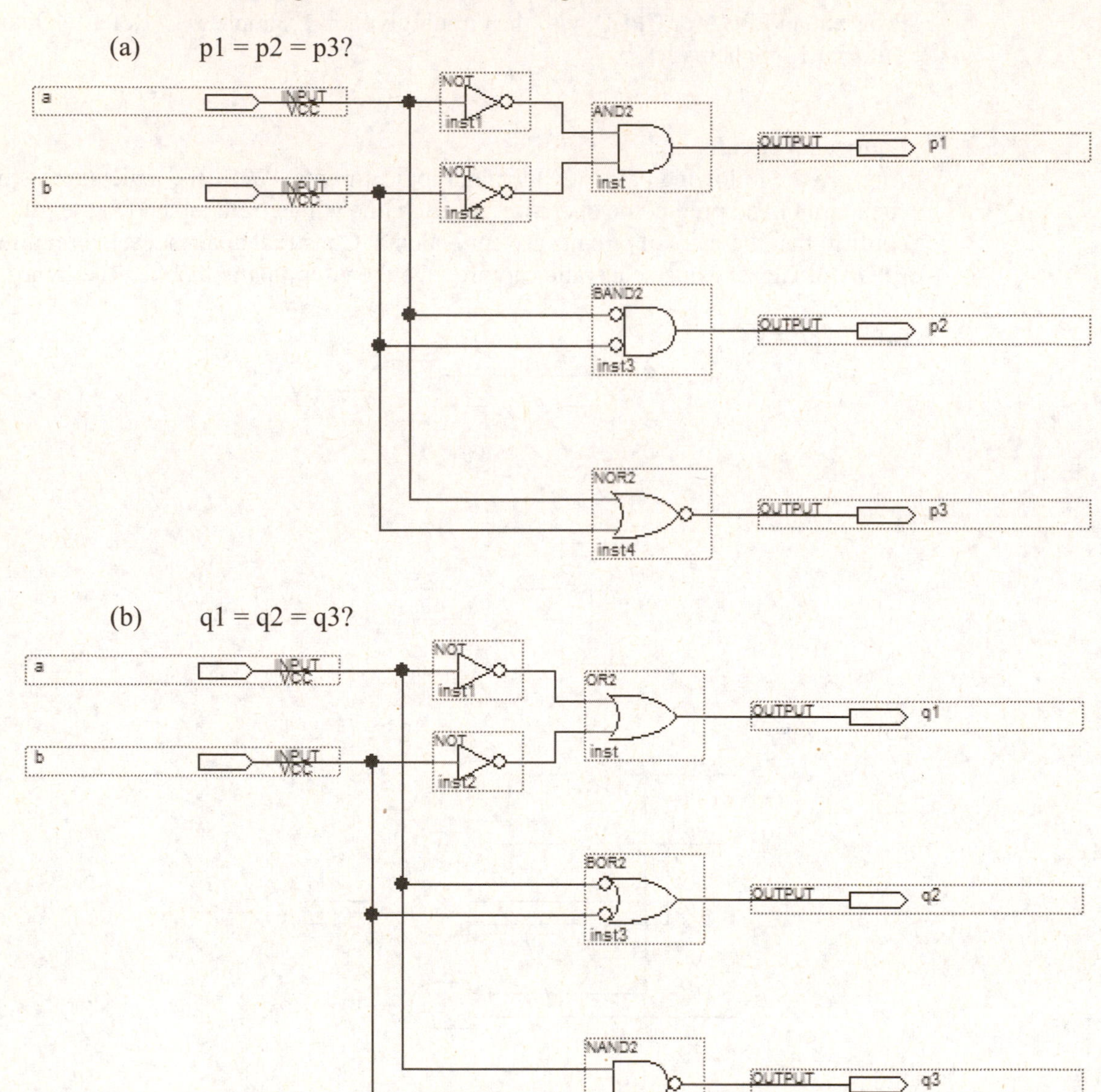

(b) q1 = q2 = q3?

3.4 2-bit adder

Analyze the following multiple-output logic circuit and predict the theoretical operation using a truth table. Construct and test the circuit in an FPGA or CPLD using schematic capture. Verify that the circuit's 3-bit output S2 S1 S0 is the result of adding the two 2-bit numbers A1 A0 plus B1 B0.

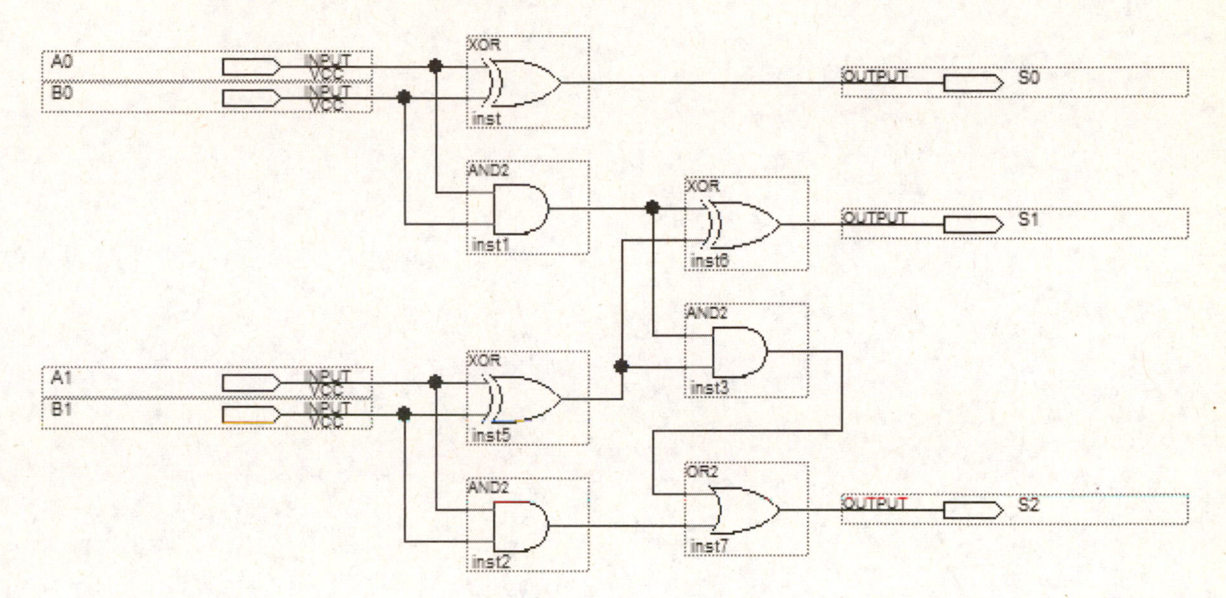

UNIT 4

DESIGN & SIMULATION OF COMBINATIONAL CIRCUITS

Objectives

- To describe given story problems in a truth table.
- To write sum-of-product (SOP) logic expressions for functions defined in truth tables.
- To design a combinational logic circuit from an SOP logic expression.
- To simulate the combinational logic circuit using Quartus II.
- To program a CPLD or FPGA using Quartus II to perform a stated task.

Combinational Circuit Design

SOP Expressions

The most commonly used format for writing logic expressions is a standard form called sum-of-product (SOP). SOP expressions can be quickly written from a truth table and they are easily implemented using a two-level (not counting inverters that may be needed) gate network. Conversely, these standard circuits that are used to implement SOP functions can be quickly analyzed just by inspection. A sum-of-product expression consists of two or more product (AND) terms that are ORed together. The SOP expression is obtained from a truth table by writing down all of the product terms (also called minterms) whose outputs are HIGH for the desired function and then ORing them together. The resultant SOP expression can be directly implemented with an AND/OR circuit design (or, alternatively, a NAND/NAND circuit).

Example 4-1 (*Tutorial 2*)

Design a combinational circuit that will indicate the majority result of 3 individuals voting.

First, define the problem in a truth table as shown in Table 4-1.

A	B	C	V
0	0	0	0
0	0	1	0
0	1	0	0
0	1	1	1
1	0	0	0
1	0	1	1
1	1	0	1
1	1	1	1

Table 4-1 Truth table for Example 4-1

The unsimplified SOP expression for Example 4-1 would be:

$$V = \overline{A}\,B\,C + A\,\overline{B}\,C + A\,B\,\overline{C} + A\,B\,C$$

We will implement this design using a CPLD or FPGA. Any PLD will have tremendously more logic resources than will be needed for this application but we want to explore further the versatility of this type of logic device and the development system software used to program it. Additionally, most digital systems will have many more functions that will be needed in an application and these other functions can also be included in the same PLD. Building a complex digital system using only SSI standard parts would require hundreds or thousands of ICs! We will also use this simple application to investigate the simulation capabilities of Quartus II. Since Quartus will simplify the logic function for us, we will not need to manually simplify the function by Karnaugh Mapping (see optional subsection starting on the next page) or some other simplification technique before entering the design. It will require a bit more drawing to create the schematic but it will save us the Karnaugh mapping step, a procedure that is outlined the next subsection. Fig. 4-1 shows our schematic for this problem.

A tutorial for this example (titled Quartus Tutorial 2 – Simulation) is available from the publisher's textbook web site www.pearsonhighered.com/electronics. Navigate to the *pdf* tutorial file and follow the step-by-step procedures. **Note that some specific details will be dependent upon the hardware that is used in your laboratory. Your lab instructor will provide this information for you.** The simulation results are shown in Fig. 4-2.

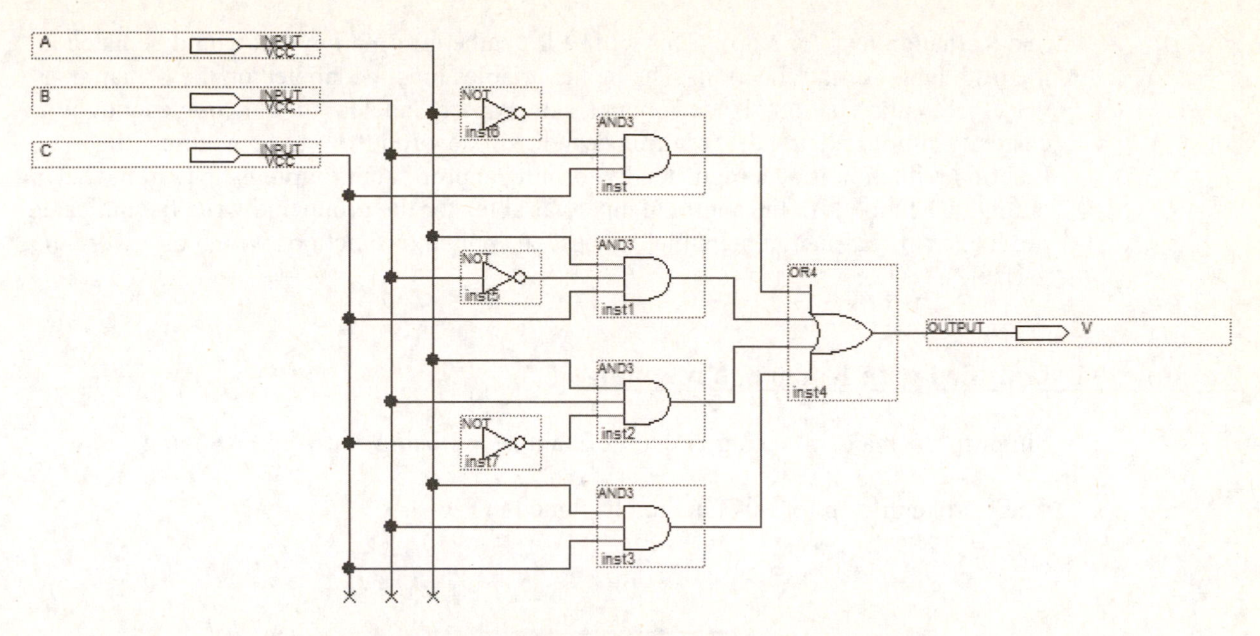

Fig. 4-1 Quartus II schematic for Example 4-1

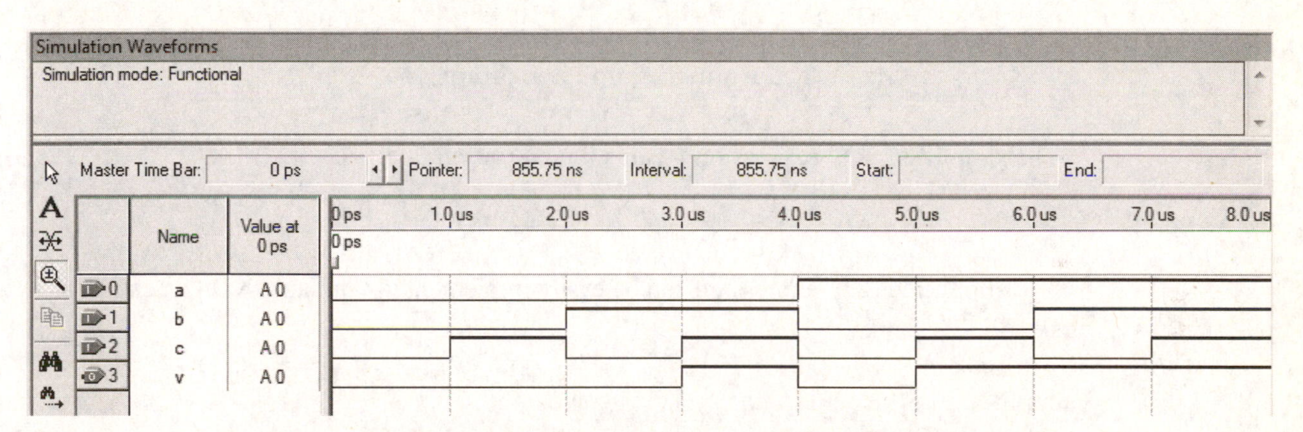

Fig. 4-2 Quartus II simulation results for Example 4-1

Karnaugh Mapping (optional)

Before logic design software such as Quartus II (and many other similar software packages) became ubiquitous in industry, the basic procedure for combinational logic circuit design was to first develop the truth table that defines the desired function and then, from the table write a simplified SOP expression. Now our logic design software will automatically perform the simplification of the desired function. The expression can still be simplified manually (by a human) using various techniques such as Boolean algebra, Karnaugh mapping, and others. Karnaugh mapping is a simple and fast procedure for reducing SOP logic expressions and thereby also reducing the implemented circuit's complexity. In Karnaugh mapping, the function is defined graphically. The relationships between the function's inputs and the output are plotted in a Karnaugh map (K map). This will be the same information that would be listed in the truth table for the function. The input variables must be labeled on the K map in a very systematic fashion. If the K map is not properly labeled, the function cannot be correctly simplified and the resulting design will be wrong. With K mapping, the function reduction is accomplished by forming appropriate groupings of 1s in the output. Then identify the common input variables for the group and write the indicated product term. Karnaugh mapping can best be applied to functions with 5 or fewer input variables.

Example 4-1 simplified with Karnaugh mapping

Simplify the majority voting circuit of Example 4-1 using Karnaugh mapping.

The Karnaugh map for this function is plotted in Fig. 4-3.

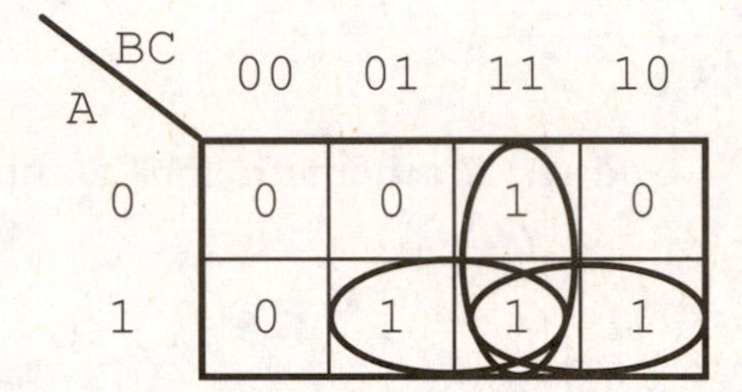

Fig. 4-3 Karnaugh map for Example 4-1

The simplified SOP expression for this function would be:

$$V = BC + AC + AB$$

The simplified SOP expression can be easily implemented with an AND/OR circuit as shown in Fig. 4-4.

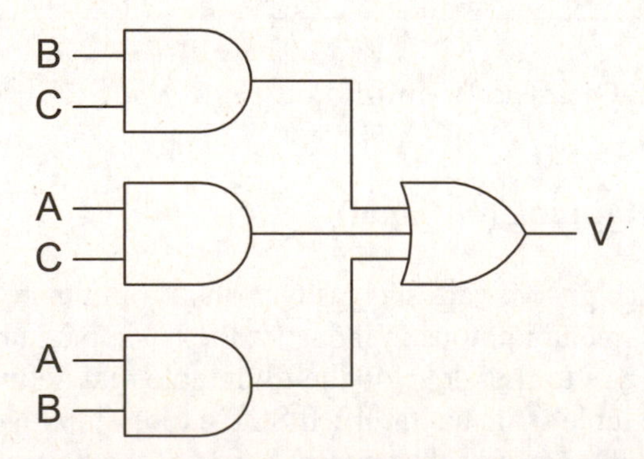

Fig. 4-4 Schematic for simplified SOP solution to Example 4-1

Example 4-2

Design a simplified logic circuit to implement the function W defined in the truth table given in Table 4-2.

A	B	C	D	W
0	0	0	0	0
0	0	0	1	1
0	0	1	0	1
0	0	1	1	1
0	1	0	0	0
0	1	0	1	0
0	1	1	0	0
0	1	1	1	0
1	0	0	0	1
1	0	0	1	1
1	0	1	0	0
1	0	1	1	0
1	1	0	0	0
1	1	0	1	0
1	1	1	0	0
1	1	1	1	0

Table 4-2 Truth table for Example 4-2

The output produced for the function W is plotted in the Karnaugh map in Fig. 4-5. Then appropriate groups of 1s are identified in the K map to create the SOP expression for W. Two different (but equivalent) simplified expressions can be obtained with K mapping. The groupings of 1s shown in the K map are represented by the first expression. The two solutions each require the same number of gates and input variables, so either simplified SOP expression may be implemented.

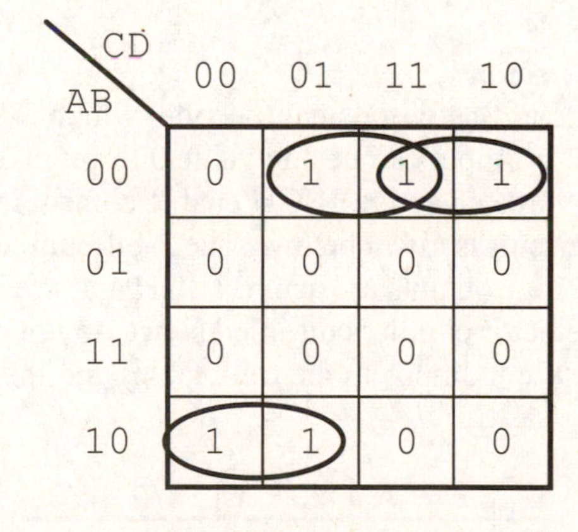

$$W = \overline{A}\,\overline{B}\,C + A\,\overline{B}\,\overline{C} + \overline{A}\,\overline{B}\,D$$

$$= \overline{A}\,\overline{B}\,C + A\,\overline{B}\,\overline{C} + \overline{B}\,\overline{C}\,D$$

Fig. 4-5 Karnaugh map and simplified expressions for Example 4-2

Laboratory Projects

Design logic circuits to be constructed with an FPGA or CPLD to perform each of the following functions. Define the problem with a truth table, write the SOP expression, and draw the schematic for design entry. Functionally simulate the circuit to verify your design, then program an FPGA/CPLD to test its operation. Be sure to follow any special instructions given by your instructor.

4.1 Majority vote
Program an FPGA or CPLD with the circuit given in Example 4-1 using the Quartus Simulation tutorial.

4.2 Two-input multiplexer
Design a circuit whose output (Y) is equivalent to one of two possible data inputs (A or B). A control input (S) selects either the data on the A input (if S is LOW) or the data on the B input (if S is HIGH) to be routed to the single output line.

4.3 Non-BCD input
Design a circuit whose output will be HIGH if the 4-bit data input is an invalid BCD number (i.e., greater than 9).

4.4 Two-bit comparator
Design a comparator circuit to compare two 2-bit numbers (A1 A0 and B1 B0). The circuit will have a HIGH output (named EQ) if the 2-bit A value is equal to the 2-bit B value.

4.5 Binary number detector
Design a logic circuit that will detect (decode) which 2-bit binary number is applied to the circuit. The 2-bit input can be any value 0 through 3 in decimal. The number detector will also have an active-HIGH enable control input named G. The truth table for this detector circuit is given below. The 2-bit number is B A (A is the LSB) and the enable is G. The four outputs are active-HIGH and are labeled Y0 through Y3. When G is LOW, the detector circuit is disabled so we do not care about the B A inputs (indicated by the X conditions in the truth table) and the outputs will all be inactive at a logic zero.

G	B	A	Y0	Y1	Y2	Y3
1	0	0	1	0	0	0
1	0	1	0	1	0	0
1	1	0	0	0	1	0
1	1	1	0	0	0	1
0	X	X	0	0	0	0

4.6 Alarm circuit

Design an alarm circuit to be used in a process control system. Temperature (T), pressure (P), flow (F), and level (L) of a fluid are each monitored by separate sensor circuits that produce a HIGH logic output signal when the following indicated physical conditions exist:

high fluid temperature
high fluid pressure
low fluid flow rate
low fluid level

The alarm circuit output (A) should be HIGH if any of the following physical conditions exist in the system:

(1) *the pressure is high when the flow rate is low*
(2) *the temperature is high when either the pressure is high or the level is low*

The alarm circuit will be connected to the outputs of the sensors, so be sure to convert the above description of the physical conditions for the alarm into the logic levels that will actually be applied to the logic circuit. The truth table should list the logic levels for the alarm inputs, not the physical conditions. List the truth table inputs (which are the sensor outputs) in the order T P F L.

4.7 Elevator control

Design an elevator control system for a large building that has 5 elevators. Four of the elevators are turned on all of the time, while the fifth is activated only if a majority of the other 4 are being used (to save energy costs). The control system will have an input for each of the 4 primary elevators to indicate that that elevator is being used (with a logic "1"). A HIGH output from the control system will activate the fifth elevator for its use.

UNIT 5

CREATING HIERARCHICAL LOGIC CIRCUITS

Objectives

- To implement hierarchical combinational logic circuits using programmable logic devices (PLDs) with Quartus II development system software.
- To simulate hierarchical combinational logic circuits in Quartus II.

Hierarchical Logic Circuits

Large designs can be more easily developed by subdividing the system into a hierarchical description of the design. With this method, a complex design can be broken down into multiple layers to make the design entry simple and convenient. The hierarchical design technique breaks a complex design into smaller, more easily defined blocks. This process can also make short work of designs that repeat the same circuit block many times to create the desired system. As seen in Figure 5-1, smaller circuit blocks at the lowest level of the project hierarchy can be combined into more complex intermediate blocks, which may then be combined into still more complex blocks and so on until we have created the complete system. In addition to using schematics, Hardware Description Languages (HDLs) may be used for design entry. Two different HDLs will be introduced in Units 6A and 6V. *A very important rule to remember when using multiple design blocks to define a system is that each design block must have a unique file name.* Example 5-1 illustrates how to apply the hierarchical design technique in Quartus II.

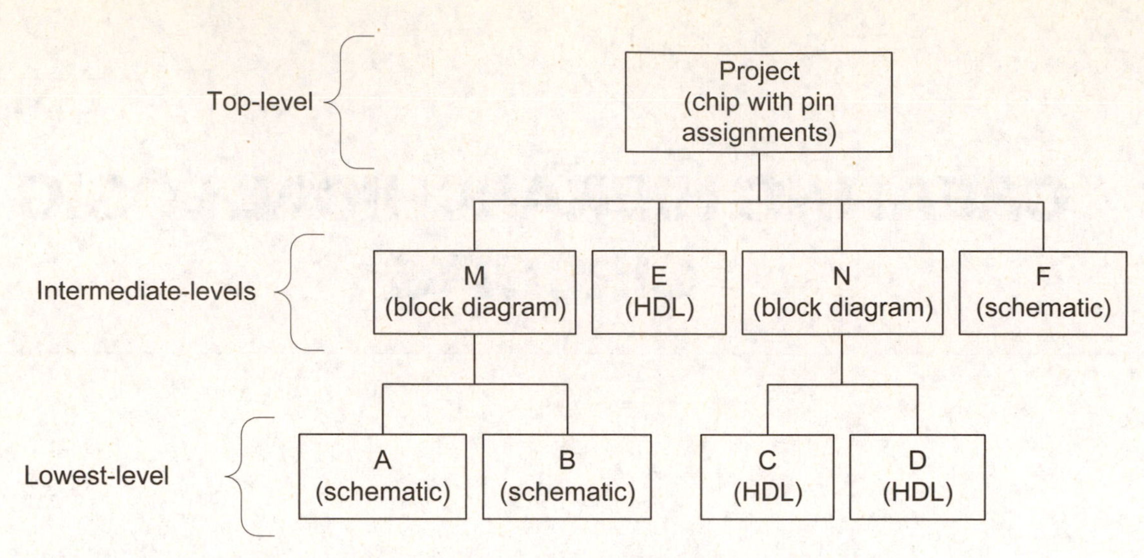

Fig. 5-1 Hierarchical design of digital systems

Example 5-1 (*Tutorial 3*)

Design a 1-out-of-4 data selector (or multiplexer) circuit using a circuit design for a 1-out-of-2 data selector. A data selector uses a control input to select a single data input (from several choices) to be routed to the output of the circuit. See Fig. 5-2 and Table 5-1. If the control SEL is LOW, then output Y = D0; otherwise Y = D1. A data selector is also called a multiplexer.

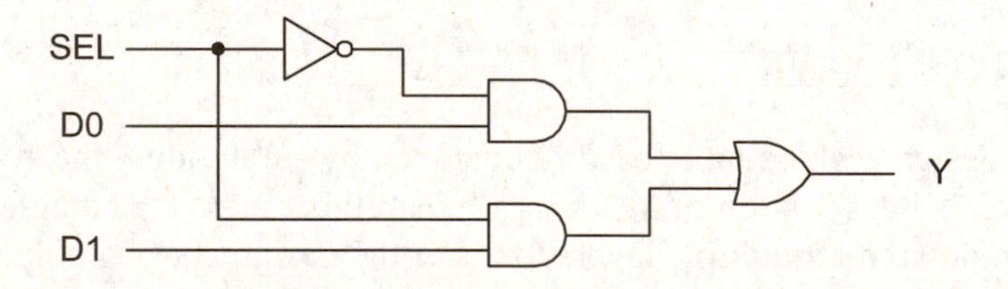

Fig. 5-2 1-out-of-2 data selector circuit for Example 5-1

SEL	Y
0	D0
1	D1

Table 5-1 Truth table for 1-out-of-2 data selector circuit

A 1-out-of-4 data selector can be constructed by combining two levels of the 1-out-of-2 data selector circuit design (see Fig. 5-3). This design can be easily implemented in a hierarchical fashion. The file Quartus Tutorial 3 – Hierarchical is available from the publisher's textbook web site www.pearsonhighered.com/electronics. Navigate to the *pdf* tutorial file and follow the step-by-step procedures.

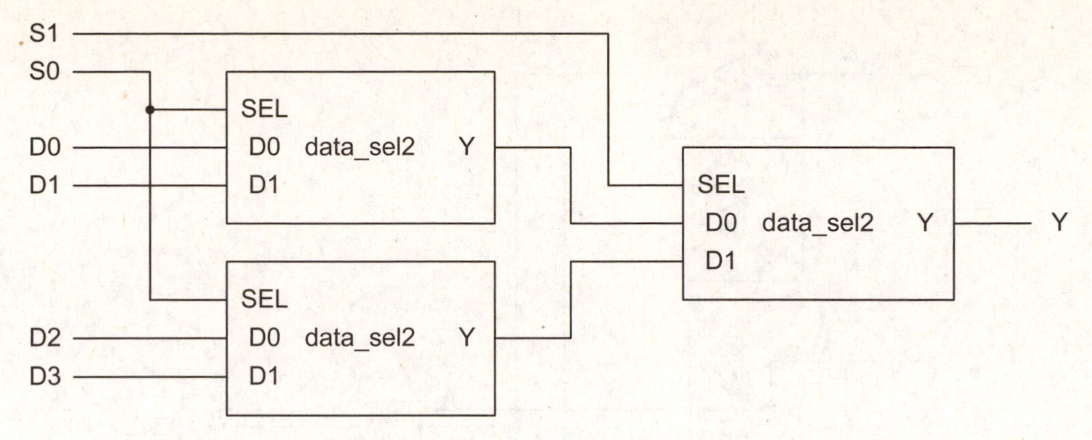

Fig. 5-3 Block diagram for 1-out-of-4 data selector circuit

Example 5-2

Design a logic circuit that will detect (or decode) which 2-bit binary number is applied to the circuit. The 2-bit binary number will have four possible input values (representing 0 through 3 in decimal) that can be applied. The number detector will also have an active-HIGH enable control input named G. The following truth table represents the desired logic function. The 2-bit number is B A (A is the LSB) and the enable is G. The four outputs are active-HIGH and are labeled Y0 through Y3. When G is LOW, the detector circuit is disabled so we do not care about the B A inputs (indicated by the X conditions in the truth table) and the outputs will all be inactive at a logic zero. We will use this design as a lower-level block in some of the lab projects in this Unit.

G	B	A	Y0	Y1	Y2	Y3
1	0	0	1	0	0	0
1	0	1	0	1	0	0
1	1	0	0	0	1	0
1	1	1	0	0	0	1
0	X	X	0	0	0	0

Note that each output is HIGH only for a single product term combination of the circuit's inputs. The schematic solution for this application is given in Fig. 5-4.

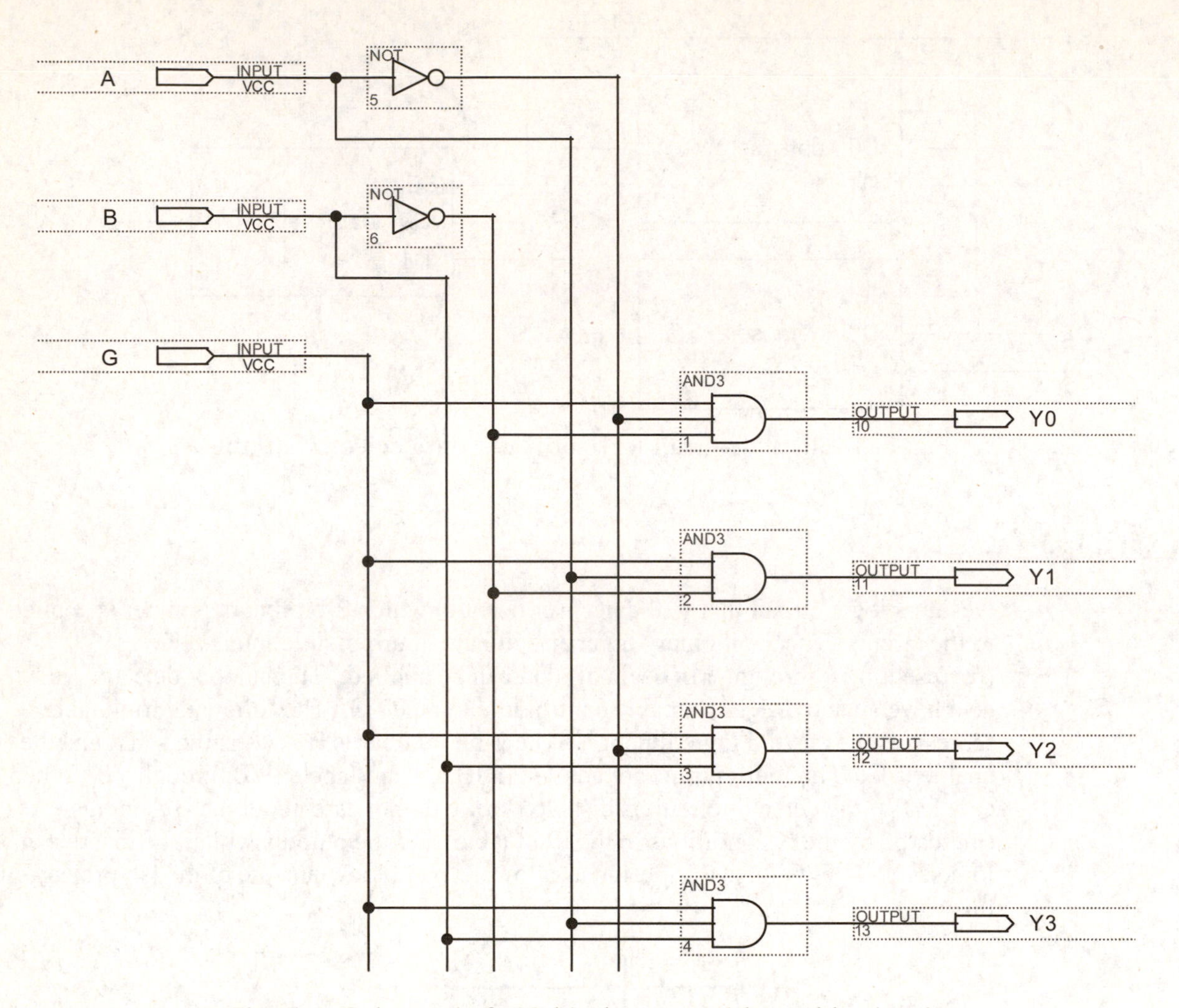

Fig. 5-4 Schematic for 2-bit detector with enable circuit

Laboratory Projects

Create, functionally simulate, and test each of the following hierarchical PLD circuits using Quartus II. Place files in the correct directory.

5.1 1-out-of-4 data selector

Test the 1-out-of-4 data selector design given in Example 5-1 (Tutorial).

S1	S0	Y
0	0	D0
0	1	D1
1	0	D2
1	1	D3

5.2 2-channel, 4-bit multiplexer

Use the 1-out-of-2 data selector given in Example 5-1 to construct a 2-channel, 4-bit multiplexer circuit that functions like the truth table below. The 2-channel, 4-bit multiplexer will select one of two 4-bit binary numbers to output from the logic circuit.

S	Y3	Y2	Y1	Y0
0	A3	A2	A1	A0
1	B3	B2	B1	B0

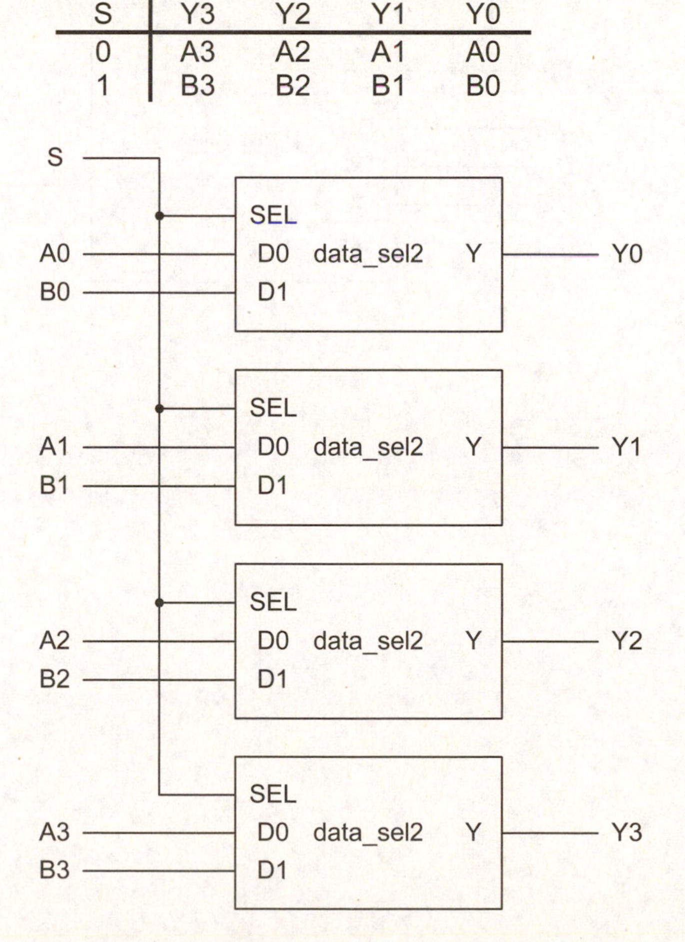

5.3 Multiplexer & decoder

Use the lower-level logic blocks described in Examples 5-1 and 5-2 to create the multiplexer and decoder circuit described in the truth table and illustrated in the diagram below. Be sure to Add Files for <u>all</u> three lower level blocks (1-out-of-2 data selector, 1-out-of-4 data selector, and 2-bit detector).

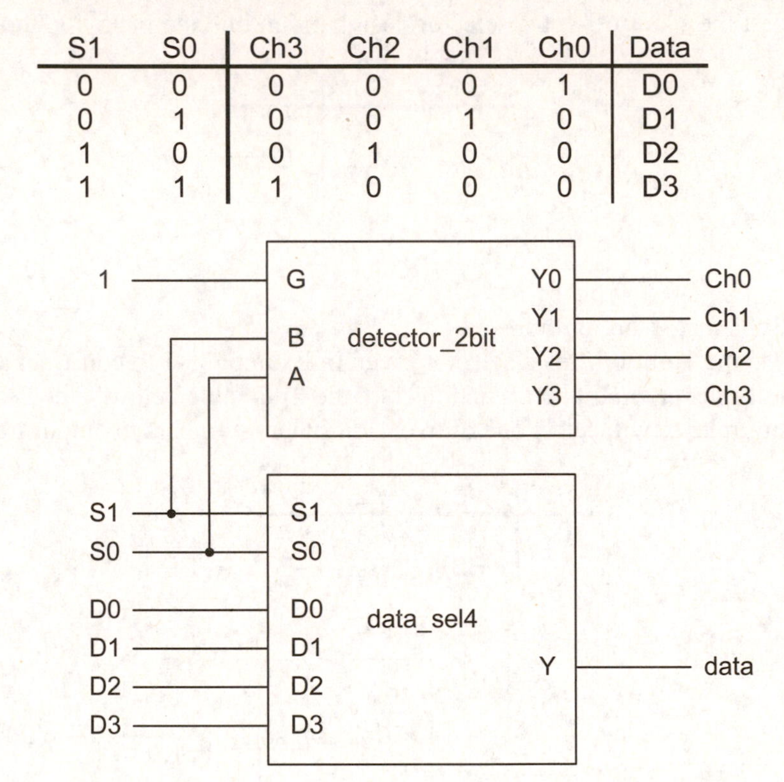

S1	S0	Ch3	Ch2	Ch1	Ch0	Data
0	0	0	0	0	1	D0
0	1	0	0	1	0	D1
1	0	0	1	0	0	D2
1	1	1	0	0	0	D3

5.4 3-bit decoder

Use the lower-level 2-bit detector block described in Example 5-2 to create a 3-bit binary number decoder described in the truth table and illustrated in the block diagram below.

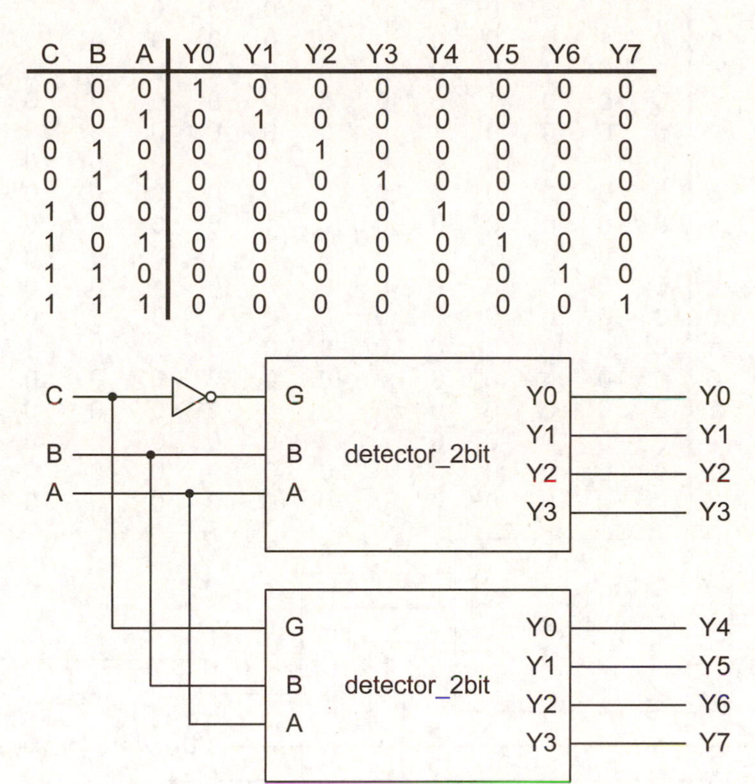

C	B	A	Y0	Y1	Y2	Y3	Y4	Y5	Y6	Y7
0	0	0	1	0	0	0	0	0	0	0
0	0	1	0	1	0	0	0	0	0	0
0	1	0	0	0	1	0	0	0	0	0
0	1	1	0	0	0	1	0	0	0	0
1	0	0	0	0	0	0	1	0	0	0
1	0	1	0	0	0	0	0	1	0	0
1	1	0	0	0	0	0	0	0	1	0
1	1	1	0	0	0	0	0	0	0	1

5.5 4-bit decoder

Use the lower-level 2-bit detector block described in Example 5-2 to create a 4-bit binary number decoder described in the truth table and illustrated in the block diagram below. The input labeled EN is an enable for the decoder.

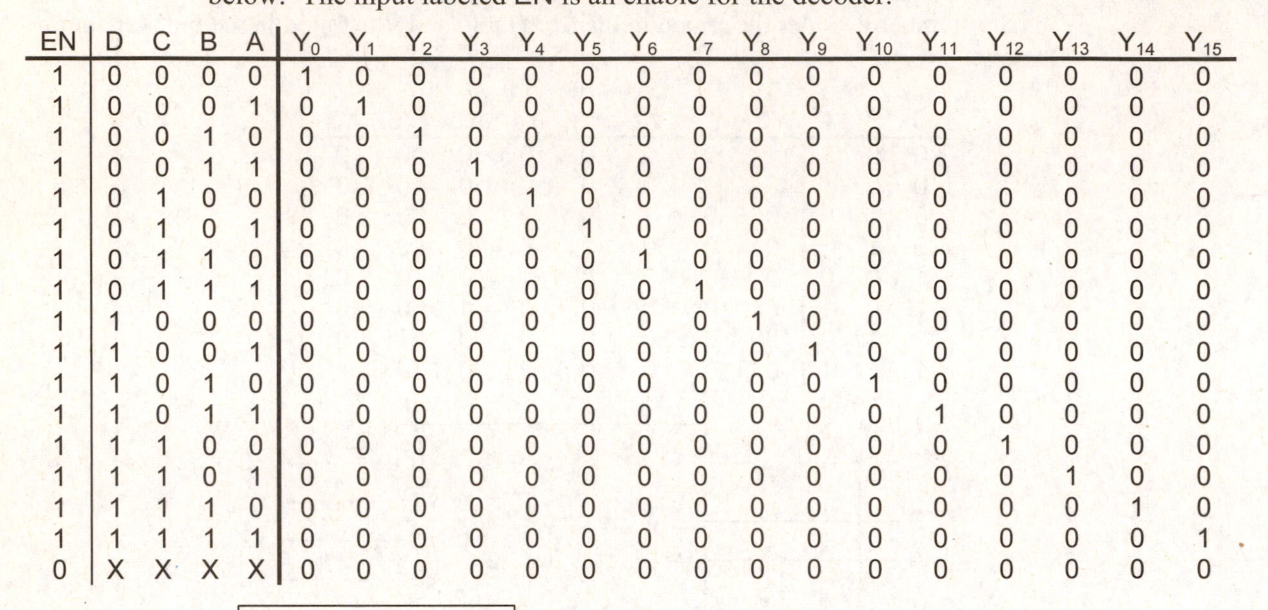

EN	D	C	B	A	Y_0	Y_1	Y_2	Y_3	Y_4	Y_5	Y_6	Y_7	Y_8	Y_9	Y_{10}	Y_{11}	Y_{12}	Y_{13}	Y_{14}	Y_{15}
1	0	0	0	0	1	0	0	0	0	0	0	0	0	0	0	0	0	0	0	0
1	0	0	0	1	0	1	0	0	0	0	0	0	0	0	0	0	0	0	0	0
1	0	0	1	0	0	0	1	0	0	0	0	0	0	0	0	0	0	0	0	0
1	0	0	1	1	0	0	0	1	0	0	0	0	0	0	0	0	0	0	0	0
1	0	1	0	0	0	0	0	0	1	0	0	0	0	0	0	0	0	0	0	0
1	0	1	0	1	0	0	0	0	0	1	0	0	0	0	0	0	0	0	0	0
1	0	1	1	0	0	0	0	0	0	0	1	0	0	0	0	0	0	0	0	0
1	0	1	1	1	0	0	0	0	0	0	0	1	0	0	0	0	0	0	0	0
1	1	0	0	0	0	0	0	0	0	0	0	0	1	0	0	0	0	0	0	0
1	1	0	0	1	0	0	0	0	0	0	0	0	0	1	0	0	0	0	0	0
1	1	0	1	0	0	0	0	0	0	0	0	0	0	0	1	0	0	0	0	0
1	1	0	1	1	0	0	0	0	0	0	0	0	0	0	0	1	0	0	0	0
1	1	1	0	0	0	0	0	0	0	0	0	0	0	0	0	0	1	0	0	0
1	1	1	0	1	0	0	0	0	0	0	0	0	0	0	0	0	0	1	0	0
1	1	1	1	0	0	0	0	0	0	0	0	0	0	0	0	0	0	0	1	0
1	1	1	1	1	0	0	0	0	0	0	0	0	0	0	0	0	0	0	0	1
0	X	X	X	X	0	0	0	0	0	0	0	0	0	0	0	0	0	0	0	0

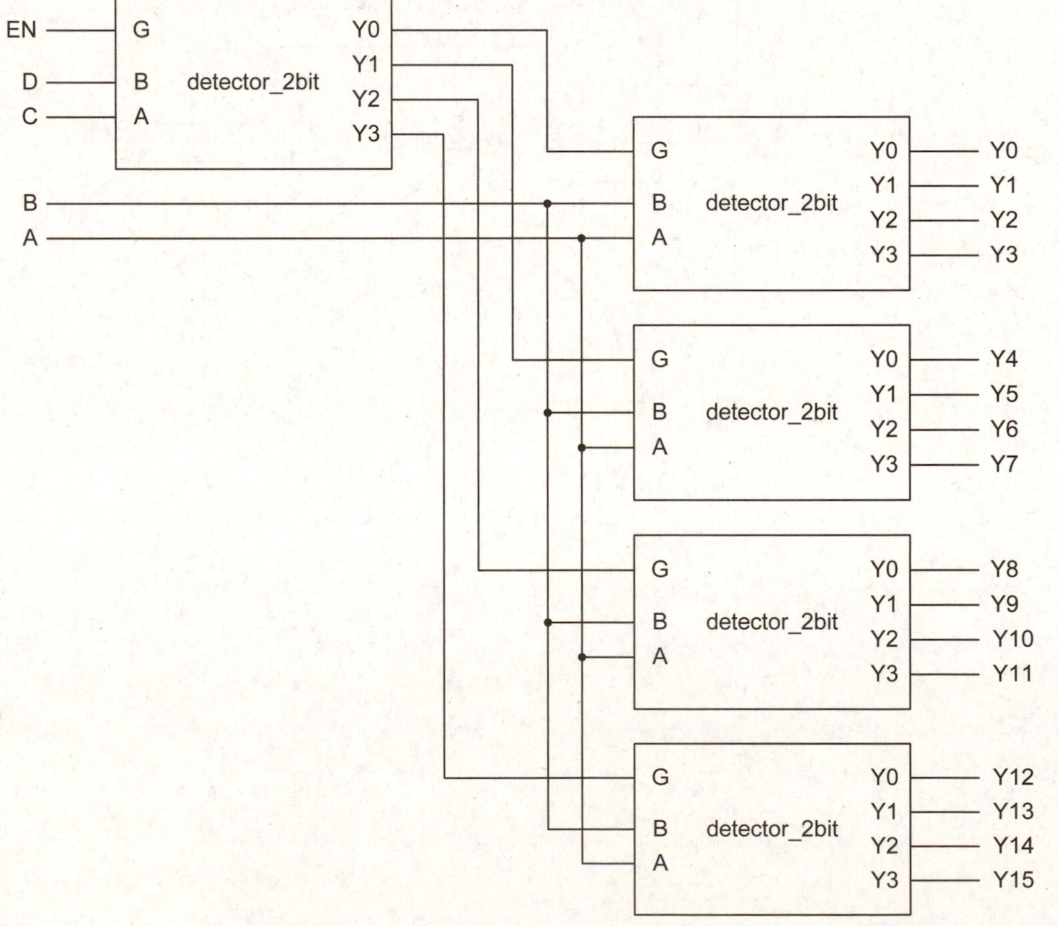

5.6 Data transmission

Create the circuit described by the truth table and block diagram below. The inputs S1 S0 control which data input D0 D1 D2 D3 will be transmitted to its corresponding output Q0 Q1 Q2 Q3. The data inputs that are not selected for transmission will be "locked out" and will have no effect on the outputs.

S1	S0	Q3	Q2	Q1	Q0
0	0	0	0	0	D0
0	1	0	0	D1	0
1	0	0	D2	0	0
1	1	D3	0	0	0

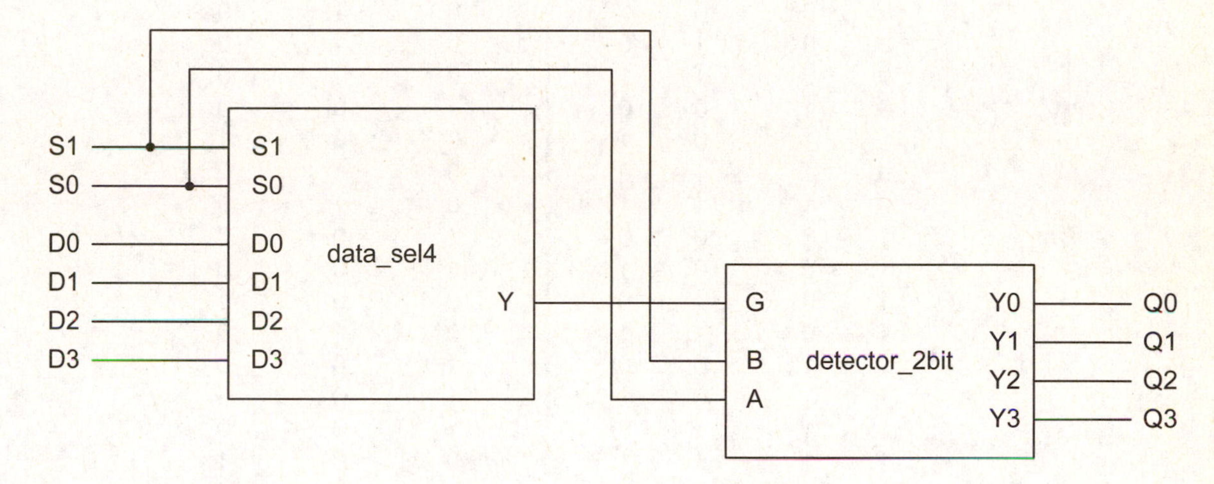

5.7 2-bit magnitude comparator

Design a circuit to compare the magnitudes of two 2-bit numbers (A1 A0 and B1 B0). The circuit will have three outputs AGTB, ALTB, and AEQB. AGTB will be HIGH to indicate that the 2-bit A value is greater than the 2-bit B value. ALTB will be HIGH to indicate that the 2-bit A value is less than the 2-bit B value. AEQB will be HIGH if both 2-bit values are equal. The truth table is shown below. Because only one of the three output functions can be HIGH at a time, we will create ALTB as a function of the other two outputs (AGTB and AEQB). See Project 5.5 for the 4-bit decoder block. Note that the decoder_4bit contains every possible product term for the 4 inputs A1 A0 B1 B0. Write the three SOP output expressions for this circuit.

A1	A0	B1	B0	AGTB	ALTB	AEQB
0	0	0	0	0	0	1
0	0	0	1	0	1	0
0	0	1	0	0	1	0
0	0	1	1	0	1	0
0	1	0	0	1	0	0
0	1	0	1	0	0	1
0	1	1	0	0	1	0
0	1	1	1	0	1	0
1	0	0	0	1	0	0
1	0	0	1	1	0	0
1	0	1	0	0	0	1
1	0	1	1	0	1	0
1	1	0	0	1	0	0
1	1	0	1	1	0	0
1	1	1	0	1	0	0
1	1	1	1	0	0	1

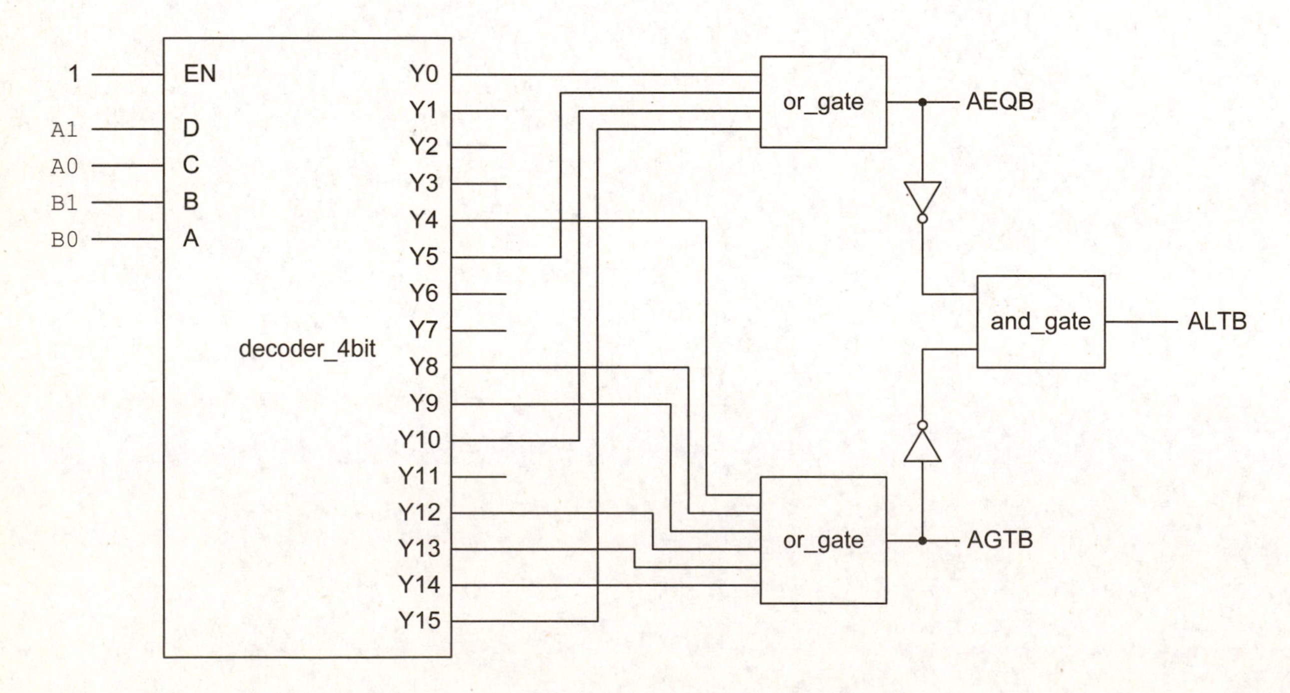

UNIT 6A

COMBINATIONAL CIRCUIT DESIGN WITH AHDL (ALTERA HARDWARE DESCRIPTION LANGUAGE)

Objectives

- To create combinational logic circuits using Altera's AHDL in Quartus II.
- To simulate AHDL combinational logic circuits in Quartus II.

Altera Hardware Description Language

Altera Hardware Description Language (AHDL) is a HIGH-level language within Quartus II. With AHDL, a text design file (*.tdf*) can be created to describe any portion of a project's hierarchy. This text-based design entry technique for a desired logic function can be used to more clearly describe the circuit's behavior than can be commonly achieved with a traditional logic schematic. The AHDL language supports various design description techniques, including Boolean expressions, truth tables, and if/then and case statements. Each TDF file contains two required sections, a Subdesign section and a Logic section. The Subdesign section is used to declare the input and output ports for the design block. The Logic section will describe the logical operation of the block. It is important to recognize that AHDL is a concurrent language. This means that all behavior specified in the Logic section is evaluated at the same time rather than sequentially as in computer languages. There are additional optional TDF sections (e.g., Title, Constant, and Define statements and Variable section) that may also be included according to the needs of the application.

A Text Design File for a simple combinational logic circuit is shown in Fig. 6-1. AHDL keywords are given in capital letters. While AHDL is not case-sensitive, it is a recommended practice to capitalize keywords for readability. Comments can be placed in the AHDL text file by either enclosing the comment within percent symbols **%** … **%** or preceding a comment with double-dashes **--** (with the comment then running to the end of the line). In this example, we have created a brief explanation of the function of this design file and identified the designer in a "header" using the percent symbol % ... % comments. Individual lines have also been commented with the descriptions that follow the double-dash -- comments. **A good design file will be liberally commented for reader clarity.**

The Subdesign section, identified by the keyword SUBDESIGN, is followed by the name of the TDF file (simple_circuit.*tdf*). Note that the file type *tdf* is omitted in the Subdesign section. The input and output signals (ports) for the subdesign are enclosed in parentheses **(...)** and are labeled with INPUT and OUTPUT for the respective port type (signal direction). Multiple port names of the same type are separated by commas, a colon separates the port names from the port type, and a semicolon ends the logical line.

The Variable section is used to identify a buried NODE in the subdesign block. Buried node signals are not available outside the subdesign block. The Logic section, which is sandwiched between BEGIN and END, describes the functionality for this block. The buried node p is the function of input a ANDed with NOT b. The subdesign output y is the logic function p ORed with input c.

```
%%%%%%%%%%%%%%%%%%%%%%%%%%%%%%%%%%%%%%%%%%%%%%%%%%%%%
% Simple circuit to illustrate AHDL design format   %
% Designer:  Greg Moss                              %
%%%%%%%%%%%%%%%%%%%%%%%%%%%%%%%%%%%%%%%%%%%%%%%%%%%%%

SUBDESIGN  simple_circuit       -- starts subdesign section
(                               -- starts list of ports
      a, b, c      :INPUT;      -- list of input ports
      y            :OUTPUT;     -- list of output ports
)                               -- ends list of ports

VARIABLE                        -- starts variable section
      p            :NODE;       -- defines a buried node

BEGIN                           -- starts logic section
      p = a & !b;               -- Boolean expression for p
      y = p # c;                -- Boolean expression for y
END;                            -- ends logic section
```

Fig. 6-1 AHDL design file for a simple circuit

AHDL text design files can be included in a project hierarchy together with other design file types (such as *.bdf* files). Breaking a design into smaller, more manageable blocks can be extremely helpful in creating larger systems. *A very important rule to remember when using multiple design blocks to define a system is that each design block must have a unique file name.* A built-in text editor within Quartus allows the user to easily create and edit the TDF files. Quartus II treats TDF files the same as other design entry files. PLD design tasks such as syntax checking, accessing on-line help, compiling, and simulation are all available when using TDF files. Fundamental language elements of AHDL are summarized in Figs. 6-2a and 6-2b.

AHDL LANGUAGE ELEMENTS

Names

- Symbolic names specify internal and external nodes and groups, constants, state machine variables, state bits, state names, and instances
- Subdesign names are user-defined names for lower-level design files and must be the same as the TDF filename
- Port names are symbolic names that identify the input or output of a logic block
- Start with an alphabet character or slash unless quoted ‘ ’
- Can contain numeric characters or underscore or dash (if quoted); cannot contain spaces
- Are not case sensitive
- May be up to 32 characters long
- Cannot contain any reserved keywords unless quoted

Reserved keywords

AND	FUNCTION	OUTPUT
ASSERT	GENERATE	PARAMETERS
BEGIN	GND	REPORT
BIDIR	HELP_ID	RETURNS
BITS	IF	SEGMENTS
BURIED	INCLUDE	SEVERITY
CASE	INPUT	STATES
CLIQUE	IS	SUBDESIGN
CONNECTED_PINS	LOG2	TABLE
CONSTANT	MACHINE	THEN
DEFAULTS	MOD	TITLE
DEFINE	NAND	TO
DESIGN	NODE	TRI_STATE_NODE
DEVICE	NOR	VARIABLE
DIV	NOT	VCC
ELSE	OF	WHEN
ELSIF	OPTIONS	WITH
END	OR	XNOR
FOR	OTHERS	XOR

Reserved identifiers

CARRY	JKFF	SRFF
CASCADE	JKFFE	SRFFE
CEIL	LATCH	TFF
DFF	LCELL	TFFE
DFFE	MCELL	TRI
EXP	MEMORY	USED
FLOOR	OPENDRN	WIRE
GLOBAL	SOFT	X

Fig. 6-2a Summary of AHDL language elements

AHDL LANGUAGE ELEMENTS

Symbols

Symbol	Description
% %	enclose comments
--	begin comments (to end of line)
( )	enclose port names in Subdesign section; enclose highest priority operations in Boolean and arithmetic expressions
[ ]	enclose the range of a group name
‘ ’	enclose quoted symbolic names
“ ”	enclose digits in nondecimal numbers
.	separate symbolic names of variables from port names
..	separate MSB from LSB in a range
;	end AHDL statements and sections
,	separate members of sequential groups and lists
:	separate symbolic names from types in declarations
=	assign values in Boolean equations; assign values to state machine states
=>	separate inputs from outputs in truth table statements
+	addition operator
–	subtraction operator
==	numeric or string equality operator
!	NOT operator
!=	not equal to operator
>	greater than comparator
>=	greater than or equal to comparator
<	less than comparator
<=	less than or equal to comparator
&	AND operator
!&	NAND operator
$	XOR operator
!$	XNOR operator
#	OR operator
!#	NOR operator

Numbers

- Default base for numbers is decimal
- Binary values (series of 0’s, 1’s, X’s) are enclosed in double quotes and prefixed with B
- Hexadecimal values (series from 0 to 9, A to F) are enclosed in double quotes and prefixed with H
- Numbers cannot be assigned to single nodes in Boolean equations; use VCC and GND

Fig. 6-2b Summary of AHDL language elements

Example 6-1 (*Tutorial 4*)

Design a magnitude comparator circuit to compare the magnitudes of two 2-bit numbers (A1 A0 and B1 B0). The circuit will have three output signals: GT, LT, and EQ. GT will be HIGH to indicate that the 2-bit A value is greater than the 2-bit B value. LT will be HIGH if the 2-bit A value is less than the 2-bit B value. EQ will be HIGH if the two 2-bit values are equal. The desired 3-output function is defined in the truth table shown in Table 6-1.

A1	A0	B1	B0	GT	LT	EQ
0	0	0	0	0	0	1
0	0	0	1	0	1	0
0	0	1	0	0	1	0
0	0	1	1	0	1	0
0	1	0	0	1	0	0
0	1	0	1	0	0	1
0	1	1	0	0	1	0
0	1	1	1	0	1	0
1	0	0	0	1	0	0
1	0	0	1	1	0	0
1	0	1	0	0	0	1
1	0	1	1	0	1	0
1	1	0	0	1	0	0
1	1	0	1	1	0	0
1	1	1	0	1	0	0
1	1	1	1	0	0	1

Table 6-1 Truth table for Example 6-1

Five different Logic section solutions are given. Each design solution will use the same Subdesign section and will produce the same desired output function. Create a design entry file by adding one of the Logic section solutions to the common Subdesign section and name the file twobit_compare.*tdf*.

The Subdesign section, shown on the next page, declares the two 2-bit inputs using AHDL group notation. A group is a collection of nodes that can be acted upon as a single unit. The group a[1..0] refers to the individual nodes a1 and a0. The group range is enclosed in brackets [ ]. The two dots separate the beginning and end of the inclusive range specified within the brackets. After a group has been defined, the entire group can then be referenced as a[ ].

```
%%%%%%%%%%%%%%%%%%%%%%%%%%%%%%%%
% 2-bit comparator             %
% Designer:  Greg Moss         %
%%%%%%%%%%%%%%%%%%%%%%%%%%%%%%%%

SUBDESIGN twobit_compare        -- declare I/O ports
(
      a[1..0], b[1..0]          :INPUT;      -- group notation
      gt, lt, eq                :OUTPUT;
)
```

The traditional approach to logic design is to develop the Boolean expressions for each output function. We would then commonly try to simplify the expressions using Karnaugh mapping or some other minimizing technique to be able to build a simpler logic circuit. While Quartus has the ability to simplify the Boolean equations for us, it will be much easier to type in simplified expressions. The simplified expressions will save a lot of typing but remember that it also takes time to perform the simplification of the logic expressions by hand! Our first AHDL solution will use simplified SOP expressions for GT and LT that have been obtained from Karnaugh mapping. Then EQ can be written as a function of GT and LT because the output functions are mutually exclusive (i.e., one, and only one, output can be HIGH at a time – see Table 6-1). The simplified equations are:

$$GT = A1\ \overline{B1} + A0\ \overline{B1}\ \overline{B0} + A1\ A0\ \overline{B0}$$

$$LT = \overline{A1}\ B1 + \overline{A0}\ B1\ B0 + \overline{A1}\ \overline{A0}\ B0$$

$$EQ = \overline{GT}\ \overline{LT}$$

These equations are entered into the Logic section of the AHDL design file for solution 1 (see next page) between the keywords BEGIN and END. Note the use of the AHDL symbols for AND, OR, and NOT logic operators. You must end each of the logic expressions with a semicolon in the TDF file. While this approach illustrates how AHDL can use the same Boolean expressions that we might use to create a schematic solution, you might also notice that the overall function of the design is not very clear! Other solution examples will show us that there are much better ways to describe the logic function in AHDL.

```
% solution 1 Logic section %

BEGIN
            -- functions defined with Boolean expressions
      gt = a1 & !b1 # a0 & !b1 & !b0 # a1 & a0 & !b0;
      lt = !a1 & b1 # !a0 & b1 & b0 # !a1 & !a0 & b0;
      eq = !gt & !lt;
%     Boolean symbols
            &  AND
            #  OR
            !  NOT       %
END;
```

The second example Logic section solution also uses logic expressions, but this time we will compare the values of the 2-bit numbers using greater than, less than, and equal to operations. Remember the [ ] notation for a group name represents all of the members of the named group. If the comparison test for an output function is true, then the corresponding output will be HIGH.

```
% solution 2 Logic section %

BEGIN
      gt = a[] > b[];           -- gt is high if a > b
      lt = a[] < b[];           -- lt is high if a < b
      eq = a[] == b[];          -- eq is high if a = b
END;
```

The third solution uses an IF/THEN statement in the Logic section to compare the values of the 2-bit numbers. When the expression following IF is true, the behavior following the keyword THEN will be applied. If the first expression is false, then the expression following ELSIF is evaluated. If the second expression is true, the behavior following the second THEN will be applied. If both of the test expressions are false, then the behavior following ELSE will be applied. Note that the IF/THEN/ELSIF statement will automatically establish a priority. The first expression that is evaluated to be true will determine the resulting behavior. Any test expressions listed after the first true one do not matter.

```
% solution 3 Logic section %

BEGIN
      IF      a[] > b[]         -- 1st expression tested
            THEN gt = VCC; lt = GND; eq = GND;
      ELSIF  a[] < b[]          -- 2nd expression tested
            THEN gt = GND; lt = VCC; eq = GND;
      ELSE        gt = GND; lt = GND; eq = VCC;
      END IF;
END;
```

The fourth solution uses a truth table format to define the three outputs. The input and output signals are given on the first line after the keyword TABLE. Multiple input and output signals are separated by commas. The inputs and outputs in the table are separated by the **=>** symbol. The table ends with END TABLE.

```
% solution 4 Logic section %

BEGIN
      TABLE        -- truth table format
            a[],    b[]         =>     gt, lt, eq;
            B"00", B"00"        =>     0,  0,  1;
            B"00", B"01"        =>     0,  1,  0;
            B"00", B"10"        =>     0,  1,  0;
            B"00", B"11"        =>     0,  1,  0;
            B"01", B"00"        =>     1,  0,  0;
            B"01", B"01"        =>     0,  0,  1;
            B"01", B"10"        =>     0,  1,  0;
            B"01", B"11"        =>     0,  1,  0;
            B"10", B"00"        =>     1,  0,  0;
            B"10", B"01"        =>     1,  0,  0;
            B"10", B"10"        =>     0,  0,  1;
            B"10", B"11"        =>     0,  1,  0;
            B"11", B"00"        =>     1,  0,  0;
            B"11", B"01"        =>     1,  0,  0;
            B"11", B"10"        =>     1,  0,  0;
            B"11", B"11"        =>     0,  0,  1;
      END TABLE;
END;
```

The fifth solution (on the next page) uses a CASE statement to define the three output functions. A CASE statement allows us to test the value of an input and, then based on that value, make an assignment to one or more output signals. With this example problem, we actually have to consider two 2-bit inputs (a[1..0] and b[1..0]) when determining which output signal should be HIGH. To take care of this dilemma, this solution includes the optional Variable section that defines a set of 4 internal nodes that are arbitrarily named input_comb[3..0]. They are described as "buried" nodes because they do not connect to any input or output ports on the subdesign block. Ports for the subdesign (which are PLD pins at the top level) must be declared in the Subdesign section. Internal bits are declared in the Variable section.

The 4 buried nodes are defined in the Logic section using a logic expression that concatenates the two 2-bit inputs. To concatenate means to join them together. The expression

```
input_combo[3..0] = (a[1..0],b[1..0]);
```

is equivalent to the set of expressions

```
input_combo3 = a1      input_combo2 = a0
input_combo1 = b1      input_combo0 = b0;
```

The signal to be tested by the CASE statement is the internal variable named input_combo[ ]. This signal name is placed between the keywords CASE and IS. All possible 4-bit input values for input_combo[ ] are listed in a set of WHEN clauses. The values for input_combo[ ] are represented in binary as indicated by the notation B"xxxx" in the list. Each input condition will produce a set of outputs that are defined by the three assignment statements that follow the **=>** symbol. The output result will, therefore, be dependent upon the input values for a[1..0] and b[1..0]. Each output assignment statement ends with a semicolon. Each of the three outputs is a single bit. Single-bit HIGH and LOW logic levels are represented in AHDL by VCC and GND, respectively.

```
% solution 5 Variable & Logic sections %

VARIABLE
      input_combo[3..0]        :NODE;       -- "buried" nodes

BEGIN
      input_combo[3..0] = (a[1..0],b[1..0]);     -- concatenate

      CASE  input_combo[]  IS        -- evaluates on input_combo
         WHEN B"0000"   =>   gt = GND; lt = GND; EQ = VCC;
         WHEN B"0001"   =>   gt = GND; lt = VCC; EQ = GND;
         WHEN B"0010"   =>   gt = GND; lt = VCC; EQ = GND;
         WHEN B"0011"   =>   gt = GND; lt = VCC; EQ = GND;
         WHEN B"0100"   =>   gt = VCC; lt = GND; EQ = GND;
         WHEN B"0101"   =>   gt = GND; lt = GND; EQ = VCC;
         WHEN B"0110"   =>   gt = GND; lt = VCC; EQ = GND;
         WHEN B"0111"   =>   gt = GND; lt = VCC; EQ = GND;
         WHEN B"1000"   =>   gt = VCC; lt = GND; EQ = GND;
         WHEN B"1001"   =>   gt = VCC; lt = GND; EQ = GND;
         WHEN B"1010"   =>   gt = GND; lt = GND; EQ = VCC;
         WHEN B"1011"   =>   gt = GND; lt = VCC; EQ = GND;
         WHEN B"1100"   =>   gt = VCC; lt = GND; EQ = GND;
         WHEN B"1101"   =>   gt = VCC; lt = GND; EQ = GND;
         WHEN B"1110"   =>   gt = VCC; lt = GND; EQ = GND;
         WHEN B"1111"   =>   gt = GND; lt = GND; EQ = VCC;
      END CASE;
END;
```

See Quartus Tutorial 4 – HDL available from the publisher's textbook web site at www.pearsonhighered.com/electronics for procedures.

Example 6-2 (*Tutorial 4*)

Design and construct a 2421-BCD-to-5421-BCD code converter. The truth table for this design is given in Table 6-2. The inputs are labeled d c b a and the outputs are labeled p q r s. Note that in this situation, we care about only 10 of the 16 possible input combinations. The other 6 input combinations are listed at the bottom of the truth table and are labeled as "invalid." Each of the invalid input conditions should be given a default output value of 1111.

Decimal	2	4	2	1	5	4	2	1	← weights
Value	d	c	b	a	p	q	r	s	← bit names
0	0	0	0	0	0	0	0	0	
1	0	0	0	1	0	0	0	1	
2	0	0	1	0	0	0	1	0	
3	0	0	1	1	0	0	1	1	
4	0	1	0	0	0	1	0	0	
5	1	0	1	1	1	0	0	0	
6	1	1	0	0	1	0	0	1	
7	1	1	0	1	1	0	1	0	
8	1	1	1	0	1	0	1	1	
9	1	1	1	1	1	1	0	0	
invalid	0	1	0	1	1	1	1	1	
invalid	0	1	1	0	1	1	1	1	
invalid	0	1	1	1	1	1	1	1	
invalid	1	0	0	0	1	1	1	1	
invalid	1	0	0	1	1	1	1	1	
invalid	1	0	1	0	1	1	1	1	

Table 6-2 Truth table for Example 6-2

Five AHDL solutions are presented here. An identical Subdesign section (shown below) is used for each solution. In the Subdesign section, the 4 input bits and 4 output bits are declared. All example solutions produce the same results.

AHDL

```
%
 Subdesign Section for all solutions to Example 6-2
 Designer:  Greg Moss
%

SUBDESIGN  code_conv
(
      d, c, b, a          :INPUT;
      p, q, r, s          :OUTPUT;
)
```

A truth table is used in the first TDF design file solution. The DEFAULTS statement is used to specify the output produced if the input combination is not listed in the table. The output bits have been grouped together by listing each output separated by commas and enclosed in parentheses. It is important to note that the bits are assigned in the order that they are listed. An AHDL truth table is defined in the Logic section between the TABLE and END TABLE lines. The first line following TABLE declares the input and output variables, which are again grouped together in the order specified. The table's inputs and outputs are separated by the **=>** symbol.

```
% solution 1 Logic section %

BEGIN
      DEFAULTS
            (p,q,r,s) = B"1111";     -- default outputs
      END DEFAULTS;

      TABLE        -- truth table defines valid BCD inputs
            (d,c,b,a)     =>     (p,q,r,s);
        %     2421 BCD             5421 BCD     %
            B"0000"      =>     B"0000";
            B"0001"      =>     B"0001";
            B"0010"      =>     B"0010";
            B"0011"      =>     B"0011";
            B"0100"      =>     B"0100";
            B"1011"      =>     B"1000";
            B"1100"      =>     B"1001";
            B"1101"      =>     B"1010";
            B"1110"      =>     B"1011";
            B"1111"      =>     B"1100";
      END TABLE;
END;
```

The second solution has declared two sets of 4-bit nodes in the VARIABLE section. These variables, codein[3..0] and codeout[3..0], are expressed as group names. The variables are defined in the logic section. The line, codein[] = (d, c, b, a), will assign the input bit d to the variable codein[3], the input bit c to the variable codein[2], the input bit b to the variable codein[1], and the input bit a to the variable codein[0]. The action of grouping the bits together with the notation (d, c, b, a) is referred to as concatenation. Likewise, the output bits p, q, r, and s are assigned to the corresponding bit in the variable codeout[3..0]. Each bit is assigned, in order, from left to right. An IF/THEN statement is used to define the desired function in the second solution. Each of the IF or ELSIF lines in the TDF file checks for a different valid input combination using the test for equality (==) operator. The appropriate output behavior for the variable codeout[] follows the corresponding THEN. If all of the tests fail (due to an invalid input code), then ELSE will produce the appropriate output behavior.

```
% solution 2  Logic section %
VARIABLE
      codein[3..0]            :NODE;
      codeout[3..0]           :NODE;

BEGIN
      codein[] = (d, c, b, a); -- concatenates inputs
      (p, q, r, s) = codeout[]; -- connect output ports

      -- select appropriate output with IF statement
      IF    codein[] == B"0000"     THEN  codeout[] = B"0000";
      ELSIF codein[] == B"0001"     THEN  codeout[] = B"0001";
      ELSIF codein[] == B"0010"     THEN  codeout[] = B"0010";
      ELSIF codein[] == B"0011"     THEN  codeout[] = B"0011";
      ELSIF codein[] == B"0100"     THEN  codeout[] = B"0100";
      ELSIF codein[] == B"1011"     THEN  codeout[] = B"1000";
      ELSIF codein[] == B"1100"     THEN  codeout[] = B"1001";
      ELSIF codein[] == B"1101"     THEN  codeout[] = B"1010";
      ELSIF codein[] == B"1110"     THEN  codeout[] = B"1011";
      ELSIF codein[] == B"1111"     THEN  codeout[] = B"1100";
      ELSE                                codeout[] = B"1111";
      END IF;
END;
```

The third solution uses a CASE statement to define the desired function. Again the variables codein[3..0] and codeout[3..0] are declared for our convenience to be able to group the inputs and outputs together. The input whose value is to be tested by the CASE statement is sandwiched between the keywords CASE and IS. The valid input codes are listed after each WHEN choice, followed by the => symbol and the desired output behavior for codeout[]. The invalid input codes are covered by the WHEN OTHERS (for any input combination not already tested) clause. All input combinations must be specified in a CASE statement.

```
% solution 3 %
VARIABLE
      codein[3..0]                  :NODE;
      codeout[3..0]                 :NODE;

BEGIN
      codein[] = (d, c, b, a); -- concatenates inputs
      (p, q, r, s) = codeout[]; -- connect output ports

      -- select appropriate output with CASE
      CASE  codein[]  IS
            WHEN B"0000"      =>    codeout[] = B"0000";
            WHEN B"0001"      =>    codeout[] = B"0001";
            WHEN B"0010"      =>    codeout[] = B"0010";
            WHEN B"0011"      =>    codeout[] = B"0011";
            WHEN B"0100"      =>    codeout[] = B"0100";
            WHEN B"1011"      =>    codeout[] = B"1000";
            WHEN B"1100"      =>    codeout[] = B"1001";
            WHEN B"1101"      =>    codeout[] = B"1010";
            WHEN B"1110"      =>    codeout[] = B"1011";
            WHEN B"1111"      =>    codeout[] = B"1100";
            WHEN OTHERS       =>    codeout[] = B"1111";
      END CASE;
END;
```

The fourth solution uses an IF/THEN statement. The first test determines if the input is less than or equal to the binary number for 4. A true result for this test should produce a 4-bit output that is the same as the input. The next test is only applied if the first test is false and will check to see if the input is greater than or equal to the binary number 1011 (which actually represents the decimal value for 5 in the 2421 input code). A true condition for the second test will produce an output that mathematically will be 3 less than the input number that is applied. If both of the tests fail (due to an invalid input code), then ELSE will produce the appropriate output of 1111.

```
% solution 4 Logic section %
VARIABLE
      codein[3..0]                  :NODE;
      codeout[3..0]                 :NODE;

BEGIN
      codein[] = (d, c, b, a); -- concatenates inputs
      (p, q, r, s) = codeout[]; -- connect output ports

   IF    codein[] <= B"0100"   THEN  codeout[] = in[];
   ELSIF codein[] >= B"1011"   THEN  codeout[] = codein[] - 3;
   ELSE                              codeout[] = B"1111";
   END IF;
END;
```

The fifth solution uses Boolean expressions derived with Karnaugh mapping. While this AHDL solution will produce the desired circuit operation, notice that the actual behavior of the design is not obvious with this technique. That is a significant drawback to this solution because the major objective in using HDL languages is to clearly represent the behavior of a logic design. Additionally, this technique requires a great deal more work by the designer! This suggests that traditional design techniques such as using Karnaugh mapping to reduce logic expressions may not be the most desirable approach when using a hardware description language for design entry.

```
% solution 5 Logic section %

BEGIN
            -- define outputs with Boolean expressions
      out3 = in3 # in2 & in0 # in2 & in1;
      out2 = !in3 & in2 # in3 & !in2 & !in1
            # in3 & !in2 & !in0 # in2 & in1 & in0;
      out1 = in1 & !in0 # !in3 & in1 # in2 & !in1 & in0
            # in3 & !in2 & !in1;
      out0 = !in3 & in0 # in3 & !in0 # in3 & !in2 & !in1
            # !in3 & in2 & in1;
END;
```

See Quartus Tutorial 4 – HDL available from the publisher's textbook web site at www.pearsonhighered.com/electronics for procedures.

Example 6-3

Design a 4-channel data selector (multiplexer). This is a standard type of logic function that we will be studying in more detail later. There will be four data inputs or channels, which we will name D0 through D3. The desired data input channel is selected with the input controls S1 and S0 and the multiplexer has an enable input named EN. Verify the design with a functional simulation.

Four different design solutions (with the same output) will be created in AHDL files. The 4-bit data input d[3..0] and 2-bit control input s[1..0] are declared using AHDL group notation in the common Subdesign section given below.

```
%
 Subdesign Section for all solutions to Example 6-3
 Designer:  Greg Moss
%

SUBDESIGN multiplexer
(
      en, d[3..0], s[1..0]          :INPUT;
      y                             :OUTPUT;
)
```

In the first solution (see next page), Boolean equations are used. A multiplexer is a common logic function, so its logic expression will become familiar to us. For now, we will give the expression as:

$$Y = (D0\ \overline{S1}\ \overline{S0} + D1\ \overline{S1}\ S0 + D2\ S1\ \overline{S0} + D3\ S1\ S0)\ EN$$

A set of buried nodes that will be used to detect (decode) the four data selection combinations is set up in the Variable section. The equations that actually detect the four possible data selection combinations are written in the Logic section. The final multiplexer output expression is also defined in the Logic section. Note that we have used the intermediate variables to make the output expression easier (for humans) to interpret – it doesn't really matter to the computer. This is an example where describing a logic function using Boolean expressions can be very useful.

```
% solution 1 Variable & Logic section %

VARIABLE
    sel[3..0]      :NODE;        -- "buried" nodes

BEGIN

    sel0 = !s1 & !s0;            -- define 4 select combinations
    sel1 = !s1 &  s0;
    sel2 =  s1 & !s0;
    sel3 =  s1 &  s0;

    -- Boolean expression to route selected input to y
    y = (d0 & sel0 # d1 & sel1 # d2 & sel2 # d3 & sel3) & en;

END;
```

The second example solution uses an IF/THEN statement. The IF statement tests to determine which input combination for s1 and s0 is applied while the en input is HIGH. The notation s[ ] represents the entire group of inputs s[1..0]. The double equal sign is the test for equality symbol. The ELSIFs produce a prioritized set of further conditions for testing. If all other tests should fail (because en is LOW), the ELSE clause gives a default output result *y* = *GND* (logic 0).

```
% solution 2 Logic section %

BEGIN
            -- IFs 1st determine desired output when enabled
      IF      s[] == 0 & en  THEN    y = d0;
      ELSIF   s[] == 1 & en  THEN    y = d1;
      ELSIF   s[] == 2 & en  THEN    y = d2;
      ELSIF   s[] == 3 & en  THEN    y = d3;
      ELSE  y = GND;                -- output if disabled
      END IF;
END;
```

The third solution is similar to the second except that it uses nested IF statements instead of a logic expression for the test condition. The outer condition (en is HIGH) must be true before it is necessary to evaluate the inner IF and determine which input combination for s1 and s0 is applied.

```
% solution 3 Logic section %

BEGIN
                                -- uses nested IFs
      IF en THEN                -- evaluate 2nd IF when 1st one true
            IF      s[] == 0  THEN    y = d0;
            ELSIF   s[] == 1  THEN    y = d1;
            ELSIF   s[] == 2  THEN    y = d2;
            ELSE                      y = d3;
            END IF;
      ELSE   y = GND;           -- disabled output
      END IF;
END;
```

The fourth solution is a modification of the third. Nested inside the IF statement that tests the condition of en is a CASE statement. Between keywords CASE and IS will be the expression s[1..0] that is to be evaluated by the CASE statement. The alternative values for the evaluated expression are listed in the WHEN clauses. Following the WHEN clauses (after the **=>** symbol) are the statements that are to be activated according to the value matched for the expression.

```
% solution 4 Logic section %

BEGIN
                                 -- CASE nested inside IF statement
      IF en THEN                 -- evaluate CASE when enable is true
            CASE s[]  IS
                  WHEN  0   =>   y = d0;
                  WHEN  1   =>   y = d1;
                  WHEN  2   =>   y = d2;
                  WHEN  3   =>   y = d3;
            END CASE;
      ELSE   y = GND;            -- output when disabled
      END IF;
END;
```

Functional simulation results are shown in Fig. 6-3. When en is HIGH, the d[3..0] input selected by s[1..0] will be routed to the y output. If en is LOW, the y output will always be LOW.

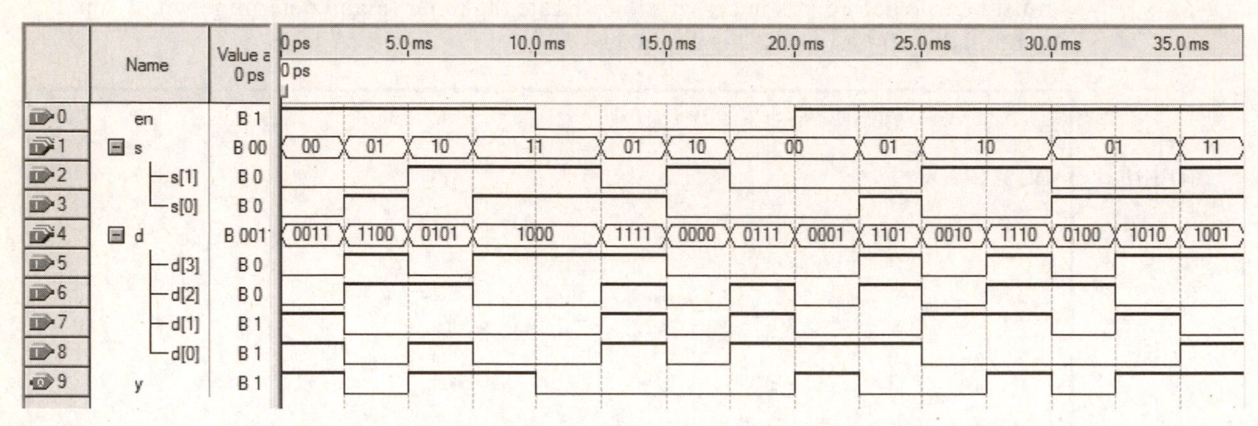

Fig. 6-3 Functional simulation results for Example 6-3

Laboratory Projects

Design PLD logic circuits for the following applications using AHDL. Provide comments in the design files and include your name in an information header. Compile the TDF file and functionally simulate your design to verify it. Program a PLD with your design and test it in the lab.

6A.1 2-bit comparator
Test the design given in Example 6-1 (Tutorial).

6A.2 Code converter
Test the design given in Example 6-2 (Tutorial).

6A.3 Modified data selector
Modify the design given in Example 6-3 to create a 2-channel data selector. The two data inputs should be named D0 and D1. The data input channel will be selected with the input control S. The data selector also needs to have an active-LOW enable named EN.

6A.4 Gray-code-to-binary conversion
Design a logic circuit that will convert a 6-bit Gray code value into its equivalent binary value. The Gray code is commonly used in shaft position encoders to decrease the possibility of errors. The Gray code is an unweighted code in which only a single bit changes from one code number to the next. For example, the Gray code value 100110 is equivalent to the binary value 111011. The Gray-to-binary code conversion algorithm is as follows:

1. *The most significant bit in the binary result is the same as the corresponding most significant (leftmost) Gray code bit.*
2. *To produce each additional binary bit, XOR the binary code bit just generated to the Gray code bit in the next adjacent position.*
3. *Repeat the process in step 2 to produce the binary result through the least significant bit.*

Write Boolean expressions for each of the binary outputs. Simulation hint: Use a Gray code count for the input waveform.

6A.5 BCD-to-binary converter

Design a logic circuit that will convert a 5-bit BCD input bcd[4..0] into its equivalent 4-bit binary value bin[3..0]. Since the output is only 4 bits long, the largest number that can be converted is the decimal value 15. This converter circuit should also have an output called err that will be HIGH if the BCD input value is not in the range of 0 through 15. Use a CASE statement to define the valid input conditions and the resultant output (bin[3..0], err) assignments. Use the WHEN OTHERS phrase to define the outputs (bin[3..0] = B"0000" and err = VCC) for any invalid input combination.

6A.6 Binary-to-BCD converter

Design a logic circuit that will convert a 5-bit binary input bin[4..0] into its equivalent 2-digit (only 6 output bits are needed) BCD value bcd[5..0]. The converter can handle numbers from 0_{10} through 31_{10}. Use a truth table to define this function.

6A.7 Tens digit detector

Design a logic circuit that will detect the equivalent tens digit value (0 through 6) for a 6-bit input number num[5..0]. The input values will range from 0 through 63 with 6 bits. Only one of the seven outputs (tens[6..0]) will be HIGH at a time, indicating the value of the tens digit for the current input number. Use IF/THEN (and ELSIF/THEN) statements for your design. IF/THEN syntax automatically creates a priority of behavior since the first IF expression that evaluates to be true will determine the behavioral statement that will be applied. Any other IF clauses will be ignored after the first one that is true. Hint: Use the <= comparator operator in the Boolean expression to be evaluated by each IF statement line, and the behavioral statement following THEN will be the appropriate assignment for the outputs (e.g., tens[ ] = B"0000001").

6A.8 Lamp display

Design a logic circuit that will light the number of lamps corresponding to the 3-bit input value that is applied (0 through 7). The inputs are named in[2..0] and the outputs are named out[7..1]. The following table describes the operation. Use a CASE statement for this design.

in2	in1	in0	out7	out6	out5	out4	out3	out2	out1
0	0	0	0	0	0	0	0	0	0
0	0	1	0	0	0	0	0	0	1
0	1	0	0	0	0	0	0	1	1
0	1	1	0	0	0	0	1	1	1
1	0	0	0	0	0	1	1	1	1
1	0	1	0	0	1	1	1	1	1
1	1	0	0	1	1	1	1	1	1
1	1	1	1	1	1	1	1	1	1

6A.9 Programmable logic unit

Design a programmable logic circuit that will perform one of four logic operations on two 4-bit inputs a[3..0] and b[3..0] when enabled by a signal named en. The desired logic function, selected by inputs s[1..0], is indicated in the following table. The 4-bit output is named f[3..0]. If the enable is LOW, controls s[1..0] do not matter and the outputs are all LOW. Each of the output bits will be a function of the corresponding a-bit and b-bit and the s1 and s0 control inputs. For example, f3 is dependent on a3, b3, s1, and s0, while f2 is dependent on a2, b2, s1, and s0. Use an IF/ELSIF statement to define this function. Hint: The behavioral expressions can be written just like the table.

en	s1	s0	Operation
1	0	0	f[ ] = a[ ] # b[ ]
1	0	1	f[ ] = a[ ] & b[ ]
1	1	0	f[ ] = a[ ] $ b[ ]
1	1	1	f[ ] = !a[ ]
0	X	X	f[ ] = B"0000"

6A.10 Number range detector

Design a logic circuit that, when enabled with an active-HIGH signal named en, will detect four different ranges of values for a 5-bit input number. The inputs are labeled num4 through num0 and the outputs are labeled range1 through range4. Range1, range2, and range3 each produce active-HIGH outputs, while range4 produces an active-LOW output signal. Write Boolean expressions for each of the outputs. Hint: Use the greater-than-or-equal-to (>=) and less-than-or-equal-to (<=) symbols in the Boolean expression to detect the desired range of input values for each output.

Outputs	Values detected
range1	4-12
range2	15-20
range3	18-24
range4	26-30

6A.11 Data switcher

Design a programmable logic circuit that will route two input data bits to the selected outputs as given in the function table below. The control inputs (s[1..0]) determine which of the two data inputs (in1 and in0) are to be routed to each of the two outputs (out1 and out0). The input en is an active-LOW enable for the data switcher. If the enable is HIGH, we don't care what input levels are on s[1..0], and the two outputs will both be LOW. Use a CASE statement inside an IF/THEN statement.

en	s1	s0	out1	out0
0	0	0	in1	in0
0	0	1	in0	in0
0	1	0	in1	in1
0	1	1	in0	in1
1	X	X	0	0

UNIT 6V

COMBINATIONAL CIRCUIT DESIGN WITH VHDL

Objectives

- To create combinational logic circuits using VHDL in Quartus II.
- To simulate VHDL combinational logic circuits in Quartus II.

VHSIC Hardware Description Language (VHDL)

VHSIC (Very High Speed Integrated Circuit) Hardware Description Language (VHDL) is a text-entry, HIGH-level language that can be used to describe complex digital circuits. VHDL files (with the filename extension .vhd) can be created to describe any portion of a project's hierarchy. This text-based design entry technique for a desired logic function can be used to more clearly describe the circuit's behavior than can be commonly achieved with a traditional logic schematic. The VHDL language supports various design description techniques including Boolean expressions and if/then and case statements. In VHDL, every logic circuit description consists of at least two design units, an entity declaration and an architecture declaration. The entity declaration is used to identify the input and output ports for the design entity. The architecture declaration will describe the logical operation of the entity. Every entity of a design must be bound to a matching architecture. Architecture declarations in VHDL can contain concurrent statements and sequential statements. All concurrent statements are executed at the same time so that there is no significance to the order in which they are listed. Sequential statements, however, are executed one after another in the order that they appear in the architecture body. It is important to note that the function specified by sequential statements (referred to as a process) will be executed concurrently with other concurrent statements.

VHDL

```
-- Simple circuit to illustrate VHDL design format
-- Designer:  Greg Moss

ENTITY simple IS                  -- starts entity declaration
      PORT                        -- port list follows
      (                           -- starts list of ports
            a     :IN BIT;        -- declare a input
            b     :IN BIT;        -- declare b input
            c     :IN BIT;        -- declare c input
            y     :OUT BIT        -- output listed
      );                          -- ends list of ports
END simple;                       -- ends entity declaration

ARCHITECTURE example OF simple IS    -- starts architecture
SIGNAL      p     :BIT; -- defines a "buried" node
BEGIN                             -- begins body of architecture
      p <= a AND NOT b;           -- assignment statement for p
      y <= p OR c;                -- assignment statement for y
END example;                      -- ends architecture
```

Fig. 6-1 VHDL design file for a simple circuit

A simple VHDL file (simple.*vhd*) for a combinational logic circuit is shown in Fig. 6-1. VHDL keywords are given in capital letters. While VHDL is not case-sensitive, it is a recommended practice to capitalize keywords for readability. Comments can be placed in the VHDL text file by preceding them with double-dashes **--** (with the comment then running to the end of the line). In this example, we have created a brief explanation of the function of this design file and identified the designer in a "header" using comments. Individual lines have also been commented with the descriptions that follow the double-dash -- comments. **A good design file will be liberally commented for reader clarity.**

The entity declaration starts with the line *ENTITY simple IS* and continues to the line *END simple*;. The name of the entity, simple, is given in both lines. The input and output ports for the entity are listed in the entity declaration and are enclosed in parentheses. Each port for the entity is identified with a name, mode (signal direction), and a type. The port names for simple.*vhd* are a, b, c, and y. A colon separates the port names from the port's mode and data type. The ports' modes are defined to be IN or OUT according to the signal directions for this entity. A port whose mode is IN can be read within the architecture declaration for this entity but cannot be assigned a value. On the other hand, a port declared to be of mode OUT can only be assigned a value within the architecture declaration and cannot be read. Each of the signals for simple.*vhd* is declared to have a data type of bit, which means that they can only have two possible values: '0' or '1.' A semicolon ends each logical line.

The architecture declaration starts with the line *ARCHITECTURE example OF simple IS* and continues to the line *END example*. The entity (simple) to which this architecture is bound is identified between the words OF and IS. VHDL permits multiple architecture declarations for an entity and, therefore, requires the architecture to be named (example, in this architecture). The keyword SIGNAL allows us to declare the name and type of buried node signals that only exist within the entity. The architecture body is sandwiched between keywords BEGIN and END and describes the functionality for this entity. Two signal assignment statements are given in the body of this architecture. Because these assignment statements are concurrent, the order in which they are listed does not matter. The buried node p is the input a ANDed with NOT b. The output port y is node p ORed with input c.

VHDL design units (.vhd files) can be included in a project hierarchy together with other design file types (such as *.bdf* files). Breaking a design into smaller, more manageable blocks can be extremely helpful in creating larger systems. *A very important rule to remember when using multiple design blocks to define a system is that each design block must have a unique file name.* A built-in text editor within Quartus II allows the user to easily create and edit the *.vhd* files. The development software treats *.vhd* files the same as other design entry files. PLD design tasks such as syntax checking, accessing on-line help, compiling, and simulation are all available when using VHDL files. Fundamental language elements of VHDL are summarized in Figs. 6-2a and 6-2b.

VHDL LANGUAGE ELEMENTS

Reserved words

ABS	FOR	PACKAGE
ACCESS	FUNCTION	PORT
AFTER	GENERATE	PROCEDURE
ALIAS	GENERIC	PROCESS
ALL	GUARDED	RANGE
AND	IF	RECORD
ARCHITECTURE	IN	REGISTER
ARRAY	INOUT	REM
ASSERT	IS	REPORT
ATTRIBUTE	LABEL	RETURN
BEGIN	LIBRARY	SELECT
BLOCK	LINKAGE	SEVERITY
BODY	LOOP	SIGNAL
BUFFER	MAP	SUBTYPE
BUS	MOD	THEN
CASE	NAND	TO
COMPONENT	NEW	TRANSPORT
CONFIGURATION	NEXT	TYPE
CONSTANT	NOR	UNITS
DISCONNECT	NOT	UNTIL
DOWNTO	NULL	USE
ELSE	OF	VARIABLE
ELSIF	ON	WAIT
END	OPEN	WHEN
ENTITY	OR	WHILE
EXIT	OTHERS	WITH
FILE	OUT	XOR

Fig. 6-2a Summary of VHDL language elements

VHDL LANGUAGE ELEMENTS

Identifiers

- An identifier is the name of an object
- Objects are named entities that can be assigned a value and have a specific data type
- Objects include signals, variables, and constants
- Must start with an alphabetic character and end with an alphabetic or a numeric character
- Can contain numeric or underscored characters and cannot contain spaces
- Are not case sensitive
- May be up to 32 characters long
- Cannot contain any reserved words

Symbols

Symbol	Description
--	begins comment (to end of line)
( )	encloses port names in entity declaration; encloses highest priority operations in Boolean and arithmetic expressions
' '	encloses scalar values
" "	encloses array values
;	ends VHDL statements and declarations
,	separates objects
:	separates object identifier names from mode and data type in declarations
<=	assigns values in signal assignment statements
:=	assigns values in variable assignment statements or to constants
=>	separates signal assignment statements from WHEN clause in CASE statements
+	addition operator
–	subtraction operator
=	equality operator
/=	inequality operator
>	greater than comparator
>=	greater than or equal to comparator
<	less than comparator
<=	less than or equal to comparator
&	concatenation operator

Synthesis Data Types

Type	Description
BIT	object can only have single-bit values of '0' or '1'
STD_LOGIC	object with multi-value logic including '0', '1', 'X', 'Z'
INTEGER	objects with whole number (decimal) values, e.g., 54, –21
BIT_VECTOR	objects with arrays of bits such as "10010110"
STD_LOGIC_VECTOR	objects with arrays of multi-value logic, e.g., "01101XX"

Fig. 6-2b Summary of VHDL language elements

Example 6-1 (*Tutorial 4*)

Design a magnitude comparator circuit to compare the magnitudes of two 2-bit numbers (A1 A0 and B1 B0). The circuit will have three output signals: GT, LT, and EQ. GT will be HIGH to indicate that the 2-bit A value is greater than the 2-bit B value. LT will be HIGH if the 2-bit A value is less than the 2-bit B value. EQ will be HIGH if the two 2-bit values are equal. The desired 3-output function is defined in the truth table shown in Table 6-1.

A1	A0	B1	B0	GT	LT	EQ
0	0	0	0	0	0	1
0	0	0	1	0	1	0
0	0	1	0	0	1	0
0	0	1	1	0	1	0
0	1	0	0	1	0	0
0	1	0	1	0	0	1
0	1	1	0	0	1	0
0	1	1	1	0	1	0
1	0	0	0	1	0	0
1	0	0	1	1	0	0
1	0	1	0	0	0	1
1	0	1	1	0	1	0
1	1	0	0	1	0	0
1	1	0	1	1	0	0
1	1	1	0	1	0	0
1	1	1	1	0	0	1

Table 6-1 Truth table for Example 6-1

Four different VHDL solutions will be given. Each one is named twobit_compare (with a *.vhd* file extension) and will produce the same desired output function. The entity declaration defines the input and output ports. Two of the solutions will use the BIT_VECTOR data type for the inputs, and the other two will use INTEGER for the data type. The BIT_VECTOR data type is only an array of bits and does not have a numerical value associated with it like an INTEGER data type does. When declaring a BIT_VECTOR data type, the range of the array is also specified. The range indicates the number of bits in the array and their order (either ascending or descending). An INTEGER data type will also have a RANGE of allowable values specified. The RANGE indicates to the compiler how many bits will be necessary to represent the declared object.

For the first solution, shown on the next page, the ports a and b are mode IN (input direction) and are two bit arrays (BIT_VECTOR data type with indexes of 1 DOWNTO 0). The bits in the array are specified to be in descending order with the word DOWNTO. The ports gt, lt, and eq are mode OUT (output direction) and are a single BIT data type. The ports for the design are enclosed in parentheses after the keyword PORT. Each of the logical lines in the entity declaration are separated by a semicolon. The last port listed does not have a semicolon separator, but there is one after the right-hand parenthesis to end the port declaration.

The traditional approach to logic design is to develop the Boolean expressions for each output function. We would then commonly try to simplify the expressions using Karnaugh mapping or some other minimizing technique to be able to build a simpler logic circuit. Karnaugh mapping was used to determine the Boolean equations for GT and LT in this solution (*ARCHITECTURE soln1 OF twobit_compare IS*). The function for EQ, on the other hand, is expressed using the XNOR function (NOT XOR in VHDL) to detect if the corresponding A and B bits are equivalent. You may recall that the XNOR function will detect when its two inputs are equivalent. The equations are:

$$GT = A1\ \overline{B1} + A0\ \overline{B1}\ \overline{B0} + A1\ A0\ \overline{B0}$$

$$LT = \overline{A1}\ B1 + \overline{A0}\ B1\ B0 + \overline{A1}\ \overline{A0}\ B0$$

$$EQ = \overline{(A1 \oplus B1)}\ \overline{(A0 \oplus B0)}$$

These equations are written as concurrent signal assignment statements in the architecture declaration of the VHDL design file between the words BEGIN and END. Each of the signal assignments ends with a semicolon. Logical operators (AND and OR) in VHDL have equal precedence. Therefore, parentheses are needed to explicitly define the order of precedence for the Boolean functions. You should note that it was not necessary to simplify the logic expressions before creating the VHDL file because Quartus will simplify them for you. The simplified expressions will save a lot of typing but remember that it also takes time to perform the simplification of the logic expressions by hand!

```
ENTITY twobit_compare IS
     PORT
     (
          a     :IN BIT_VECTOR (1 DOWNTO 0);
          b     :IN BIT_VECTOR (1 DOWNTO 0);
                -- a & b inputs are each 2-bit arrays
          gt    :OUT BIT;
          lt    :OUT BIT;
          eq    :OUT BIT
                -- each output is a single bit
     );
END twobit_compare;

ARCHITECTURE soln1 OF twobit_compare IS
BEGIN
          -- Boolean expressions for each output
     gt <= (a(1) AND NOT b(1))
          OR (a(0) AND NOT b(1) AND NOT b(0))
          OR (a(1) AND a(0) AND NOT b(0));

     lt <= (NOT a(1) AND b(1))
          OR (NOT a(0) AND b(1) AND b(0))
          OR (NOT a(1) AND NOT a(0) AND b(0));

     eq <= NOT (a(1) XOR b(1))
          AND NOT (a(0) XOR b(0));
END soln1;
```

In the second solution, the two input ports are both declared to be INTEGER data types with a RANGE of possible values from 0 TO 3. Therefore each of the two input objects a and b will require two bits in the design. The architecture declaration (*ARCHITECTURE soln2 OF twobit_compare IS*) uses an IF/THEN statement to compare the values of the 2-bit numbers. In VHDL this is described as using a behavioral style to define an entity's architecture. The IF/THEN statement is a sequential statement and must be placed within a PROCESS. The signals given in parentheses after the word PROCESS are in the sensitivity list for the process. If any of the signals in a sensitivity list change, then the process will be invoked and the behavior will be evaluated to determine the output for the defined function. The actual process behavior is placed between another BEGIN and END PROCESS keyword set. An IF/THEN statement defines an order or priority in its evaluation. If the expression following IF is true, the behavior following the word THEN will be applied. If the first expression is false, then the expression following ELSIF is evaluated. If the second expression is true, the behavior following the second THEN will be applied. If both of the test expressions are false, then the behavior following ELSE will be applied. The order of the tests in an IF/THEN/ELSIF statement will automatically establish a priority. The first expression that is evaluated to be true will determine the resulting behavior. Any other test expressions after the first true one do not matter. The IF statement is terminated with *END IF;*.

```
ENTITY twobit_compare IS
      PORT
      (
            a     :IN INTEGER RANGE 0 TO 3;
            b     :IN INTEGER RANGE 0 TO 3;
            -- integer inputs will have numerical values
            gt    :OUT BIT;
            lt    :OUT BIT;
            eq    :OUT BIT
      );
END twobit_compare;

ARCHITECTURE soln2 OF twobit_compare IS
BEGIN
      PROCESS (a, b)    -- a & b are in sensitivity list
      BEGIN
      -- 1st IF that tests true makes signal assignments
            IF (a > b) THEN
                  gt <= '1';
                  lt <= '0';
                  eq <= '0';
            ELSIF (a < b) THEN
                  gt <= '0';
                  lt <= '1';
                  eq <= '0';
            ELSE
                  gt <= '0';
                  lt <= '0';
                  eq <= '1';
            END IF;
      END PROCESS;
END soln2;
```

The third VHDL solution (*ARCHITECTURE soln3 OF twobit_compare IS*) uses another form of sequential statement, the CASE statement. The CASE statement will select a behavior for the design unit based on the value of a given expression. A CASE statement is used within a PROCESS. When a signal in the sensitivity list changes, the PROCESS will be invoked. This solution actually has nested CASE statements. The outer CASE statement evaluates the value of the INTEGER input signal a (*CASE a IS*), and the inner CASE statement evaluates the INTEGER value of b (*CASE b IS*). It does not matter which signal is evaluated by the inner CASE or the outer CASE. It also does not matter in what order the values are checked by the WHEN clause of the CASE statement, but all possible values must be covered. Unlike the IF/THEN/ELSE statement, there is no priority for the conditions contained within the CASE statement. The desired action for each WHEN and value choice is given after the symbol =>. The CASE statement chooses one and only one of the alternative signal assignments according to the current value of the signal being evaluated. A CASE statement is terminated with *END CASE;*.

```
ENTITY`twobit_compare IS
      PORT
      (
         a  :IN INTEGER RANGE 0 TO 3;
         b  :IN INTEGER RANGE 0 TO 3;
         -- a & b will have numerical values as integers
         gt :OUT BIT;
         lt :OUT BIT;
         eq :OUT BIT
      );
END twobit_compare;

ARCHITECTURE soln3 OF twobit_compare IS
BEGIN
   PROCESS (a, b)     -- change on a or b invokes process
   BEGIN

      CASE a IS              -- 1st case evaluates a value

         WHEN 0   =>
            CASE b IS        -- 2nd case evaluates b value
               WHEN 0   =>
                  gt <= '0';  lt <= '0';  eq <= '1';
               WHEN 1   =>
                  gt <= '0';  lt <= '1';  eq <= '0';
               WHEN 2   =>
                  gt <= '0';  lt <= '1';  eq <= '0';
               WHEN 3   =>
                  gt <= '0';  lt <= '1';  eq <= '0';
            END CASE;

         WHEN 1   =>
            CASE b IS
               WHEN 0   =>
                  gt <= '1';  lt <= '0';  eq <= '0';
               WHEN 1   =>
                  gt <= '0';  lt <= '0';  eq <= '1';
               WHEN 2   =>
                  gt <= '0';  lt <= '1';  eq <= '0';
               WHEN 3   =>
                  gt <= '0';  lt <= '1';  eq <= '0';
            END CASE;
```

(continued)

(continued)

```
            WHEN 2    =>
               CASE b IS
                  WHEN 0    =>
                     gt <= '1';  lt <= '0';  eq <= '0';
                  WHEN 1    =>
                     gt <= '1';  lt <= '0';  eq <= '0';
                  WHEN 2    =>
                     gt <= '0';  lt <= '0';  eq <= '1';
                  WHEN 3    =>
                     gt <= '0';  lt <= '1';  eq <= '0';
               END CASE;

            WHEN 3    =>
               CASE b IS
                  WHEN 0    =>
                     gt <= '1';  lt <= '0';  eq <= '0';
                  WHEN 1    =>
                     gt <= '1';  lt <= '0';  eq <= '0';
                  WHEN 2    =>
                     gt <= '1';  lt <= '0';  eq <= '0';
                  WHEN 3    =>
                     gt <= '0';  lt <= '0';  eq <= '1';
               END CASE;
         END CASE;
   END PROCESS;
END soln3;
```

The fourth solution (*ARCHITECTURE soln4 OF twobit_compare IS*) will use a selected signal assignment statement. This is a concurrent signal assignment statement that is used to select and assign a value to a specified signal from a list of alternatives. Because SELECT is a concurrent signal assignment statement, it is not placed within a process. The input ports a and b are again given the data type BIT_VECTOR since the architecture declaration does not associate a numerical value to the signals. Two buried node signals named input and output are created in the architecture declaration and are given the data type BIT_VECTOR with array sizes of 4 bits and 3 bits, respectively. The signal input is assigned to be the concatenation of the two input ports a and b. The symbol & is used to indicate that the bits for the signals should be concatenated together. This then gives us a 4-bit value to represent each of the 16 possible input signal combinations, like in a truth table. The SELECT statement starts with the phrase *WITH input SELECT*. The expression to be evaluated by the SELECT statement is placed between the keywords WITH and SELECT in that phrase. The signal assignment statement then follows the WITH … SELECT phrase, with all of the desired signal assignment values to be selected by the set of possible values for the evaluated expression given after each WHEN. You must specify all possible conditions for the evaluated expression. There is no priority implied by a selected signal assignment statement. The bits assigned for the signal output correspond to the individual port output bits gt, lt, and eq. The output port signals are assigned to the respective bit in the array called output.

```
ENTITY twobit_compare IS
      PORT
      (
            a     :IN BIT_VECTOR (1 DOWNTO 0);
            b     :IN BIT_VECTOR (1 DOWNTO 0);
                  -- bit vectors do not have a value
            gt    :OUT BIT;
            lt    :OUT BIT;
            eq    :OUT BIT
      );
END twobit_compare;

ARCHITECTURE soln4 OF twobit_compare IS
SIGNAL      input       :BIT_VECTOR (3 DOWNTO 0);
SIGNAL      output      :BIT_VECTOR (2 DOWNTO 0);
      -- these signals are created for our convenience

BEGIN
      input <= a & b;         -- concatenates input bits

      WITH input SELECT       -- input selects output
            output      <=      "001" WHEN "0000",
                                "010" WHEN "0001",
                                "010" WHEN "0010",
                                "010" WHEN "0011",
                                "100" WHEN "0100",
                                "001" WHEN "0101",
                                "010" WHEN "0110",
                                "010" WHEN "0111",
                                "100" WHEN "1000",
                                "100" WHEN "1001",
                                "001" WHEN "1010",
                                "010" WHEN "1011",
                                "100" WHEN "1100",
                                "100" WHEN "1101",
                                "100" WHEN "1110",
                                "001" WHEN "1111";

      gt <= output(2);        -- connect array bits to ports
      lt <= output(1);
      eq <= output(0);

END soln4;
```

See Quartus Tutorial 4 – HDL available from the publisher's textbook web site at www.pearsonhighered.com/electronics for procedures.

Example 6-2 (*Tutorial 4*)

Design and construct a 2421-BCD-to-5421-BCD code converter. The truth table for this design is given in Table 6-2. The inputs are labeled d c b a and the outputs are p q r s. Note that in this situation, we care about only 10 of the 16 possible input combinations. The other 6 input combinations are listed at the bottom of the truth table and are labeled as "invalid." Each of the invalid input conditions should be given a default output value of 1111.

Decimal	2	4	2	1	5	4	2	1	← weights
Value	d	c	b	a	p	q	r	s	← bit names
0	0	0	0	0	0	0	0	0	
1	0	0	0	1	0	0	0	1	
2	0	0	1	0	0	0	1	0	
3	0	0	1	1	0	0	1	1	
4	0	1	0	0	0	1	0	0	
5	1	0	1	1	1	0	0	0	
6	1	1	0	0	1	0	0	1	
7	1	1	0	1	1	0	1	0	
8	1	1	1	0	1	0	1	1	
9	1	1	1	1	1	1	0	0	
invalid	0	1	0	1	1	1	1	1	
invalid	0	1	1	0	1	1	1	1	
invalid	0	1	1	1	1	1	1	1	
invalid	1	0	0	0	1	1	1	1	
invalid	1	0	0	1	1	1	1	1	
invalid	1	0	1	0	1	1	1	1	

Table 6-2 Truth table for Example 6-2

Four VHDL solutions, each producing the same results, are given here. In each, the 4 input ports (d, c, b, a) and 4 output ports (p, q, r, s) are declared. All signals are declared to be a BIT data type.

A selected signal assignment statement is used in the first VHDL design file solution (*ARCHITECTURE solution1 OF code IS*). Two arrays (data type BIT_VECTOR) of bits named input and output are declared with SIGNAL and are, therefore, visible within the architecture for the entity named code. The signal assignment statement for input concatenates the four input bits together so that they can be treated as a unit. Each of the output port signals is assigned a specific bit in the array named output. The SELECT statement is a concurrent signal assignment statement that will select the correct value to assign to output based on the current set of bits for input. All possible values of the expression to be evaluated (input) must be covered in a SELECT statement. The phrase WHEN OTHERS is used to cover any values that are not explicitly named in the WHEN list. This will take care of any invalid bit pattern that may be applied to input. The order of the choices for the signal being evaluated does not matter because there is no priority implied in a selected signal assignment statement, but all possible values must be accounted for.

```
ENTITY code IS
      PORT
      (
            d, c, b, a          :IN BIT;
            p, q, r, s          :OUT BIT
      );
END code;

ARCHITECTURE solution1 OF code IS
SIGNAL      input         :BIT_VECTOR (3 DOWNTO 0);
SIGNAL      output        :BIT_VECTOR (3 DOWNTO 0);

BEGIN
input <= d & c & b & a;      -- concatenation of input bits

p <= output(3);              -- connect array bit to output
q <= output(2);
r <= output(1);
s <= output(0);

WITH input SELECT            -- input selects output pattern
      output    <=     "0000" WHEN "0000",
                       "0001" WHEN "0001",
                       "0010" WHEN "0010",
                       "0011" WHEN "0011",
                       "0100" WHEN "0100",
                       "1000" WHEN "1011",
                       "1001" WHEN "1100",
                       "1010" WHEN "1101",
                       "1011" WHEN "1110",
                       "1100" WHEN "1111",
                       "1111" WHEN OTHERS;
END solution1;
```

The second solution (*ARCHITECTURE solution2 OF code IS*) uses an IF/THEN statement. Like the first solution, the individual input and output port signals are grouped into arrays named input and output. Because IF/THEN is a sequential statement, it must be placed within a PROCESS. The PROCESS will be invoked whenever the BIT_VECTOR signal in its sensitivity list (input) changes. Each of the IF or ELSIF lines tests for a different valid input combination using the equal to operator. The appropriate output behavior follows the corresponding THEN. If all of the tests should fail (due to an invalid input code), then ELSE will produce the appropriate output behavior. After the PROCESS is evaluated, the signals are all concurrently updated.

```
ENTITY code IS
      PORT
      (
            d, c, b, a          :IN BIT;
            p, q, r, s          :OUT BIT
      );
END code;

ARCHITECTURE solution2 OF code IS
SIGNAL      input       :BIT_VECTOR (3 DOWNTO 0);
SIGNAL      output      :BIT_VECTOR (3 DOWNTO 0);
BEGIN
      input <= d & c & b & a;     -- concatenates inputs
      p <= output(3);             -- output array to ports
      q <= output(2);
      r <= output(1);
      s <= output(0);

      PROCESS (input)    -- change on input invokes process
      BEGIN
          IF    input = "0000"   THEN   output <= "0000";
          ELSIF input = "0001"   THEN   output <= "0001";
          ELSIF input = "0010"   THEN   output <= "0010";
          ELSIF input = "0011"   THEN   output <= "0011";
          ELSIF input = "0100"   THEN   output <= "0100";
          ELSIF input = "1011"   THEN   output <= "1000";
          ELSIF input = "1100"   THEN   output <= "1001";
          ELSIF input = "1101"   THEN   output <= "1010";
          ELSIF input = "1110"   THEN   output <= "1011";
          ELSIF input = "1111"   THEN   output <= "1100";
          ELSE                          output <= "1111";
          END IF;
      END PROCESS;
END solution2;
```

The third solution (*ARCHITECTURE solution3 OF code IS*) uses a CASE statement to define the desired function. The same two buried arrays are created, and signal assignments are made. CASE is a sequential statement and is, therefore, placed inside a PROCESS. If the array of bits named input changes, the PROCESS will be invoked. The signal (input) whose value is to be tested is sandwiched between the keywords CASE and IS. The valid input codes are listed after each WHEN choice, followed by the => symbol and the appropriate signal assignment behavior for output. The invalid input codes are covered by the WHEN OTHERS (for any input combination not already tested) clause. The order of the choices for the signal being evaluated does not matter, but all possible values must be accounted for.

```
ENTITY code IS
      PORT
      (
            d, c, b, a          :IN BIT;
            p, q, r, s          :OUT BIT
      );
END code;

ARCHITECTURE solution3 OF code IS
SIGNAL      input        :BIT_VECTOR (3 DOWNTO 0);
SIGNAL      output       :BIT_VECTOR (3 DOWNTO 0);
BEGIN
      input <= d & c & b & a;    -- concatenates inputs
      p <= output(3);            -- connects output ports
      q <= output(2);
      r <= output(1);
      s <= output(0);

      PROCESS (input)    -- change on input invokes process
      BEGIN
            CASE input IS
                  WHEN "0000"     =>    output <= "0000";
                  WHEN "0001"     =>    output <= "0001";
                  WHEN "0010"     =>    output <= "0010";
                  WHEN "0011"     =>    output <= "0011";
                  WHEN "0100"     =>    output <= "0100";
                  WHEN "1011"     =>    output <= "1000";
                  WHEN "1100"     =>    output <= "1001";
                  WHEN "1101"     =>    output <= "1010";
                  WHEN "1110"     =>    output <= "1011";
                  WHEN "1111"     =>    output <= "1100";
                  WHEN OTHERS     =>    output <= "1111";
            END CASE;
      END PROCESS;
END solution3;
```

The fourth solution (*ARCHITECTURE solution4 OF code IS*) uses Boolean expressions derived with Karnaugh mapping for the output port signal assignments. Unlike the previous three solutions, we do not need to group the input and output bits into arrays. Four single-bit signals (nd, nc, nb, and na) are created within the architecture for this design. The signal assignments for these buried nodes are simply the NOTing of each of the corresponding input port signal. This makes it somewhat easier to write the signal assignment expressions for the output ports. Because signal assignment statements are concurrent statements, the order in which they are listed in the architecture declaration for the design does not matter. Logical operators (AND and OR) in VHDL have equal precedence. Therefore, parentheses are needed to explicitly define the order of precedence for the Boolean functions. While this VHDL solution will produce the desired circuit operation, notice that the circuit behavior is not obvious with this technique. The major objective in using HDL languages is to clearly represent the behavior of a logic design. Additionally, this technique requires a great deal more work by the designer! This may suggest that traditional design techniques such as using Karnaugh mapping to reduce logic expressions may not be the most desirable approach when using a hardware description language for design entry.

VHDL

```
ENTITY code IS
      PORT
      (
            d, c, b, a        :IN BIT;
            p, q, r, s        :OUT BIT
      );
END code;

ARCHITECTURE solution4 OF code IS
SIGNAL nd, nc, nb, na         :BIT; -- buried nodes
BEGIN
      nd <= NOT d;
      nc <= NOT c;
      nb <= NOT b;
      na <= NOT a;

      p <= d OR (c AND a) OR (c AND b);
      q <= (nd AND c) OR (d AND nc AND nb)
            OR (d AND nc AND na) OR (c AND b AND a);
      r <= (b AND na) OR (nd AND b) OR (c AND nb AND a)
            OR (d AND nc AND nb);
      s <= (nd AND a) OR (d AND na) OR (d AND nc AND nb)
            OR (nd AND c AND b);
END solution4;
```

See Quartus Tutorial 4 – HDL available from the publisher's textbook web site at www.pearsonhighered.com/electronics for procedures.

Example 6-3

Design a 4-channel data selector (multiplexer). This is a standard type of logic function that we will be studying in more detail later. There will be four data inputs or channels, which we will name D0 through D3. The desired data input channel is selected with the input controls S1 and S0 and the multiplexer has an enable input named EN.

Five different design solutions will be created in VHDL files. All solutions will use the same ENTITY declaration. The 4-bit data input d and 2-bit control input s are declared using BIT_VECTOR arrays.

The first solution will be based on the general logic expression for a multiplexer. A multiplexer is a common logic function, so its logic expression will become familiar to us. For now, we will give the expression as:

$$Y = (D0\ \overline{S1}\ \overline{S0} + D1\ \overline{S1}\ S0 + D2\ S1\ \overline{S0} + D3\ S1\ S0)\ EN$$

In the first solution (*ARCHITECTURE boolean OF multiplexer IS*), two different variations of signal assignment statements will be used. A set of buried nodes, declared by SIGNAL sel, will be used to detect (decode) the four data selection combinations. One of the four bits for BIT_VECTOR sel will be HIGH to indicate which of the four possible bit combinations for input port s is applied. The four data selection combinations are detected using conditional signal assignment statements (WHEN … ELSE). A Boolean expression to be tested follows the keyword WHEN. If that expression is true, then the signal assignment before the WHEN will be applied. If the expression is false, then the signal assignment will be the value after the keyword ELSE. Priority is implied by the ordering of expressions in conditional signal assignments. The signal assignment for the output port y is a Boolean expression derived from the function expression above and the appropriate sel bit. Logical operators (AND and OR) in VHDL have equal precedence. Therefore, parentheses are needed to explicitly define the order of precedence for the Boolean functions.

```
ENTITY multiplexer IS
      PORT ( en    :IN BIT;
             d     :IN BIT_VECTOR (3 DOWNTO 0);
             s     :IN BIT_VECTOR (1 DOWNTO 0);
             y     :OUT BIT );
END multiplexer;

ARCHITECTURE boolean OF multiplexer IS
SIGNAL sel         :BIT_VECTOR (0 TO 3);
BEGIN
      sel(0) <= '1' WHEN (s = "00") ELSE '0';    -- conditional
      sel(1) <= '1' WHEN (s = "01") ELSE '0';    -- signal
      sel(2) <= '1' WHEN (s = "10") ELSE '0';    -- assignment
      sel(3) <= '1' WHEN (s = "11") ELSE '0';    -- statements

      y <= ((d(0) AND sel(0))         -- Boolean expression
         OR (d(1) AND sel(1))
         OR (d(2) AND sel(2))
         OR (d(3) AND sel(3))) AND en;
END boolean;
```

The second example solution (*ARCHITECTURE ifand OF multiplexer IS*) will use an IF/THEN statement. IF statements are sequential statements and are therefore contained in a PROCESS. The sensitivity list for the PROCESS must include all input signals because this is a combinational function that will be affected by any input change. The IF statement uses a compound test to determine the input combination for s if the en input is simultaneously HIGH. The appropriate signal assignment for y is given after THEN. The ELSIFs produce a prioritized set of further conditions for testing. If all other tests should fail (because en is LOW), the ELSE clause gives a default output signal assignment for y.

```
ENTITY multiplexer IS
      PORT ( en    :IN BIT;
             d     :IN BIT_VECTOR (3 DOWNTO 0);
             s     :IN BIT_VECTOR (1 DOWNTO 0);
             y     :OUT BIT );
END multiplexer;

ARCHITECTURE ifand OF multiplexer IS
BEGIN
      PROCESS (en, d, s)      -- process with sensitivity list
      BEGIN
            IF    (s = "00" AND en = '1') THEN  y <= d(0);
            ELSIF (s = "01" AND en = '1') THEN  y <= d(1);
            ELSIF (s = "10" AND en = '1') THEN  y <= d(2);
            ELSIF (s = "11" AND en = '1') THEN  y <= d(3);
            ELSE        y <= '0';         -- disabled output
            END IF;
      END PROCESS;
END ifand;
```

The third solution (*ARCHITECTURE nestif OF multiplexer IS*) is similar to the second except that it uses nested IF statements. The outer condition (en is HIGH) must be true before it is necessary to evaluate the inner IF and determine the input combination for s. Note that both the inner and outer IFs must have a corresponding END IF;.

```
ENTITY multiplexer IS
      PORT ( en    :IN BIT;
             d     :IN BIT_VECTOR (3 DOWNTO 0);
             s     :IN BIT_VECTOR (1 DOWNTO 0);
             y     :OUT BIT );
END multiplexer;

ARCHITECTURE nestif OF multiplexer IS
BEGIN
      PROCESS (en, d, s)       -- process with sensitivity list
      BEGIN
            IF (en = '1') THEN                   -- nested IFs
                  IF    (s = "00") THEN    y <= d(0);
                  ELSIF (s = "01") THEN    y <= d(1);
                  ELSIF (s = "10") THEN    y <= d(2);
                  ELSE                     y <= d(3);
                  END IF;
            ELSE        y <= '0';          -- disabled output
            END IF;
      END PROCESS;
END nestif;
```

The fourth solution (*ARCHITECTURE ifcase OF multiplexer IS*) is a modification of the third. Nested inside the IF statement that tests the condition of en is a CASE statement. The signal to be evaluated by the CASE statement is s. The alternative values for s are listed after the keyword WHEN. Following the WHEN clauses (after the => symbol) are the different signal assignment statements that are to be activated according to the value for s. A PROCESS is used because IF statements and CASE statements are sequential. Each of the input signals is in the PROCESS sensitivity list because a change in any one of the inputs could cause a change in the output y.

```
ENTITY multiplexer IS
      PORT ( en    :IN BIT;
             d     :IN BIT_VECTOR (3 DOWNTO 0);
             s     :IN BIT_VECTOR (1 DOWNTO 0);
             y     :OUT BIT );
END multiplexer;

ARCHITECTURE ifcase OF multiplexer IS
BEGIN
      PROCESS (en, d, s)      -- invoke on changes
      BEGIN
            IF (en = '1') THEN      -- en high?
                  CASE s IS         -- nested CASE
                        WHEN "00"   =>    y <= d(0);
                        WHEN "01"   =>    y <= d(1);
                        WHEN "10"   =>    y <= d(2);
                        WHEN "11"   =>    y <= d(3);
                  END CASE;
            ELSE                  -- otherwise disabled output
                                          y <= '0';
            END IF;
      END PROCESS;
END ifcase;
```

The fifth solution (*ARCHITECTURE assign OF multiplexer IS*) uses the WHEN ELSE of a conditional signal assignment statement to select the appropriate d input (out of four choices) for the y signal assignment statement. The Boolean expression (given after the keyword WHEN) tests for both a specific value for s and that the circuit is enabled (en = 1). According to which expression is true, the signal assignment before the appropriate WHEN will be applied. If all expressions are false (because they are disabled), then the signal assignment will be the value after the final ELSE.

```
ENTITY multiplexer IS
      PORT ( en    :IN BIT;
             d     :IN BIT_VECTOR (3 DOWNTO 0);
             s     :IN BIT_VECTOR (1 DOWNTO 0);
             y     :OUT BIT );
END multiplexer;

ARCHITECTURE assign OF multiplexer IS
BEGIN
            -- conditional signal assignment statement
      y <=  d(0)  WHEN s = "00" AND en = '1' ELSE
            d(1)  WHEN s = "01" AND en = '1' ELSE
            d(2)  WHEN s = "10" AND en = '1' ELSE
            d(3)  WHEN s = "11" AND en = '1' ELSE
            '0';        -- disabled
END assign;
```

Functional simulation results are shown in Fig. 6-3. When en is HIGH, the d[3..0] input selected by s[1..0] will be routed to the y output. If en is LOW, the y output will always be LOW.

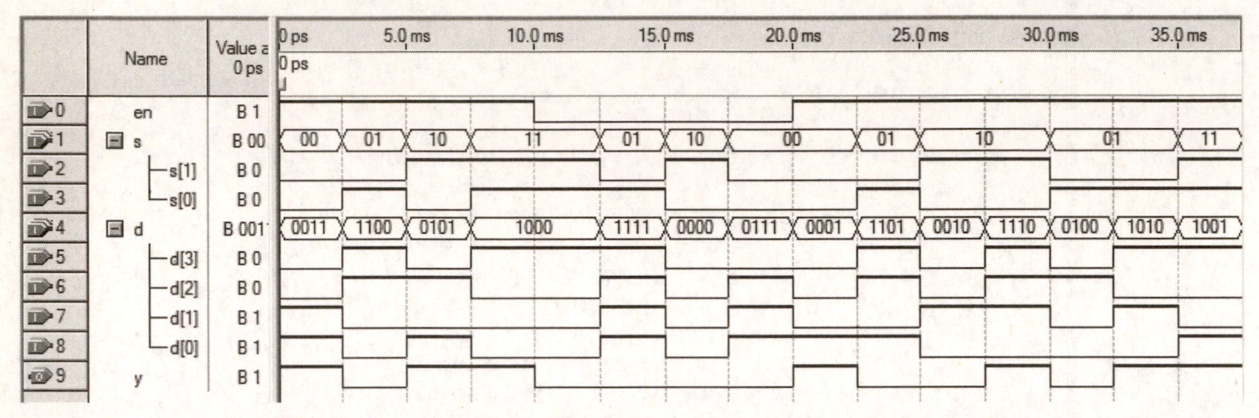

Fig. 6-3 Functional simulation results for Example 6-3

Laboratory Projects

Design PLD logic circuits for the following applications using VHDL. Provide comments in the design files and include your name in an information header. Compile the text file and functionally simulate your design to verify it. Program a PLD with your design and test it in the lab.

6V.1 2-bit comparator
Test the design given in Example 6-1 (Tutorial).

6V.2 Code converter
Test the design given in Example 6-2 (Tutorial).

6V.3 Modified data selector
Modify the design given in Example 6-3 to create a 2-channel data selector. The two data inputs should be named D0 and D1. The data input channel will be selected with the input control S. The data selector also needs to have an active-LOW enable named EN.

6V.4 Gray-code-to-binary conversion
Design a logic circuit that will convert a 6-bit Gray code value into its equivalent binary value. The Gray code is commonly used in shaft position encoders to decrease the possibility of errors. The Gray code is an unweighted code in which only a single bit changes from one code number to the next. For example, the Gray code value 100110 is equivalent to the binary value 111011. The Gray-to-binary code conversion algorithm is as follows:

1. *The most significant bit in the binary result is the same as the corresponding most significant (leftmost) Gray code bit.*
2. *To produce each additional binary bit, XOR the binary code bit just generated to the Gray code bit in the next adjacent position.*
3. *Repeat the process in step 2 to produce the binary result through the least significant bit.*

Because the output ports cannot be read to produce the next binary bit, you will need to create buried signal nodes and write the Boolean expressions for each of the binary results and then assign the buried nodes to the output port. Simulation hint: Use a Gray code count for the input waveform.

6V.5 BCD-to-binary converter

Design a logic circuit that will convert a 5-bit BCD input named bcd into its equivalent 4-bit binary value bin. Since the output is only 4 bits long, the largest number that can be converted is the decimal value 15. This converter circuit should also have an output called err that will be HIGH if the BCD input value is not in the range of 0 through 15.

6V.6 Binary-to-BCD converter

Design a logic circuit that will convert a 5-bit binary input bin into its equivalent 2-digit (only 6 output bits are needed) BCD value bcd. The converter can handle numbers from 0_{10} through 31_{10}.

6V.7 Tens digit detector

Design a logic circuit that will detect the equivalent tens digit value (0 through 6) for a 6-bit input number num. The input values will range from 0 through 63 with 6 bits. Only one of the seven outputs (tens) will be HIGH at a time, indicating the value of the tens digit for the current input number. Use IF/THEN (and ELSIF/THEN) statements for your design. IF/THEN syntax automatically creates a priority of behavior since the first IF expression that evaluates to be true will determine the behavioral statement that will be applied. Any other IF clauses will be ignored after the first one that is true. Hint: Use the <= comparator operator in the Boolean expression to be evaluated by each IF statement line, and the behavioral statement following THEN will be the appropriate assignment for the outputs (e.g., tens <= "0000001").

6V.8 Lamp display

Design a logic circuit that will light the number of lamps corresponding to the 3-bit input value that is applied (0 through 7). The inputs are named num and the outputs are named lites. The following table describes the operation. Use a CASE statement for this design.

num2	num1	num0	lites7	lites6	lites5	lites4	lites3	lites2	lites1
0	0	0	0	0	0	0	0	0	0
0	0	1	0	0	0	0	0	0	1
0	1	0	0	0	0	0	0	1	1
0	1	1	0	0	0	0	1	1	1
1	0	0	0	0	0	1	1	1	1
1	0	1	0	0	1	1	1	1	1
1	1	0	0	1	1	1	1	1	1
1	1	1	1	1	1	1	1	1	1

6V.9 Programmable logic unit

Design a programmable logic circuit that will perform one of four logic operations on two 4-bit input arrays named a and b when enabled by a signal named en. The desired logic function, selected by a 2-bit input array named s, is indicated in the following table. The 4-bit output array is named f. If the enable is LOW, the control s does not matter and the outputs are all LOW. Each of the output bits will be a function of the corresponding a-bit and b-bit and the s1 and s0 control inputs. For example, f3 is dependent on a3, b3, s1, and s0, while f2 is dependent on a2, b2, s1, and s0. Use an IF/ELSIF statement to define this function. Hint: The signal assignments can be written just like the table.

en	s1	s0	Array Logic Operation
1	0	0	f <= a OR b
1	0	1	f <= a AND b
1	1	0	f <= a XOR b
1	1	1	f <= NOT a
0	X	X	f <= "0000"

6V.10 Number range detector

Design a logic circuit that, when enabled with an active-HIGH signal named en, will detect four different ranges of values for a 5-bit input number (see table below). The inputs are labeled num4 through num0 and the outputs are labeled values1 through values4. Values1, values2, and values3 each produce active-HIGH outputs, while values4 produces an active-LOW output signal. Use IF/ELSIF to test the 5-bit input for the desired value ranges. Hint: Use the greater-than-or-equal-to (>=) and less-than-or-equal-to (<=) symbols in a Boolean expression to detect the desired range of input values for each output result.

Outputs	Values detected
values1	4–12
values2	15–20
values3	21–24
values4	26–30

6V.11 Data switcher

Design a programmable logic circuit that will route two input data bits to the selected outputs as given in the function table below. The control input, a 2-bit array named sel, determines which of the two data inputs (input1 and input0) are to be routed to each of the two outputs (output1 and output0). The input en is an active-LOW enable for the data switcher. If the enable is HIGH, we don't care what input levels are on sel, and the two outputs will both be LOW. Use a CASE statement inside an IF/THEN statement.

en	sel1	sel0	output1	output0
0	0	0	input1	input0
0	0	1	input0	input0
0	1	0	input1	input1
0	1	1	input0	input1
1	X	X	0	0

UNIT 7

TESTING FLIP-FLOPS AND SEQUENTIAL CIRCUIT APPLICATIONS

Objectives

- To test the operation of SR and D latches.
- To test the operation of D and JK flip-flops.
- To test basic counter and register circuits.

Flip-Flops and Latches

Combinational logic circuits have outputs that are dependent only on the current inputs to the circuit. Sequential circuits, on the other hand, are dependent not only on the current inputs but also on the prior circuit conditions. Sequential circuits contain memory elements that allow them to utilize the prior circuit conditions in determining what circuit output will be produced next. These memory elements consist of flip-flops and latches that can each store one bit of data. There are various types of flip-flops and latches, including SR, D, and JK. Flip-flops are generally devices that are edge-triggered by a clock input control while latches may be enabled by a control signal that must be at a specific logic level. The trigger or enable input signals are used to control the timing of the action of a flip-flop or latch. Several flip-flops or latches can be connected together in specific circuit configurations to construct various types of registers and counters. The storage category of primitives in the Quartus II library has various types of positive-edge-triggered flip-flops (DFF, JKFF, SRFF, and TFF) and latches. The flip-flop device names that end with E have an additional enable control that may be useful with certain designs. See Quartus Notes available from the publisher's textbook web site at www.pearsonhighered.com/electronics.

Example 7-1

Simulate the SR latch circuit shown in Fig. 7-1 with the given input waveforms.

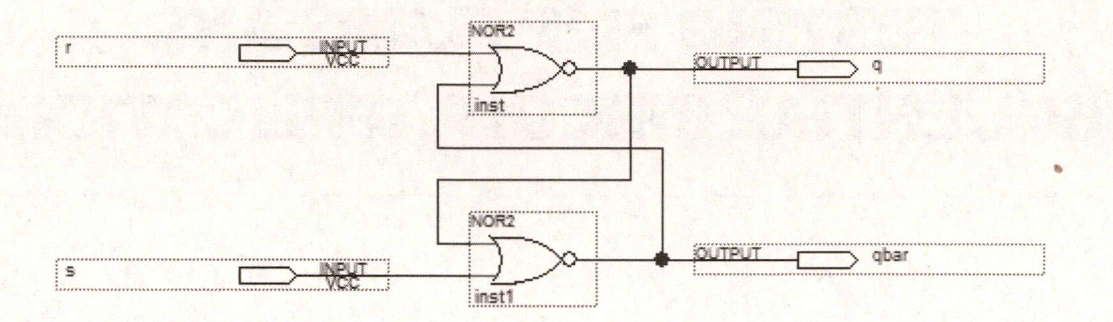

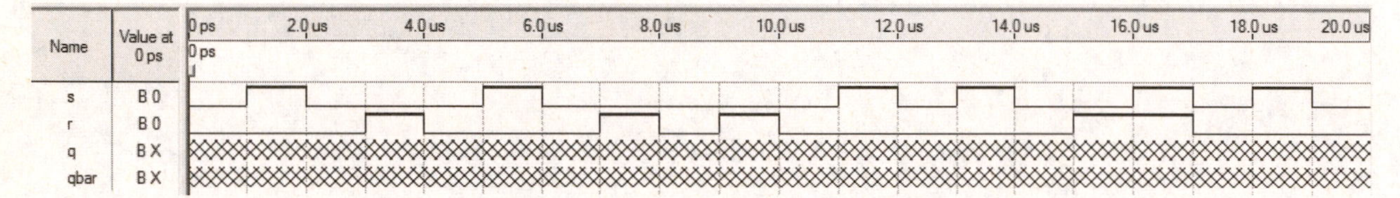

Fig. 7-1 Unclocked SR latch and input conditions

The SR latch shown in Fig. 7-1 will function as indicated in Table 7-1. The specified input waveforms will produce the results shown in Fig. 7-2. The sections of the output waveforms that are shown in a cross-hatch cannot be determined. The time period that is shaded (by me for this figure) in the simulation is the result of applying an invalid input condition (S = R = 1).

S	R	Q	$\overline{Q}$	Command
0	0	Q	$\overline{Q}$	no change
0	1	0	1	Reset
1	0	1	0	Set
1	1	0	0	*Invalid*

Table 7-1 Truth table for active-HIGH input SR latch

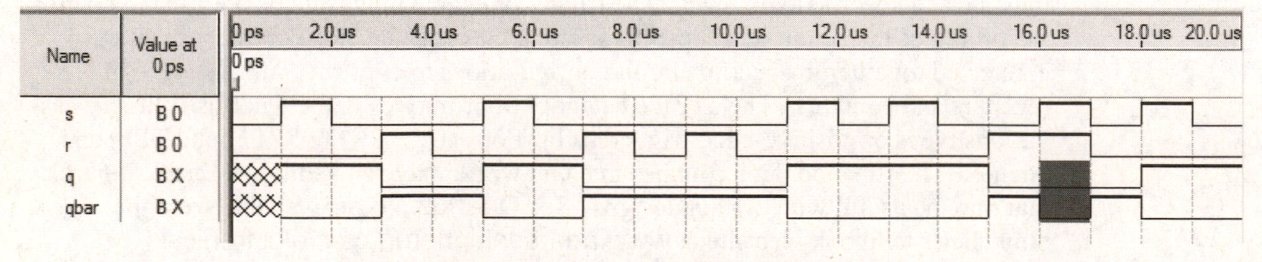

Fig. 7-2 Simulation results for Example 7-1

Example 7-2

Simulate the D flip-flop circuit shown in Fig. 7-3 with the given input waveforms. Note that Quartus will automatically make the PRN and CLRN inputs on the D flip-flop inactive if they are left unconnected in the schematic.

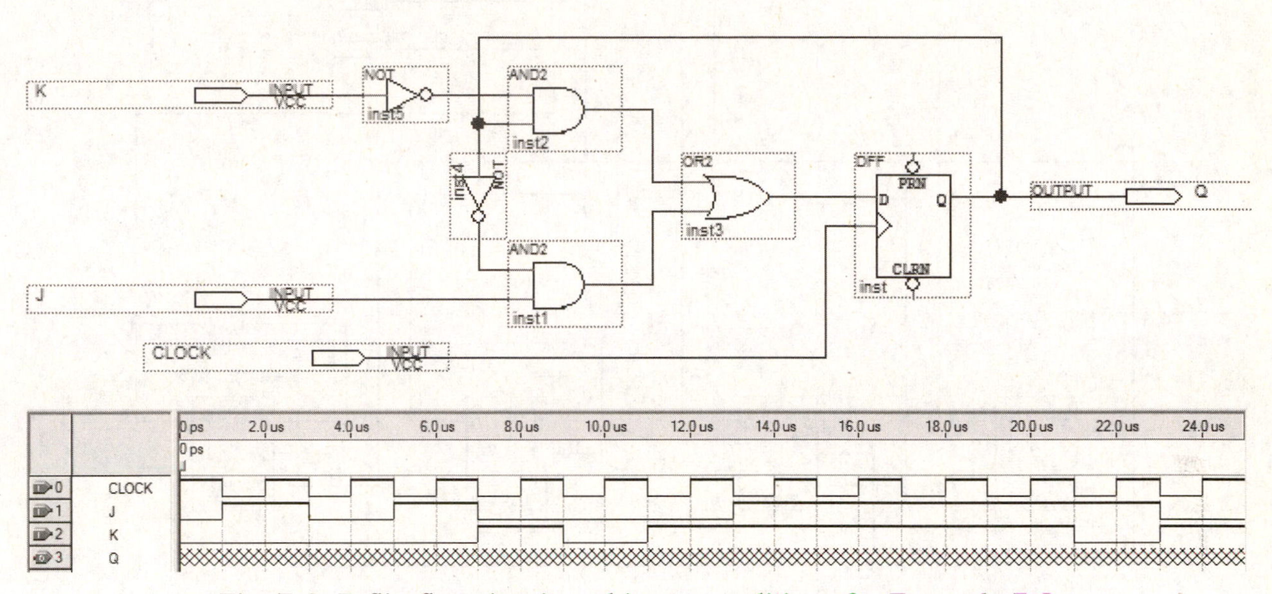

Fig. 7-3 D flip-flop circuit and input conditions for Example 7-2

The simulation results are shown in Fig. 7-4. These results show that the D flip-flop circuit will operate the same as a JK flip-flop.

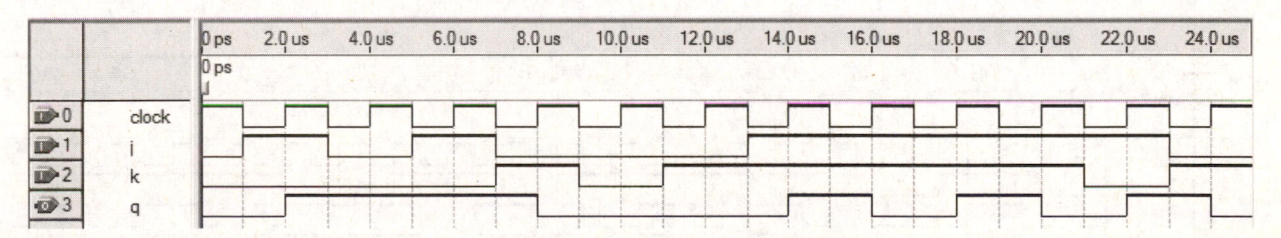

Fig. 7-4 Timing diagram for Example 7-2

Example 7-3

Prove that each of the four flip-flops in Fig. 7-5 will have the same function. What function is performed by the flip-flops when t is LOW? What function is performed when t is HIGH?

For the functional simulation shown in Fig. 7-6, the flip-flop circuit is clocked several times with t LOW and again with t HIGH. These results show that each flip-flop in the circuit has the same operation. Each flip-flop will toggle on the ↑-edge of clock when t = 1 and will hold its data when t = 0.

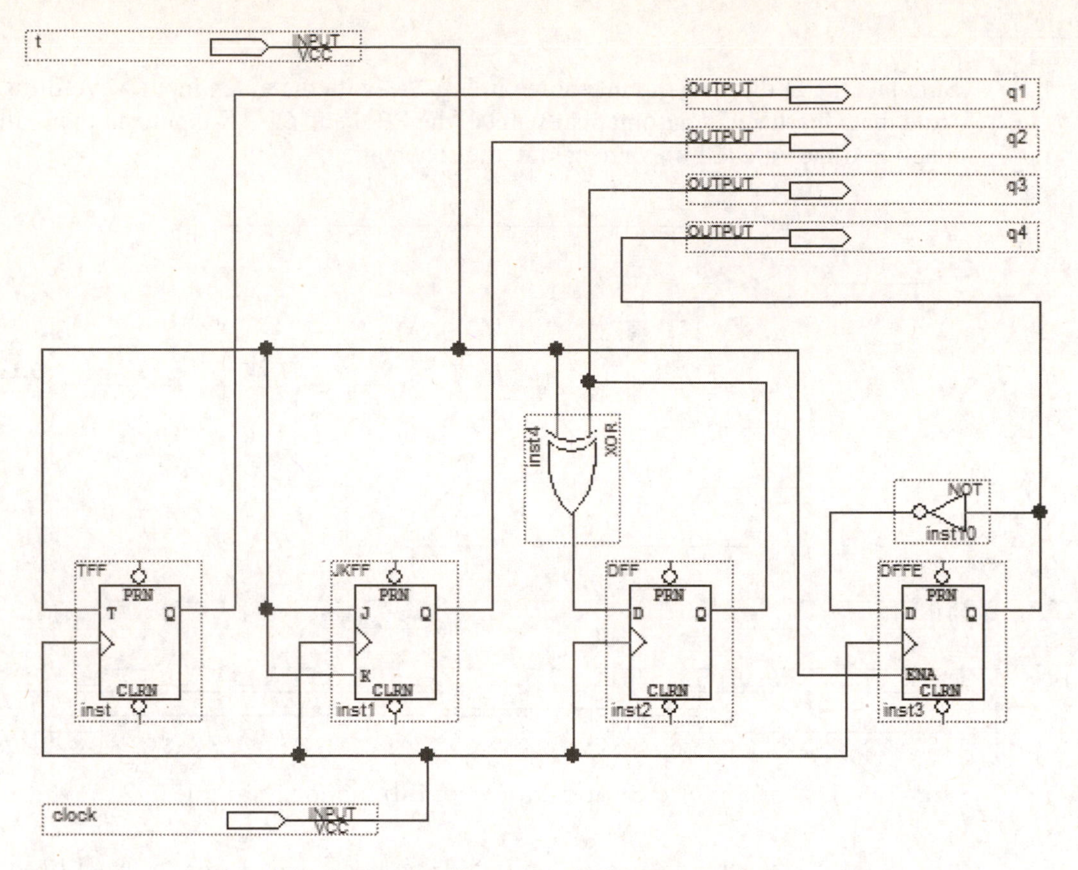

Fig. 7-5 Flip-flop circuit for Example 7-3

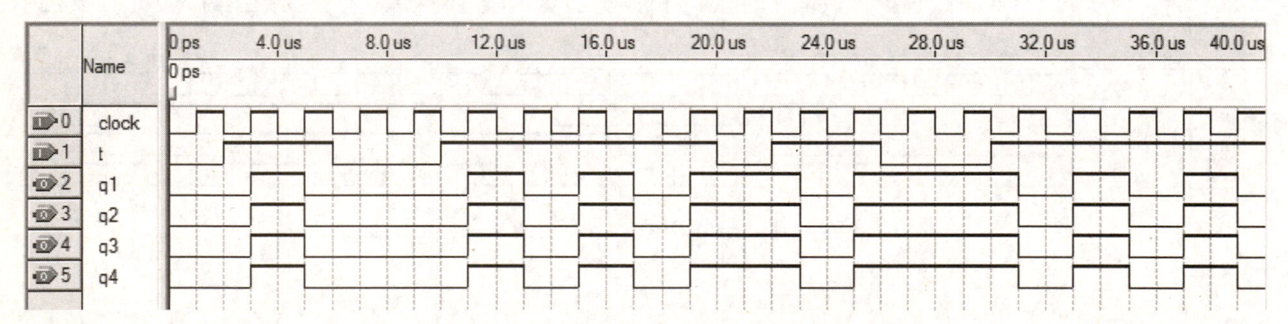

Fig. 7-6 Simulation results for Example 7-3

Laboratory Projects

Use schematic capture to create the following logic circuits. Functionally simulate the circuits and test by programming a PLD.

7.1 NAND SR latch

Construct and test a NAND SR latch (see schematic below). Use two logic switches for the control inputs (Set and Reset) and monitor both latch outputs (Q and Qbar) with separate lamps. Test the latch under each of the four possible input conditions. How do you set the latch? What is the condition of the latch's two outputs (Q and Qbar) when the latch is set? How do you reset the latch? What is the condition of the latch's outputs when the latch is reset? How do you control the latch so that it will hold a bit of data? Why are the outputs named Q and Qbar? What happens to a NAND latch in the invalid state? Are the latch's control inputs active-HIGH or active-LOW? Describe the operation of the latch.

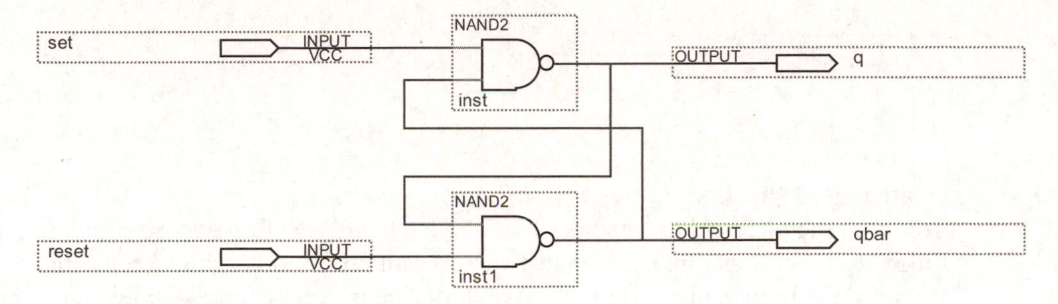

7.2 D latch

Construct and test a D latch using the NAND latch in Project 7.1. Use a logic switch for the Data input and another for the Enable. Monitor both latch outputs (Q and Qbar) with separate lamps. What does the latch do when the enable is HIGH and the data input is changed? What does the latch do when the enable is LOW and the data input is changed? Is the latch's Enable active-HIGH or active-LOW? How do you store data in the D latch? What part of the D latch circuit actually stores the data and which part provides the data steering (or control) of the latch? Describe the operation of the latch.

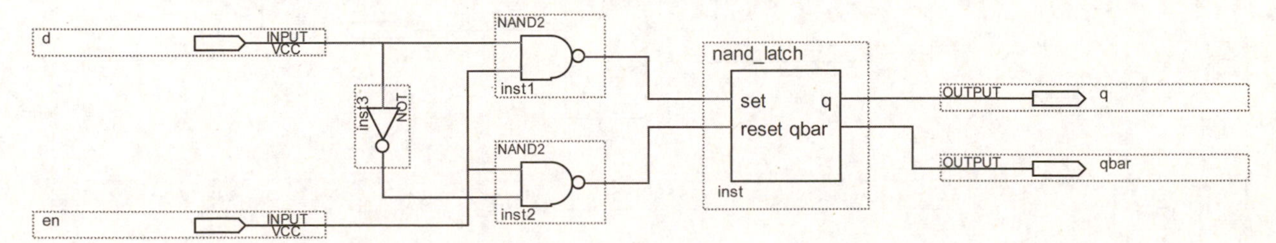

7.3 JK flip-flop

Test the operation (both synchronous and asynchronous) of a JK flip-flop. Connect the flip-flop's J, K, PRE, and CLR control inputs to logic switches, the clock to a pushbutton, and the Q output to a lamp. Which control inputs are synchronous? Determine how to synchronously control the flip-flop to hold data, set, reset, and toggle. Which control inputs are asynchronous? Determine how to asynchronously store a zero or a one in the flip-flop. Why are the controls classified as synchronous or asynchronous? Which set of controls (synchronous or asynchronous) have priority? What is the function of the clock? Give the truth table for this JK flip-flop.

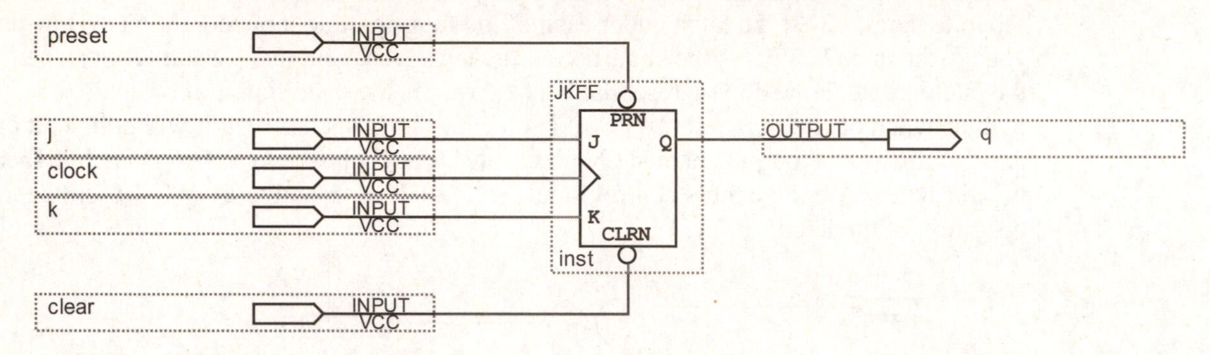

7.4 D latch vs. D flip-flop

Compare the difference in operation of a D latch and a D flip-flop by simulating the following circuit using the given test vectors and programming the circuit into a PLD. Why are the two results different? Which device is level-enabled and which is edge-triggered? On which level or edge does the device read new data? What happens at any other time besides the appropriate level or edge for the respective device?

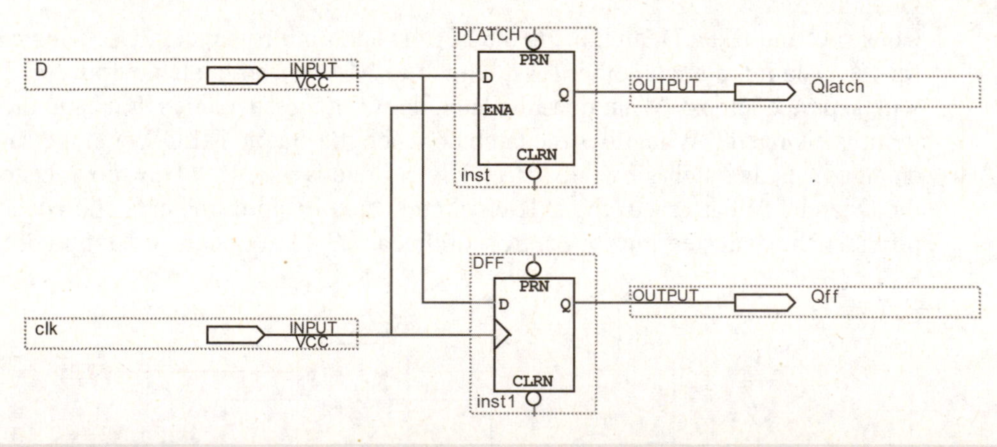

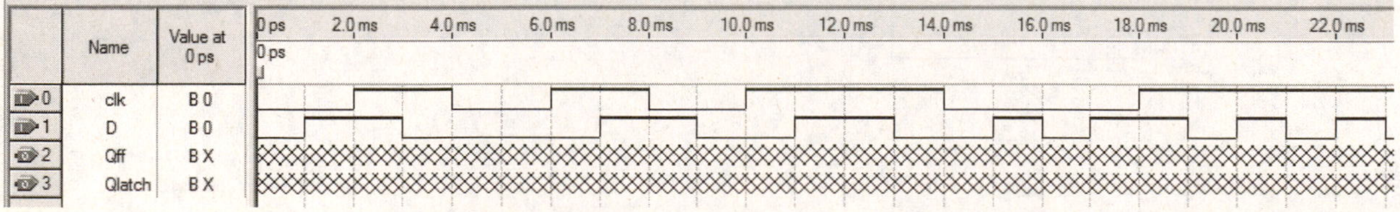

7.5 Binary counters

There are four binary counter circuits, each using a different type of flip-flop, given on the next 2 pages. Construct and test one of the 3-bit counters. Use a debounced pushbutton to manually clock the counter and a logic switch to control the enable input. Monitor the counter's output (QC QB QA) on 3 lamps. Assume that QC is the most significant bit and QA is the least significant bit. What is the count sequence produced by this circuit? Is this an up-counter or a down-counter? What is the counter's modulus? Why is this counter described as a "synchronous" counter? What does the enable input do? What is the active-level for enable? Describe what circuit conditions will cause each flip-flop to toggle.

7.6 Counter timing diagram

Use an oscilloscope to observe the input and output waveforms (4 signals: CLOCK, QA, QB, QC) for the 3-bit counter created in Laboratory Project 7.5. A good display on the oscilloscope will be obtained if the clock frequency is increased to at least 10 kHz. **Remember that the clock signal must be compatible with your PLD – check with your lab instructor.** A 4-channel scope will provide the best display because all 4 waveforms can be viewed simultaneously, but a dual-channel scope can also be used by carefully swapping signals. Arrange the scope display in the order of decreasing signal frequency, with the highest frequency signal at the top of the screen and the lowest frequency signal at the bottom. **Trigger the oscilloscope on the lowest frequency signal.** Which signal appears to be the lowest frequency? Why? Carefully and neatly draw the waveforms to show the timing relationships between each of the counter outputs and the clock signal.

7.7 Frequency division

Use a frequency counter to measure the actual signal frequencies of the clock input and QA, QB, and QC outputs for the 3-bit counter in Laboratory Project 7.5. What is the frequency relationship between each of these signals?

7.8 4-bit binary counter

Modify the 3-bit counter in Project 7.5 to create a 4-bit binary counter. What do you need to add to the circuit besides another flip-flop? What inputs will be needed for this new gate? Observe the counter outputs on 4 lamps while clocking with a low frequency clock. What is the count sequence produced by the new circuit? What is the new counter's modulus?

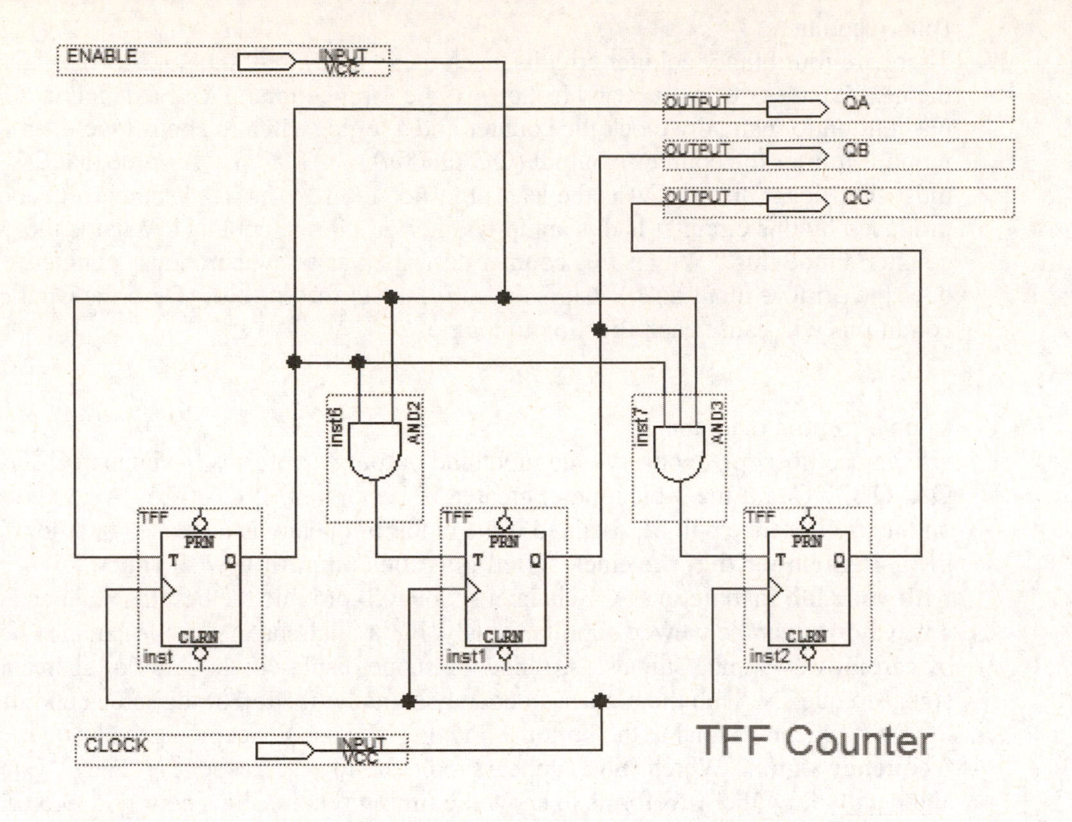
ENABLE
INPUT
VCC
OUTPUT
QA
OUTPUT
QB
OUTPUT
QC
inst6
AND2
inst7
AND3
TFF
PRN
T
Q
CLRN
inst
inst1
inst2
CLOCK
TFF Counter

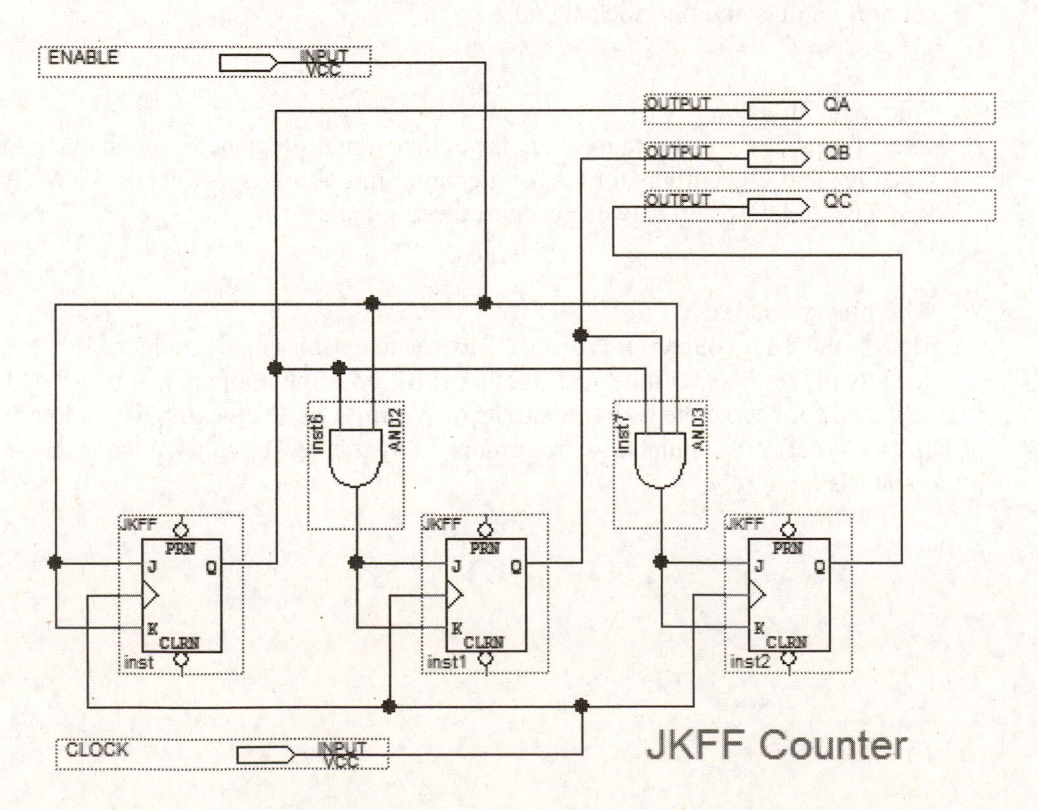
ENABLE
INPUT
VCC
OUTPUT
QA
OUTPUT
QB
OUTPUT
QC
inst6
AND2
inst7
AND3
JKFF
PRN
J
K
Q
CLRN
inst
inst1
inst2
CLOCK
JKFF Counter

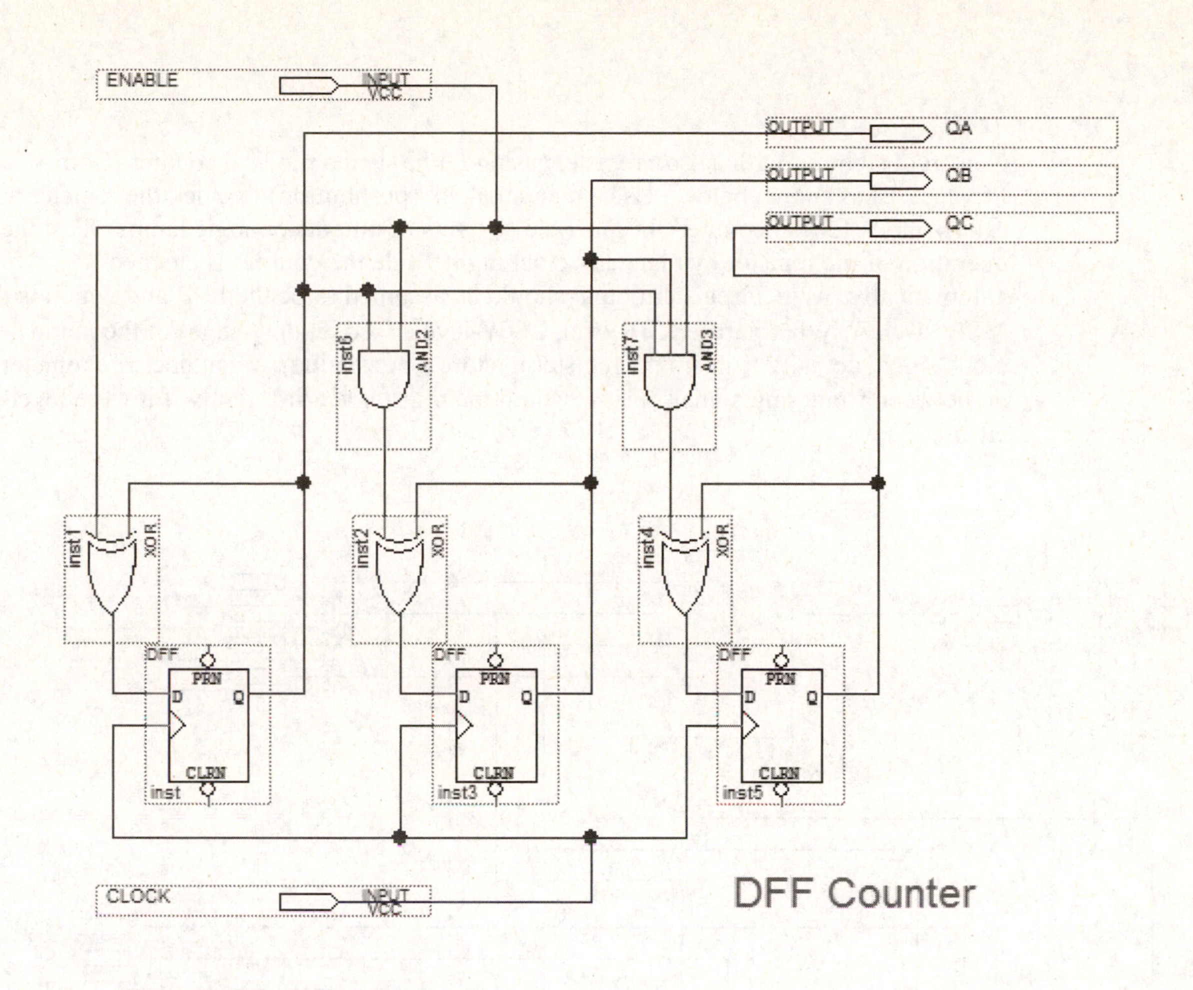

DFF Counter

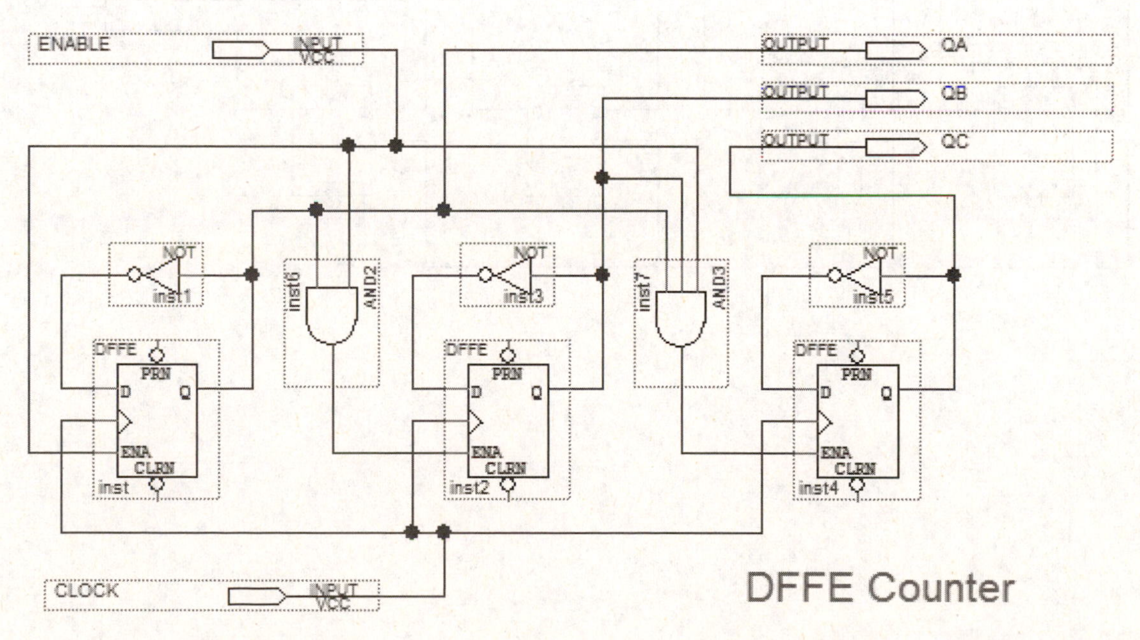

DFFE Counter

7.9 4-bit register

Construct a circuit with a 4-bit register (using DFFs) and a mod-16 counter (from Project 7.8) as shown below. Use a manual clock (pushbutton) to clock the register and a 1-Hz clock for the counter. Display the two sets of outputs on logic lamps. Test the operation of the register by manually clocking it while the counter is clocked automatically. Which register output should be assumed to be the LSB and which is the MSB? Why? What part (HIGH-level, LOW-level, ↑-edge, or ↓-edge) of the manual clock signal actually triggers the register to store a new value? What does the register do between triggering signals? How would the register act differently if it were level-enabled?

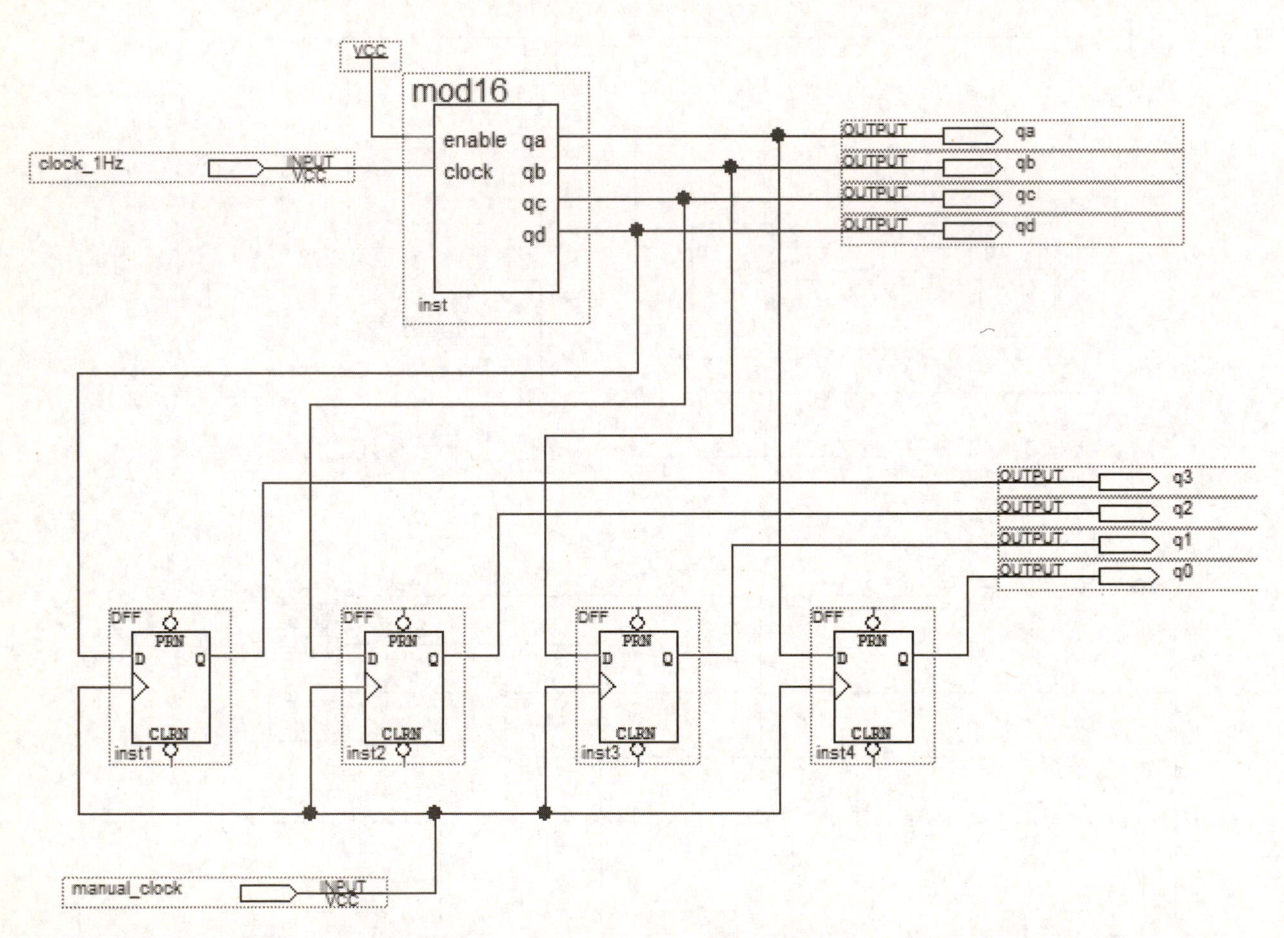

7.10 LPM_FF megafunction

Use the Quartus MegaWizard to create and test a 4-bit register that is implemented with the LPM_FF megafunction.

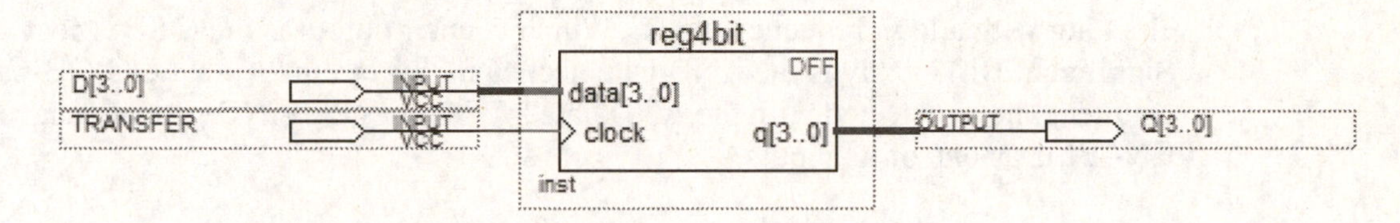

7.11 Binary counter

Create a mod-16 binary counter by designing the feedback circuit to connect to a 4-bit register. Use LPM_FF for the register. Use an HDL to create the feedback circuit, a combinational circuit "look_up_table" that will provide the next-state for each present-state in the register.

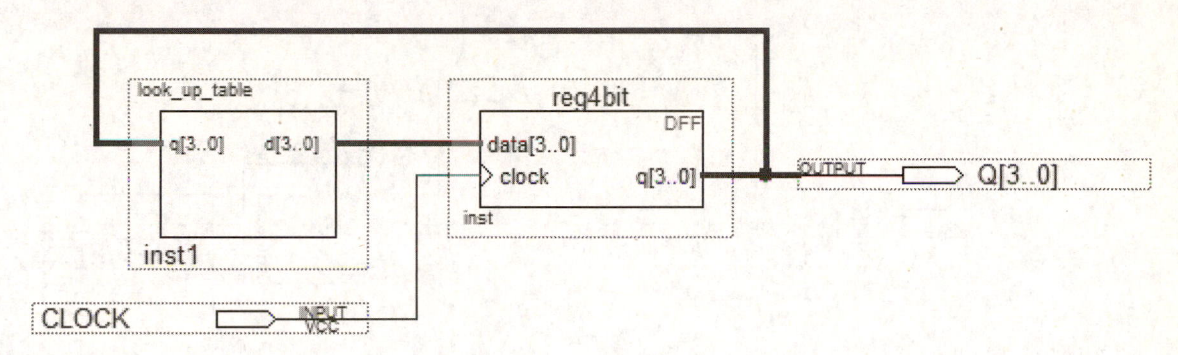

7.12 Cascading counters

Connect two mod-8 counter blocks (from Laboratory Project 7.5) together to create a mod-64 binary counter. Don't forget to create a symbol for the mod-8 counter. This circuit will be a hierarchical design. Observe the counter outputs on 6 lamps while clocking with a low frequency clock. Which counter output is the MSB (Most Significant Bit)? Why is the AND gate needed in this circuit? What QC QB QA counter state is detected with the AND gate? Why does the AND gate also have ENABLE as one of its inputs?

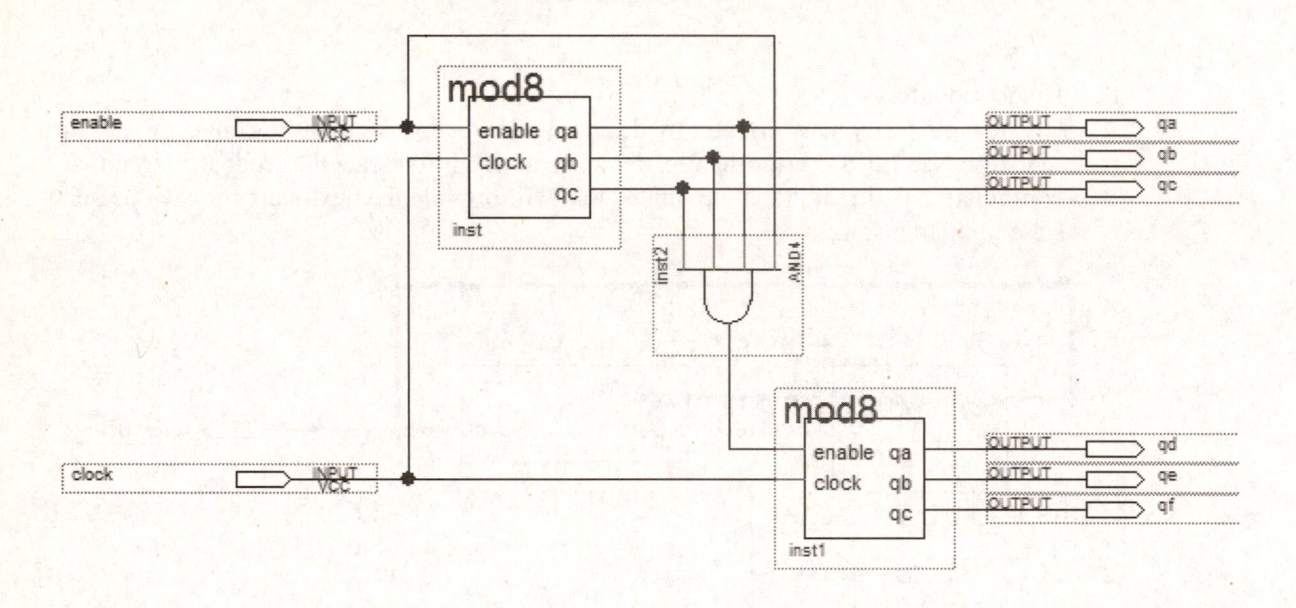

7.13 Debouncing a logic switch

(a) Try to manually clock a 4-bit binary counter with a simple digital switch (see the following schematic). Connect the "Simple Digital Switch Output" to the clock input for the counter. **Check with your lab instructor to determine what voltage should be applied for the +V in the circuit to be compatible with the counter used. Make sure that circuit grounds are connected together!** Note that the switch circuit used for part (a) is actually a portion of the circuit to be used for part (b). The objective is for the counter to produce a binary sequence that increments by one count each time the logic switch is flipped back and forth. Does the counter seem to count correctly? What is the counter actually doing each time the switch is cycled? What is switch bounce? How does switch bounce affect a digital circuit?

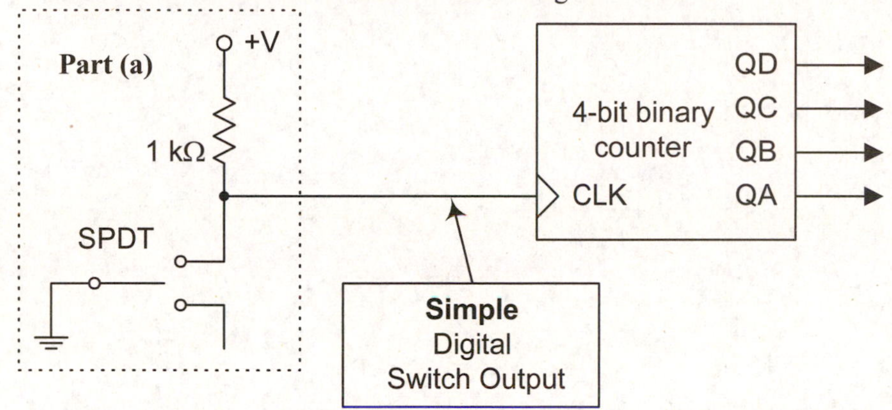

(b) Add a NAND SR latch to eliminate the switch's contact bounce as shown in the schematic below. Now use the "Debounced Digital Switch Output" to clock the counter again. Does the counter now operate correctly? What is the sequence of states for the counter's outputs? Describe the operation of the debouncing circuit. What command does the SR latch receive when the switch bounces open? How does the latch eliminate the contact bounce of the switch? Why are the two resistors called "pull-up" resistors?

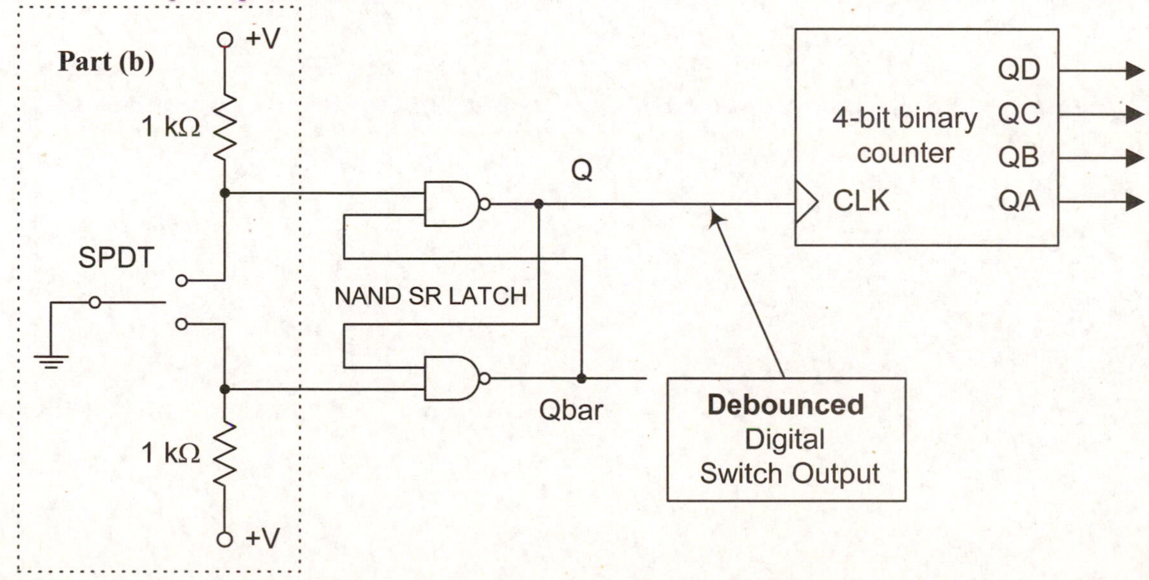

UNIT 8

TIMING AND WAVESHAPING CIRCUITS

Objectives

- To convert analog signals into logic-compatible signals using Schmitt trigger devices.
- To design and construct astable multivibrators that produce specified square waveforms.
- To design and construct monostable multivibrators that produce specified time delay patterns.

Suggested Parts			
7404	7414	74221	555
Resistors: 1.0, 3.3, 8.2, 27, 33, 47, 62, 68, 72, 82 kΩ			Capacitors: 0.001, 0.0047, 0.01, 0.1, 10 μF
Potentiometer: 10 kΩ (10-turn)			Diode: 1N4001

Schmitt Triggers

If a slow-changing signal is applied to the input of a logic device, the device will often produce an output signal that oscillates as the input slowly transitions between HIGH and LOW logic levels. A Schmitt trigger is a waveshaping device used to convert signals in which the voltage is slowly changing (and, therefore, incompatible with logic devices) to a signal of the same frequency but compatible with the signal transition times of logic devices. Schmitt trigger devices have an input hysteresis because they have two different specific triggering points. A positive-going threshold voltage and a negative-going threshold voltage will switch the Schmitt trigger's output between the two possible logic levels. The 7414 IC is a hex inverter chip with Schmitt trigger inputs.

One-Shots or Monostable Multivibrators

A one-shot is a timing device in which the output is triggered into a quasi-stable state and then returns to its stable state. The length of the quasi-stable state is usually controlled by an external resistor and capacitor. One-shots are typically used to produce delays in control signals for digital systems. There are two types of one-shots: retriggerable and nonretriggerable. The 74221 IC (see Fig. 8-1) contains two independent, nonretriggerable one-shots. The one-shot can be triggered by either a positive-edge on the B input or a negative-edge on the A input. The Q output will go HIGH after the one-shot is triggered and will return to a LOW at the end of the delay time. The length of delay is controlled by the resistor and capacitor. The active-LOW CLR input will immediately terminate the delay and return the one-shot to its stable state.

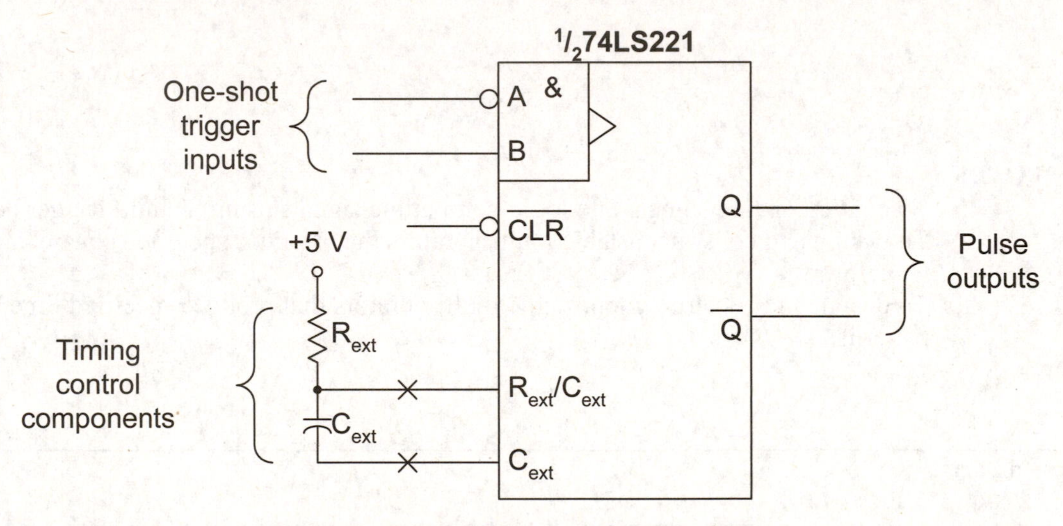

Fig. 8-1 A 74LS221 one-shot IC chip

Example 8-1

Determine the delay that will be produced when the 74LS221 one-shot shown in Fig. 8-1 is triggered if R_{ext} = 47 kΩ, C_{ext} = 0.033 μf, and the CLR input is disabled (HIGH).

The formula for the time delay is:

$t_w = 0.693\ R_{ext}\ C_{ext} = 0.693\ (47\ k\Omega)\ (0.033\ \mu f) = 1.075\ ms$

Clocks or Astable Multivibrators

An astable or free-running multivibrator has an output that continually switches back and forth between the two states, producing a square wave signal. This type of signal is often used as a clock signal to control (or trigger) synchronous circuits. The 555 IC timer is a device that can be used to produce a clock signal whose frequency and duty cycle are dependent on two external timing resistors and a timing capacitor (see Fig. 8-2).

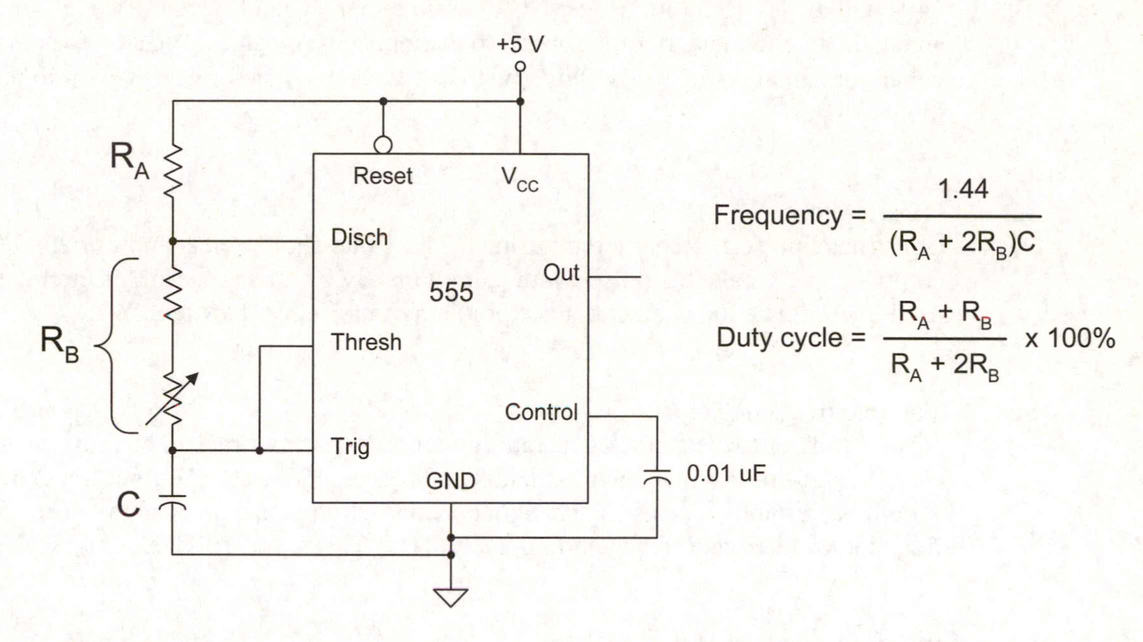

Fig. 8-2 A 555 astable multivibrator with variable frequency output

Example 8-2

Determine the output frequency range that will be produced for the 555 clock circuit shown in Fig. 8-2 if $R_A = 1\ k\Omega$, R_B consists of a fixed 10-kΩ resistor in series with a 100-kΩ potentiometer, and $C = 0.01\ \mu f$. Also determine the duty cycle range for the output signal.

The lowest output frequency will be obtained when R_B is at its maximum value (10 kΩ + 100 kΩ = 110 kΩ). The resultant frequency is 452.5 Hz with a duty cycle of 50.2%. The highest frequency will be obtained when the potentiometer is at its minimum value (0 Ω) so that $R_B = 10\ k\Omega$. This will produce an output frequency of 4762 Hz with a duty cycle of 52.4%.

Laboratory Projects

Design timing and waveshaping circuits for the following applications. Breadboard and verify your designs.

8.1 Schmitt trigger waveshaper
Use an oscilloscope to compare the output waveforms produced by a NOT gate in a 7404 and a 7414 (Schmitt trigger NOT) when either a triangle or sine waveform from a signal or function generator is applied to the inputs of the gates. Be sure to adjust the generator signal using the oscilloscope to $0 \leq V_{in} \leq +5$ V before applying it to the gate inputs.

8.2 Pulse stretcher
Construct a pulse stretcher circuit using a 74221 one-shot. Select appropriate components to make the pulse width approximately 0.5 s. How can this circuit also be used to eliminate the contact bounce problem of mechanical switches?

8.3 Variable frequency clock
Design and construct a clock generator circuit using a 555 timer. The duty cycle of the clock waveform should be approximately 50 percent. The clock output frequency should be variable in steps (by changing the 555 timing capacitor value). The 3 clock frequency values should be approximately 1 Hz, 1 kHz, and 10 kHz.

8.4 Waveform generator
Design and construct a waveform generator that will produce the waveforms given in the timing diagram below. Use the 555 timer to generate one of the waveforms, and then use that waveform to trigger the two one-shots in the 74221 to produce the other two waveforms.

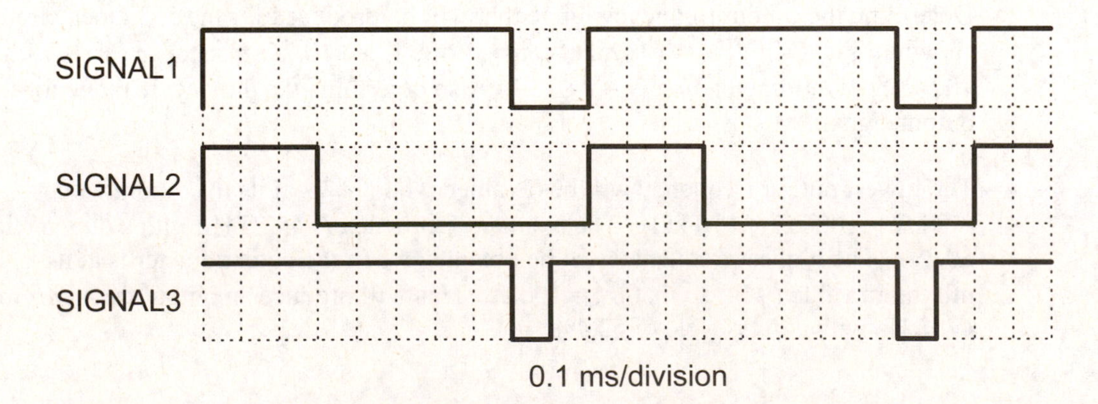

8.5 Variable duty-cycle square wave generator

Design a variable duty-cycle square wave generator using a 555 timer. The output frequency should be 10 kHz and the duty cycle should vary from 15% to 85%. Use a 10-turn, 10-kΩ potentiometer to control the duty cycle. The diodes in the modified astable multivibrator circuit below allow the R_B resistance to be bypassed during the charge cycle. The charge and discharge time equations for this circuit are dependent upon the forward voltage drop (V_F) for the diodes (which may be less than the assumed typical 0.7v).

$$t_L = K \times R_B \times C$$

$$t_H = K \times R_A \times C$$

V_F	K
0.7v	1.00
0.6v	0.94
0.5v	0.89

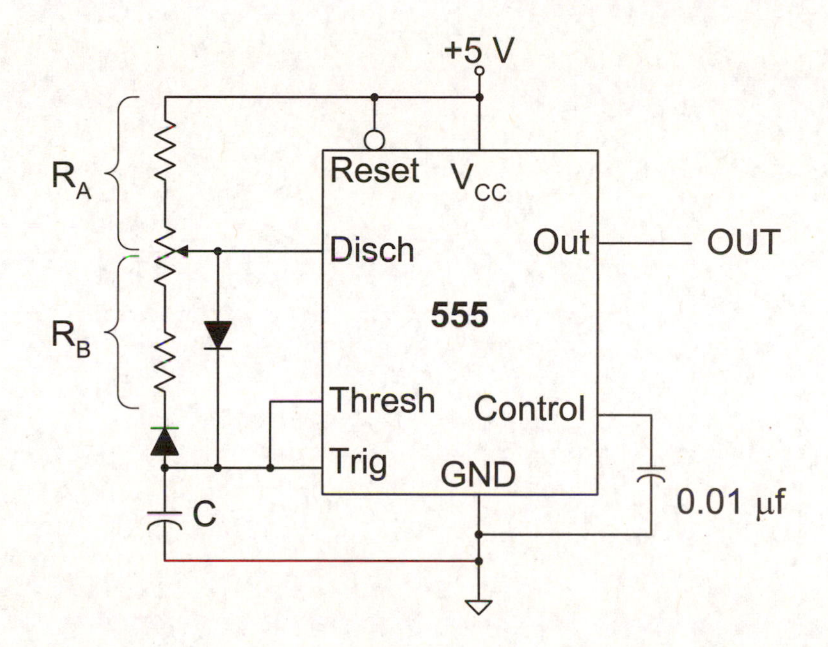

8.6 Delayed pulse

Design a timing circuit that will output a positive pulse that is 0.22 ms wide. The output pulse is delayed and does not start until 0.33 ms after being triggered by a negative-edge signal.

8.7 Clock generator using one-shots

Design a 100-Hz clock generator using the two one-shots contained in a 74221. Use a 0.1 µF timing capacitor for each one-shot. Select a timing resistor for each one-shot. Hint: Let the timeout of each one-shot trigger the other.

UNIT 9

ARITHMETIC CIRCUIT APPLICATIONS

Objective

- To create digital applications using arithmetic circuits.

Adder Circuits

The primary building blocks in adder circuits are half adders and full adders. The basic difference between a half and a full adder is that a full adder has an additional input (3 inputs total) that allows a carry input to be handled. Both half and full adders generate a sum and a carry output. These building blocks can be implemented a number of ways using various logic gates.

Due to their versatility and usefulness, parallel adders have been available as integrated circuit devices. The 74LS283 (shown in Fig. 9-1) is an example of a 4-bit parallel binary adder chip. This MSI chip can add two 4-bit binary numbers (A4 A3 A2 A1 and B4 B3 B2 B1) together. The adder chip outputs a 5-bit sum (C4 Σ4 Σ3 Σ2 Σ1). There is also a carry input (C0) available that can be used to connect multiple 74283 chips together for larger adder applications. An 8-bit parallel adder constructed with two 74283 chips is illustrated in Fig 9-2. The inputs A8 through A1 and B8 through B1 are added together to produce the 9-bit sum labeled S9 through S1. Parallel binary adders can also be used in many arithmetic applications besides just simple addition. See Quartus Notes available from the publisher's textbook web site at www.pearsonhighered.com/electronics.

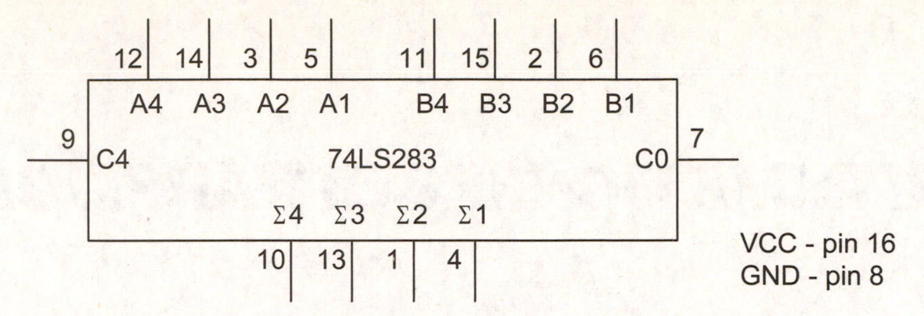

Fig. 9-1 A 74LS283 4-bit parallel binary adder chip

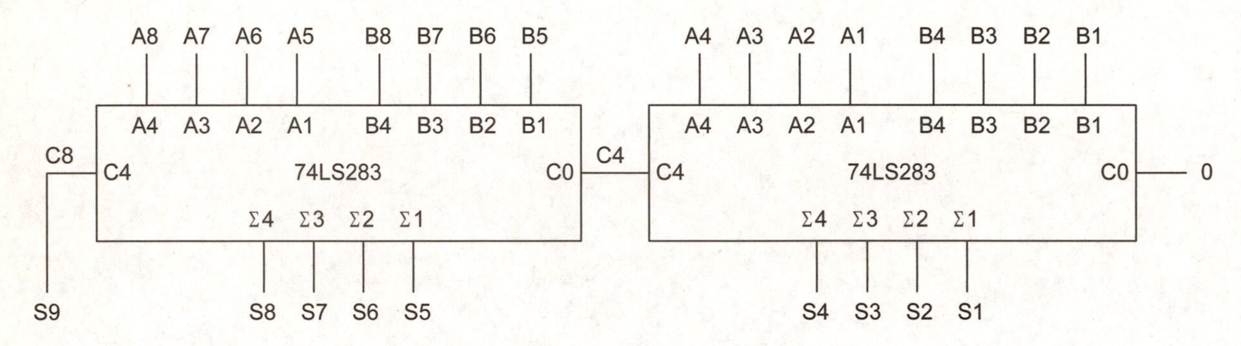

Fig. 9-2 An 8-bit parallel binary adder circuit using two 74283 chips

Example 9-1

Create an 8-bit parallel adder using Quartus II.

Quartus II provides several options for designers to use in creating our desired logic function. We will illustrate 4 equivalent solutions for this 8-bit adder circuit. Two solutions will apply schematic capture design entry techniques and two will use the HDL languages presented in this lab manual.

The first schematic capture solution (see Fig. 9-3) uses built-in library functions (found in the maxplus2 library) that mimic standard digital IC devices. Many of these functions are named after the same standard parts that they imitate. This solution, therefore, merely uses the same 74283 function that is suggested in Fig. 9-2. The inputs and outputs are drawn as buses in the *bdf* schematic. Two 8-bit input ports (one for the set of A inputs and one for the set of B inputs) and one 9-bit output port (for the set of S outputs) are needed. The individual signals are split from the buses for connection to the appropriate input and output ports on the 74283 function symbols. Note that the CIN for the least significant 4-bit adder (the one at the top in the schematic) is connected to GND. The symbols for constant HIGH and LOW logic inputs are named VCC and GND, respectively.

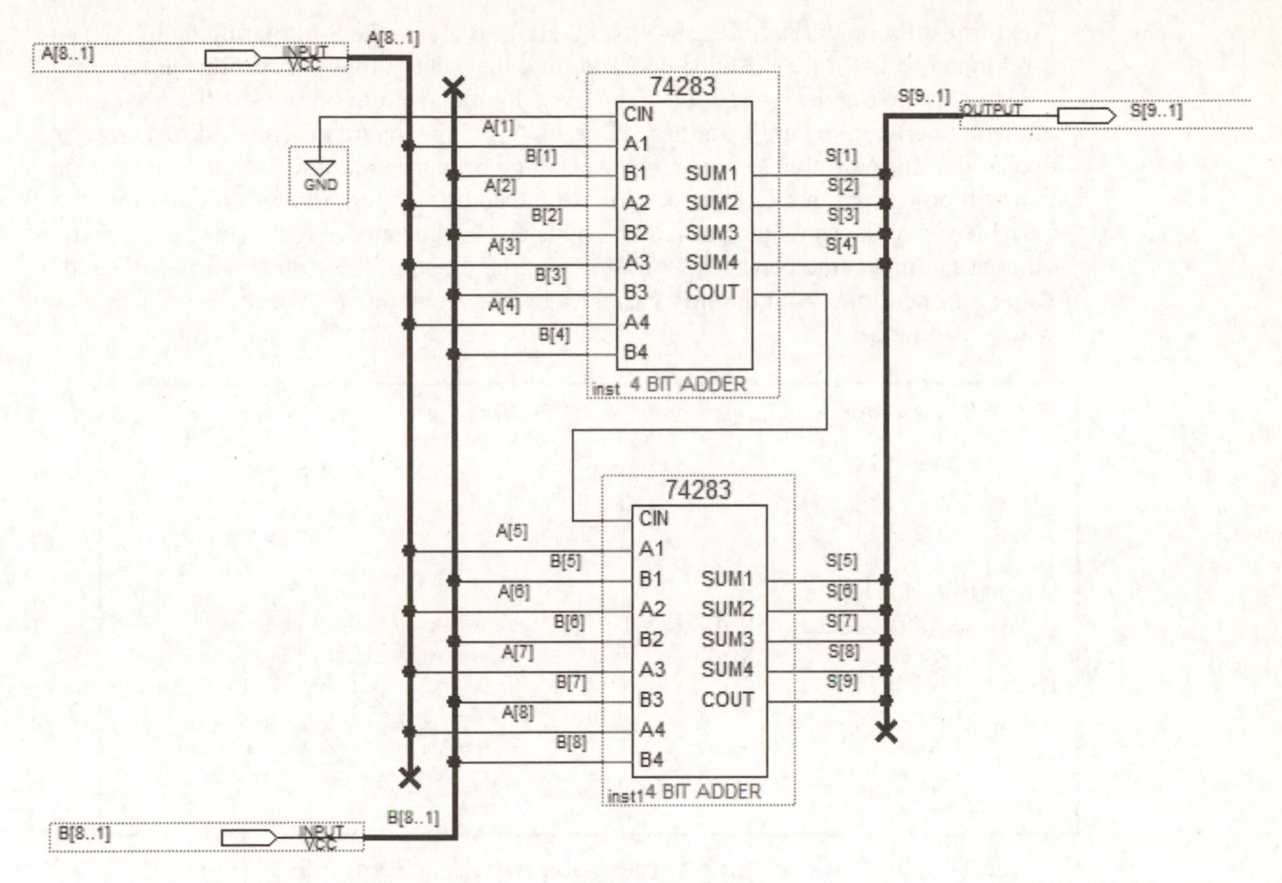

Fig. 9-3 Quartus solution one: 74283 schematic for 8-bit parallel binary adder

Our second solution shown in Fig. 9-4 is another *bdf* file, but this time we will use a more flexible Quartus function library. These are referred to as LPMs (Library of Parameterized Modules) or megafunctions. Our desired function can be found in the Arithmetic megafunction folder and is called LPM_ADD_SUB. The specific block functionality is selected from a set of available features and parameters using the MegaWizard. Our megafunction solution only needs to add the two 8-bit operands and produce the 8-bit sum and carry out.

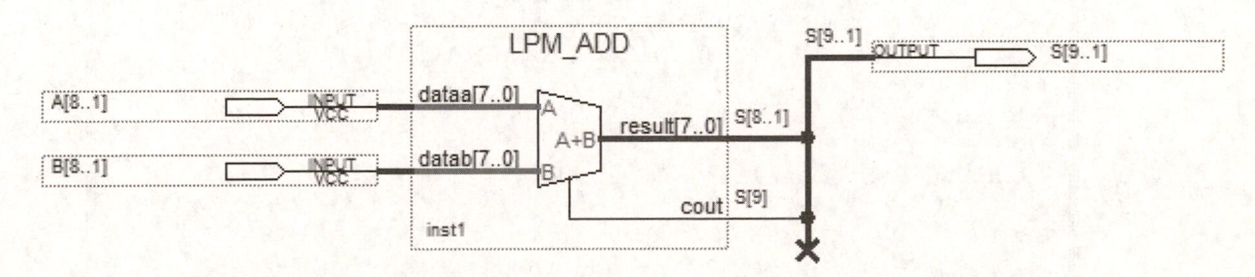

Fig. 9-4 Quartus solution two: LPM block diagram for Example 9-1

The third solution given in Fig. 9-5 uses AHDL to create the 8-bit parallel adder. The 8-bit input ports for the A and B operands and the 9-bit output port for the sum are declared in the Subdesign section. The most significant sum bit will be the final carry out generated by the parallel adder. Nine-bit "dummy" operands (aa and bb) are declared in the Variable section. This was done because the logic assignment statement that will be written in the Logic section for the addition operation will require that the variables have the same group size on both sides of the equals sign. There are also two other assignment statements in the Logic section. Each of these statements will set the value for one of the 9-bit dummy variables by concatenating one of the 8-bit data inputs with a leading zero.

```
SUBDESIGN  adderA
(
      a[8..1]             :INPUT;     -- 8-bit inputs
      b[8..1]             :INPUT;
      s[9..1]             :OUTPUT;    -- 9-bit output
)
VARIABLE
      aa[9..1]      :NODE;            -- create 9-bit
      bb[9..1]      :NODE;            -- dummy variables
BEGIN
      aa[9..1] = (GND,a[8..1]);       -- expand operands
      bb[9..1] = (GND,b[8..1]);       -- by concatenating
      s[9..1] = aa[9..1] + bb[9..1];        -- add operands
END;
```

Fig. 9-5 Quartus solution three using AHDL for Example 9-1

The fourth solution given in Fig. 9-6 uses VHDL to create the 8-bit parallel adder. The input ports for the A and B operands and an output port for the sum are declared as integer data types. An integer data type is used so that we can do an arithmetic operation with these objects. The architecture only contains the assignment statement to perform the addition.

```
ENTITY  adderV  IS
PORT (
      a           :IN INTEGER RANGE 0 TO 255;
      b           :IN INTEGER RANGE 0 TO 255;
      s           :OUT INTEGER RANGE 0 TO 511
);
END adderV;

ARCHITECTURE  parallel  OF  adderV  IS

BEGIN
      s <= a + b;        -- add the inputs
END parallel;
```

Fig. 9-6 Quartus solution four using VHDL for Example 9-1

Sample test vectors and simulation results for each of these design solutions are shown in Fig. 9-7. The simulation indicates that each circuit appears to function correctly.

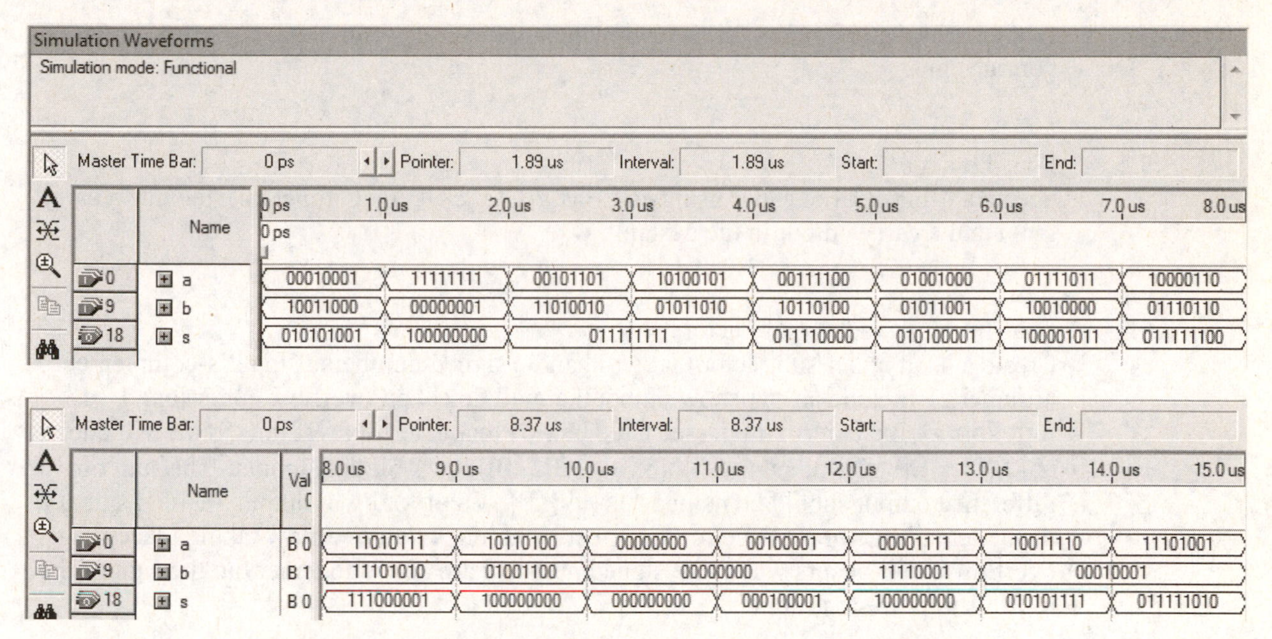

Fig. 9-7 Simulation results for 8-bit parallel adders

Laboratory Projects

Design arithmetic circuits for the following applications. Simulate, construct, and test your designs.

9.1 Four-bit adder
Create a 4-bit parallel adder using an LPM_ADD_SUB megafunction. Include a carry input and a carry output in the design.

9.2 Four-bit binary adder/subtractor
Create a 4-bit adder/subtractor that will handle signed numbers using 2's-complement arithmetic. Include an overflow output that will detect an overflow condition. Use function for the control input (see table below) that determines if the circuit will add (A[3..0] + B[3..0]) or subtract (A[3..0] – B[3..0]) the 4-bit data inputs. The sum or difference output should be named result[3..0]. Overflow will output a HIGH signal if the adder/subtractor produces an incorrect result due to an overflow taking place. An overflow result occurs when like-signed numbers are added together but the sign of the result is not the same sign.

(a) Use an LPM. Use the MegaWizard to select the specified features.

(b) Use an HDL. When subtracting, the sign of the B input will be changed.

function	Arithmetic operation
1	Result = A + B (addition)
0	Result = A – B (subtraction)

9.3 BCD adder (one digit)

Design an adder circuit that will output the sum of two 4-bit BCD numbers. The two BCD inputs are p[3..0] and q[3..0]. The carry input for the BCD adder stage is carry in. The BCD digit sum is sum[3..0], and carry out is the BCD carry output. The first parallel adder performs an initial binary addition of the 4-bit numbers. The "sum>9 detector" block determines if the **5-bit** intermediate sum produced by the first adder (binary carry, z[3..0]) is greater than 9. If so, the binary sum will be corrected by adding six (0110_2) with the second parallel adder and carry out will be HIGH. If the intermediate sum is less than or equal to 9, carry out is zero and will make no correction to the binary sum by adding 0000. Hint: An IF statement to check binary carry and z[3..0] in an HDL file is an easy design solution for the "sum> 9 detector" block. Note: Use this same circuit for each digit if adding multiple-digit BCD numbers.

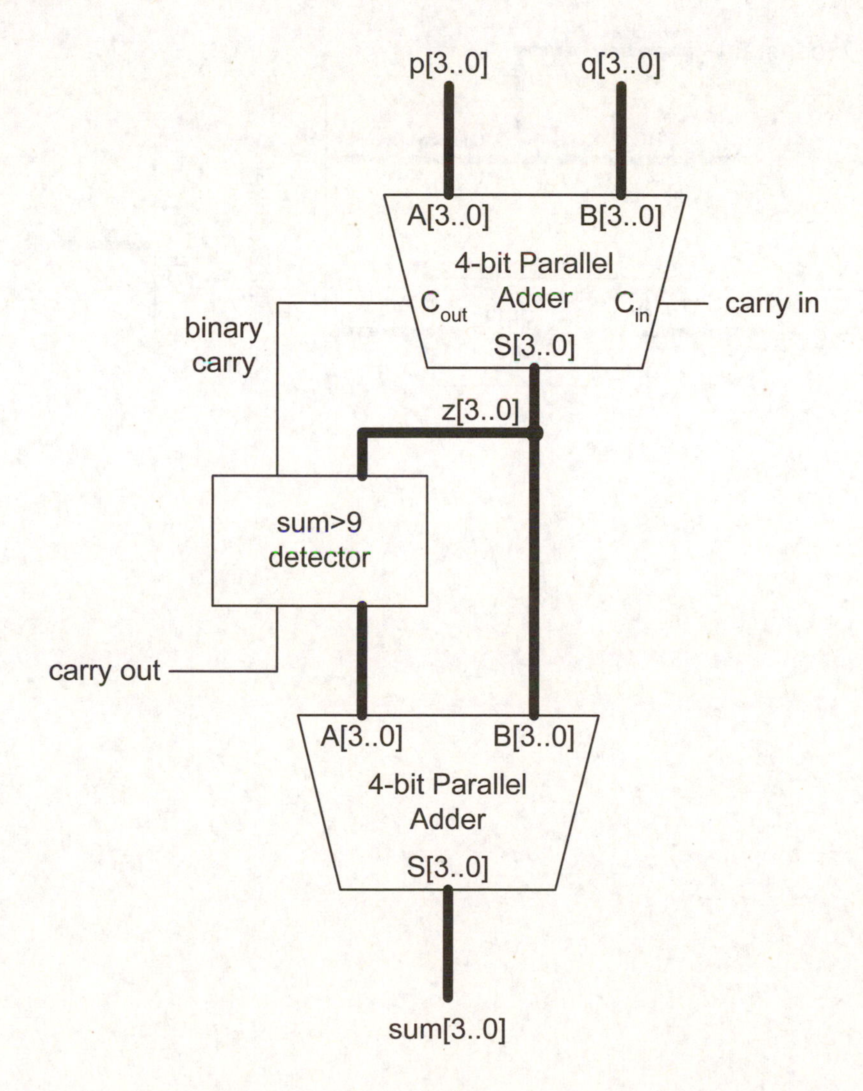

9.4 BCD to binary converter

Design a 2-digit BCD-to-binary converter. The inputs are the BCD tens digit (tens[3..0]) and ones digit (ones[3..0]). The output is the equivalent 7-bit binary value (binary[6..0]). The largest BCD input is 99, which should produce a binary output of 1100011. The design "trick" is to determine the binary equivalent for the weighted value of the tens digit that is applied, which is then added to the ones digit to produce the binary equivalent. For example, if the tens digit is 5, this would represent a value of 50 and the equivalent 7-bit binary value for 50 is 0110010. The function of the "tens digit circuit" then is to determine the value of the tens digit input and output a 7-bit binary number that is equal to 10 times the digit's input value. Note that we also need to insert leading zeros for the ones digit input to create a 7-bit input number to the parallel adder.

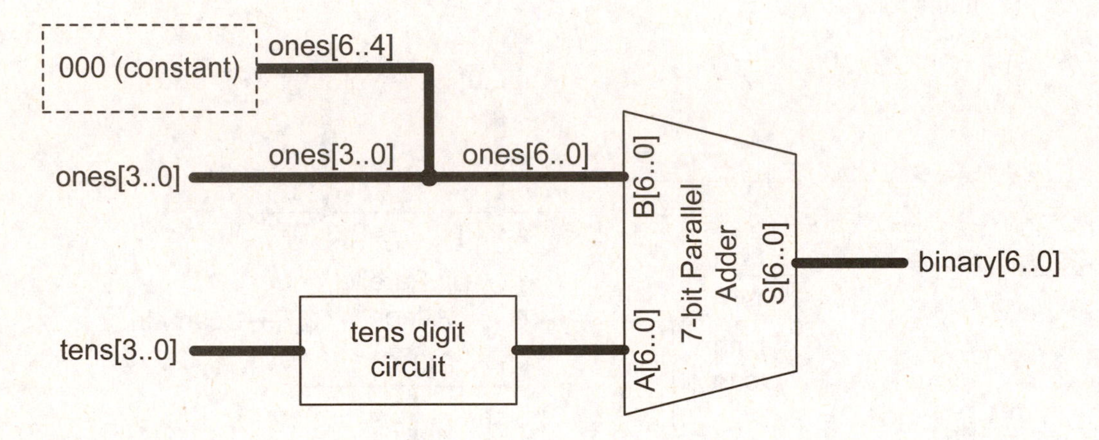

9.5 Binary to BCD converter

Design a 6-bit binary-to-BCD converter. The input is labeled binary[5..0] and the output is labeled bcd[6..0]. The largest input will be the binary number 111111, which is equal to 63_{10}. The output will contain two digits, with the tens digit equal to bcd[6..4] and the ones digit equal to bcd[3..0]. The design "trick" is to determine the decade range of the applied binary input and then output an appropriate adjustment value (see table below) that should be added to the binary input. The resultant sum will be the equivalent BCD value. For example, if the binary input is 100011 (35_{10}), we should add 010010 (18_{10}). The binary result is 0110101, which will be read as two BCD digits 011 and 0101 or 35_{10}. Note that the carry out from the adder will be the MSB of the output.

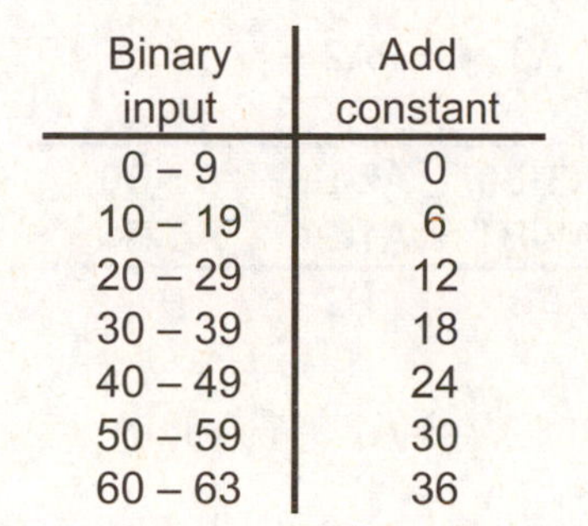

Binary input	Add constant
0 – 9	0
10 – 19	6
20 – 29	12
30 – 39	18
40 – 49	24
50 – 59	30
60 – 63	36

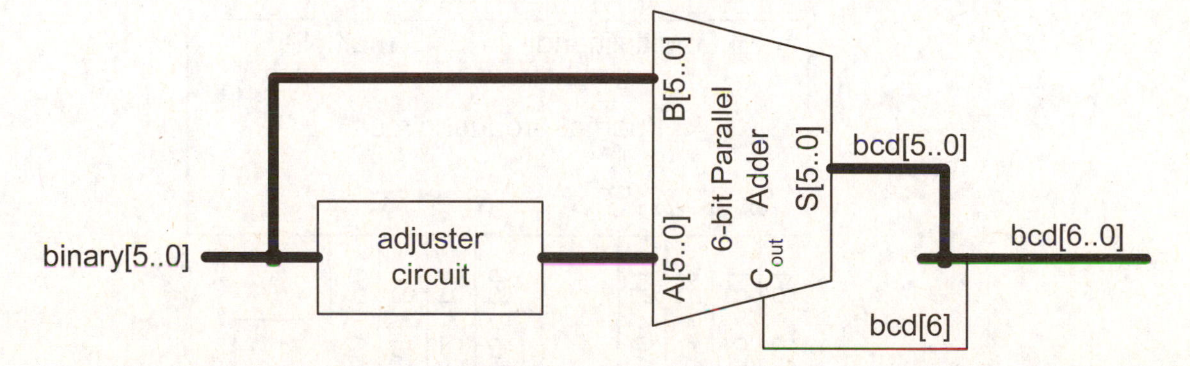

9.6 Up/down counter

Create an 8-bit, up/down counter that functions as shown in the following table. Use a parallel adder/subtractor LPM to add 1 to or subtract 1 from the current-state, which is stored in a register constructed with 8 D flip-flops-with-enables (Quartus storage primitive named dffe). The circuit will essentially "calculate" the counter's next-state. Define the adder/subtractor's b[7..0] input to be the constant 1 and connect the a[7..0] input to the register's q[7..0] output. Manually clock with a debounced pushbutton.

Enable	Up_down	Operation
0	0	hold
0	1	hold
1	0	Count down
1	1	Count up

9.7 Binary multiplier

Design an unsigned binary multiplier using a parallel adder. The multiplier circuit should handle 4-bit multiplicands (A3 A2 A1 A0) and 2-bit multipliers (B1 B0) to produce 6-bit products (P5 P4 P3 P2 P1 P0). The partial products are created by ANDing the appropriate <u>pair</u> of input bits. These partial products are then added together just like you learned to do in grade school. Note: Using signal busses is convenient in this project but it can be a little tricky to connect the signals to the blocks. You may find it convenient to insert a logic buffer named wire (a buffer component located in the primitive library) in some of the signal lines. This device will allow you to name each end of the wire with different signal names. Otherwise, you will not be able to label the single wire with both names.

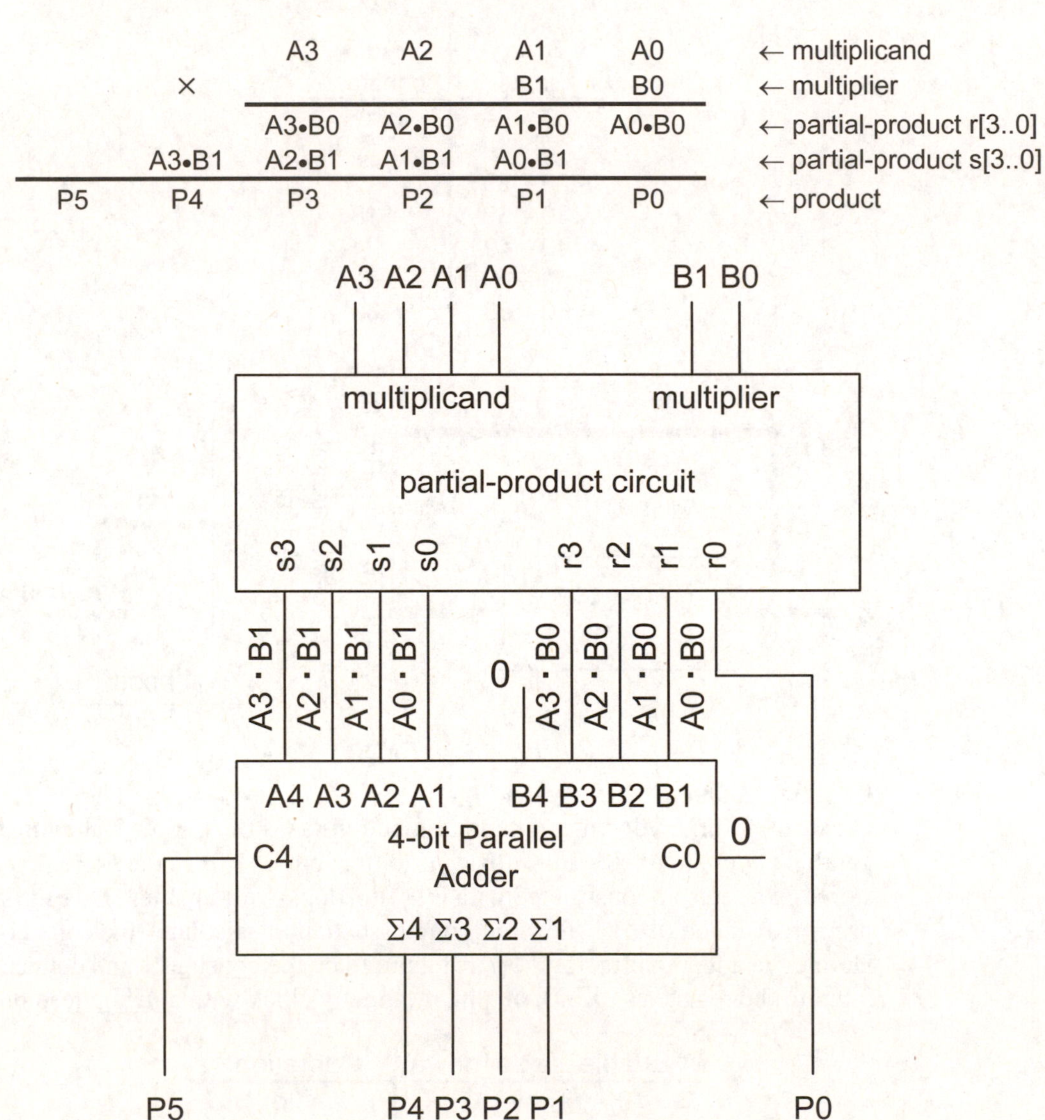

		A3	A2	A1	A0	← multiplicand
	×			B1	B0	← multiplier
		A3•B0	A2•B0	A1•B0	A0•B0	← partial-product r[3..0]
	A3•B1	A2•B1	A1•B1	A0•B1		← partial-product s[3..0]
P5	P4	P3	P2	P1	P0	← product

9.8 Multiplier megafunction

Create a multiplier using the LPM_MULT megafunction that will produce the product of two, unsigned, 4-bit numbers.

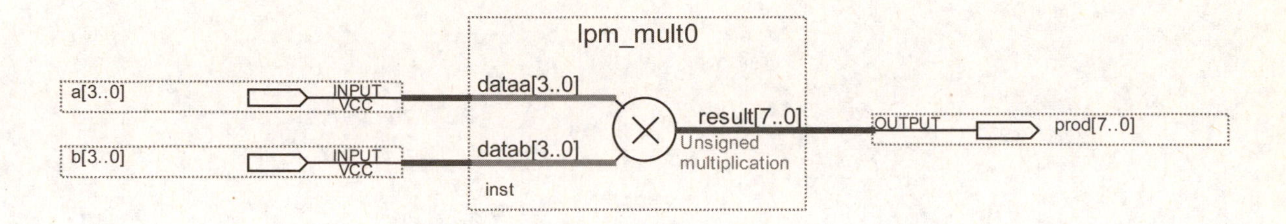

UNIT 10

ANALYZING AND TESTING SYNCHRONOUS COUNTERS

Objectives

- To analyze synchronous counter circuits constructed with JK and D flip-flops and predict their theoretical operation.
- To construct and test the operation of synchronous counter circuits using PLDs.
- To simulate synchronous counter circuits with Quartus II.

Synchronous Counters

Synchronous or parallel counters are triggered by a common clocking signal applied to each flip-flop. Because of this clocking arrangement, all flip-flops react to their individual synchronous control inputs at the same time. The count sequence depends on the control signals input to each flip-flop.

Sequential Circuits in PLDs

Sequential circuits can be easily created in PLDs from schematics using Quartus II. The primitives library contains storage devices including DFF, JKFF, SRFF, and TFF flip-flops and latches. See Quartus Notes available from the publisher's textbook web site at www.pearsonhighered.com/electronics.

Example 10-1

Analyze the synchronous counter circuit given in Fig. 10-1 to determine its count sequence. Draw the state transition diagram (include all 8 possible states) for the counter. Simulate the counter in Quartus. What is the counter's modulus? Quartus will automatically make the PRN and CLRN inputs on the JK flip-flops inactive if they are left unconnected in the schematic. If the circuit is to be constructed with standard logic devices, you should disable the PRN and CLRN inputs by tying them HIGH.

To analyze the counter, the circuit excitation (or present state/next state) table given in Table 10-1 is produced. For our analysis, we have assumed that the counter starts at state 000. The analysis indicates that the counter is a mod-5 counter (see Fig. 10-2). The simulation results are shown in Fig. 10-3.

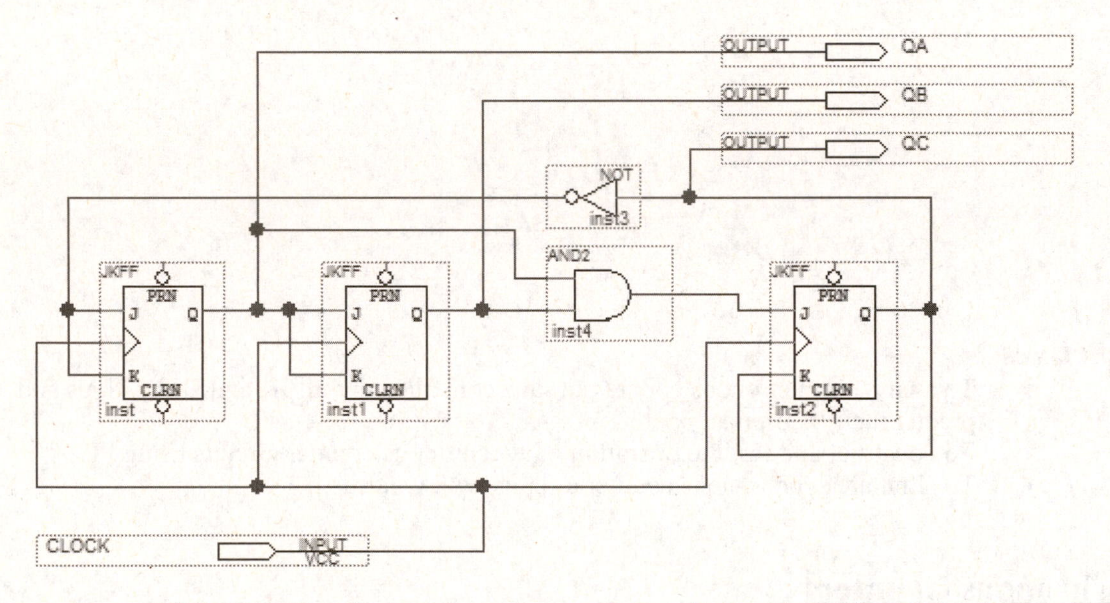

Fig. 10-1 Synchronous counter schematic for Example 10-1

CLOCK	Present State QC QB QA	J_{QC}	K_{QC}	J_{QB}	K_{QB}	J_{QA}	K_{QA}	Next State QC QB QA
0	0 0 0	0	0	0	0	1	1	0 0 1
1	0 0 1	0	0	1	1	1	1	0 1 0
2	0 1 0	0	0	0	0	1	1	0 1 1
3	0 1 1	1	0	1	1	1	1	1 0 0
4	1 0 0	0	1	0	0	0	0	0 0 0
	1 0 1	0	1	1	1	0	0	0 1 1
	1 1 0	0	1	0	0	0	0	0 1 0
	1 1 1	1	1	1	1	0	0	0 0 1

Table 10-1 Complete present state/next state table for Example 10-1

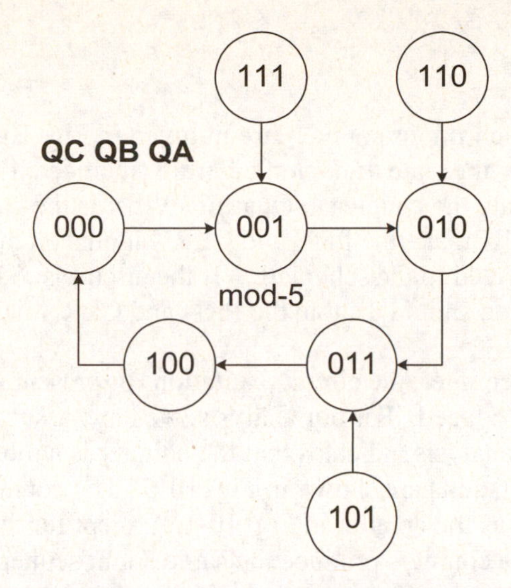

Fig. 10-2 State transition diagram for synchronous counter in Example 10-1

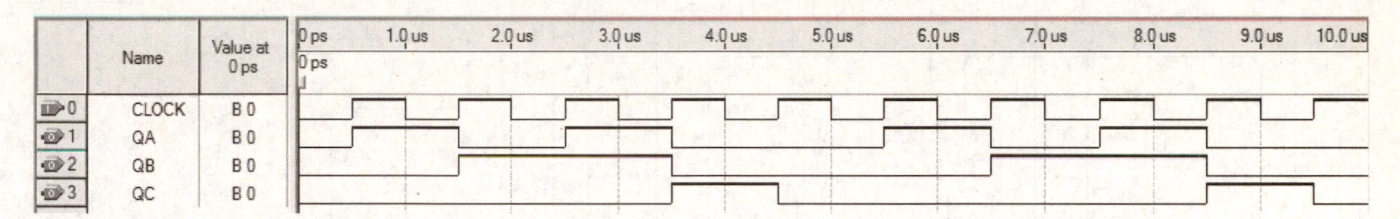

Fig. 10-3 Simulation results for synchronous counter in Example 10-1

Example 10-2

Analyze the synchronous counter circuit given in Fig. 10-4 to determine its count sequence. Draw the state transition diagram (include all 8 possible states) for the counter. Simulate the counter in Quartus. What is the counter's modulus? Quartus will automatically make the PRN and CLRN inputs on the D flip-flops inactive if they are left unconnected in the schematic. If the circuit is to be constructed with standard logic devices, you should disable the PRN and CLRN inputs by tying them HIGH.

To analyze the counter, the circuit excitation (or present state/next state) table given in Table 10-2 is produced. For our analysis, we have assumed that the counter starts at state 000. The analysis indicates that the counter is a mod-5 counter (see Fig. 10-5). The simulation results are shown in Fig. 10-6. The counter in Fig. 10-4 has the same count sequence as the counter in Fig. 10-1. Except for the states that are not normally encountered, the counters produce the same count sequence.

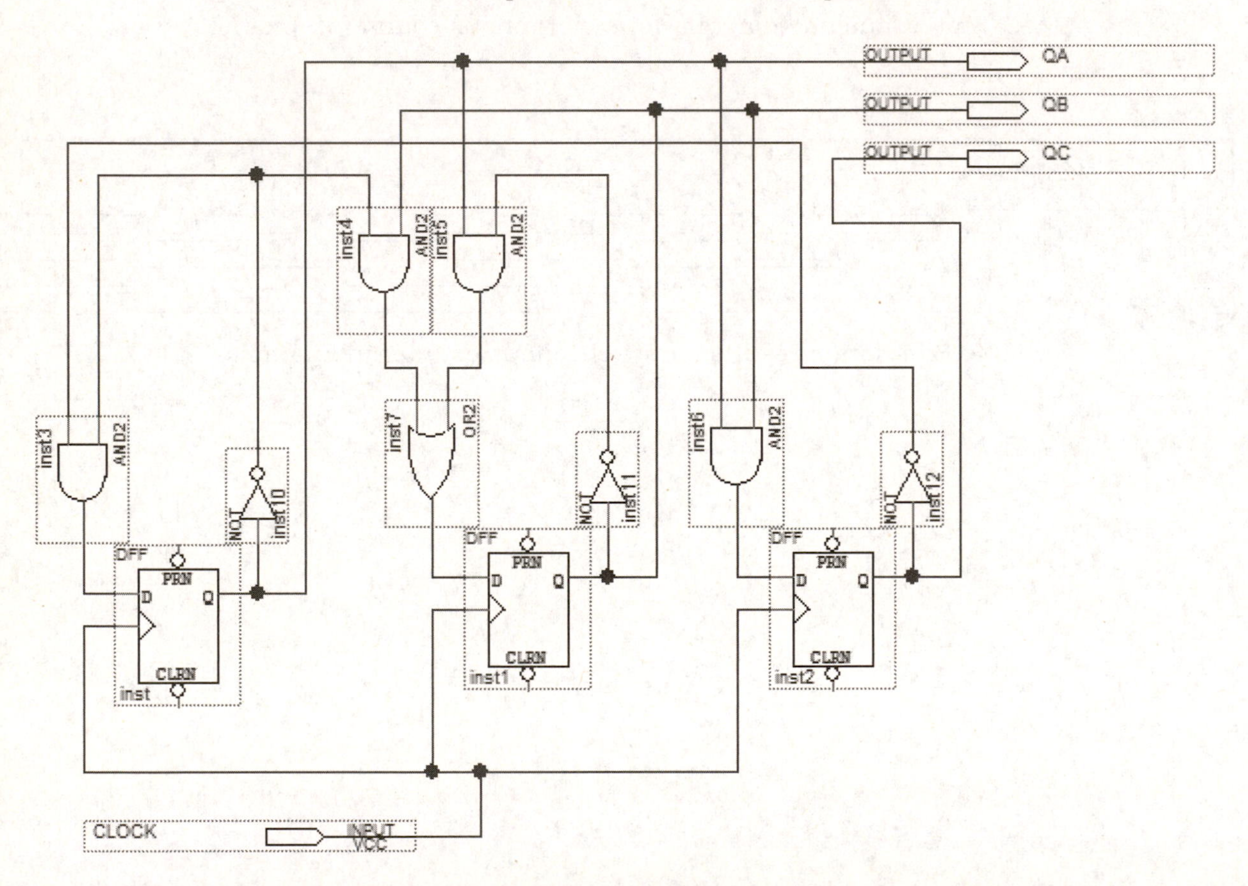

Fig. 10-4 Synchronous counter schematic for Example 10-2

CLOCK	Present State QC QB QA	D_{QC}	D_{QB}	D_{QA}	Next State QC QB QA
0	0 0 0	0	0	1	0 0 1
1	0 0 1	0	1	0	0 1 0
2	0 1 0	0	1	1	0 1 1
3	0 1 1	1	0	0	1 0 0
4	1 0 0	0	0	0	0 0 0
	1 0 1	0	1	0	0 1 0
	1 1 0	0	1	0	0 1 0
	1 1 1	1	0	0	1 0 0

Table 10-2 Complete present state/next state table for Example 10-2

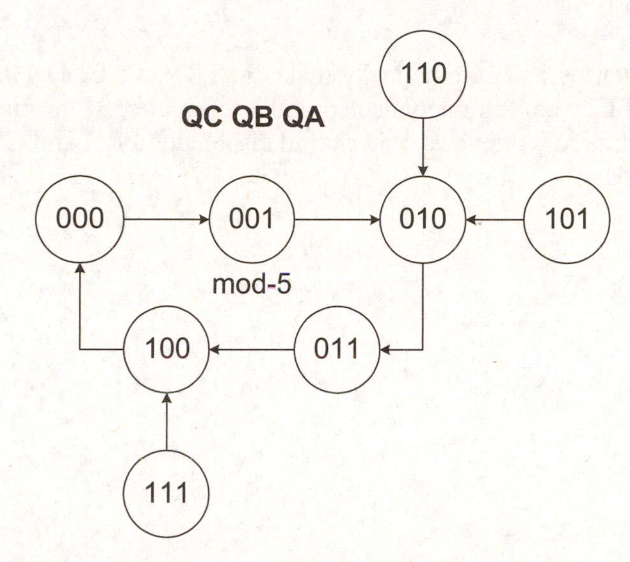

Fig. 10-5 State transition diagram for synchronous counter in Example 10-2

Fig. 10-6 Simulation results for synchronous counter in Example 10-2

Laboratory Projects

Analyze the following recycling, synchronous counter circuits. For each counter, determine the count sequence and the counter's modulus. Sketch its timing diagram (include clock and counter output waveforms).

Construct and test the operation of the counter circuits with a manual clock (use a debounced pushbutton). Connect lights to the counter outputs. Record the test results.

Then remove the manual clock and replace it with a 10 kHz **compatible** clock signal (you may need to change the pin assignment). With a 4-channel oscilloscope connected to the counter outputs (you may need to change the pin assignments), compare the output waveforms for each of the counters with your theoretical prediction. Demonstrate the counter operation on the oscilloscope to your lab instructor. Simulate each circuit.

Notes: Quartus will automatically make the PRN and CLRN inputs on the flip-flops inactive if they are left unconnected in the schematic. If the circuit is to be constructed with standard logic devices, you should disable the PRN and CLRN inputs by tying them HIGH.

10.1 (a) JK counter 1

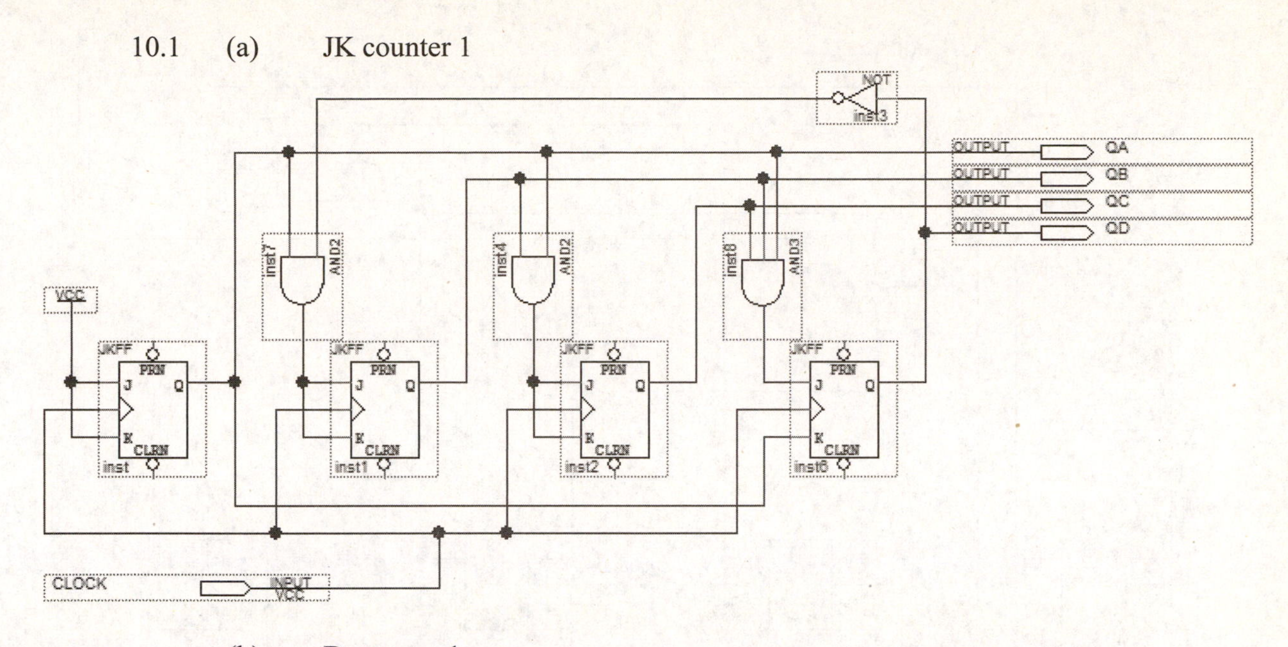

(b) D counter 1

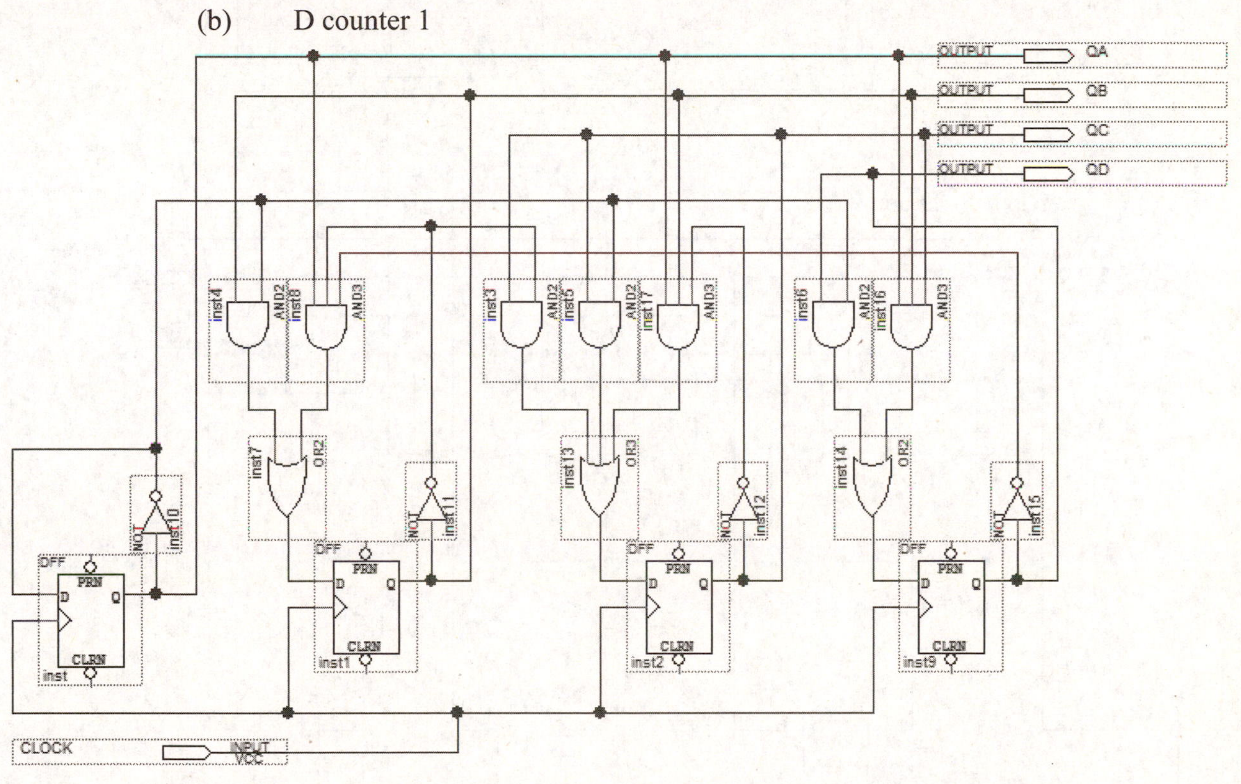

10.2 (a) JK counter 2

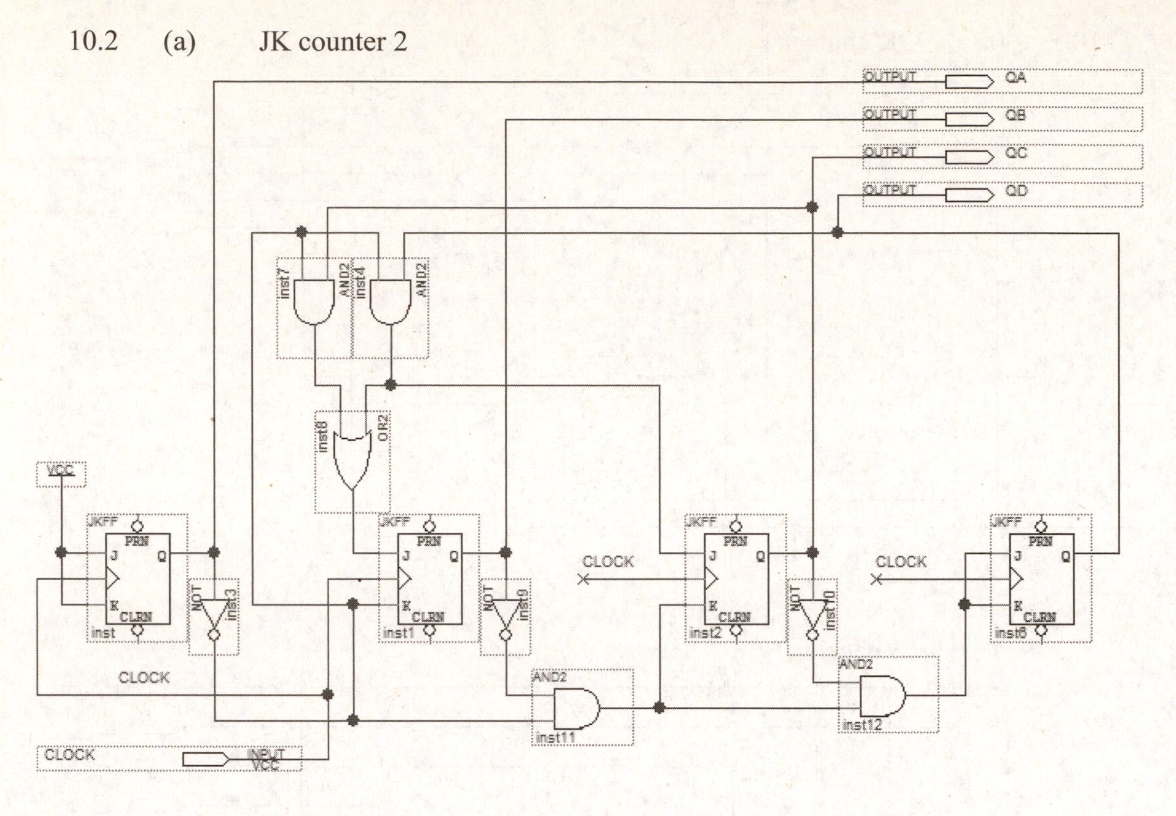

(b) D counter 2

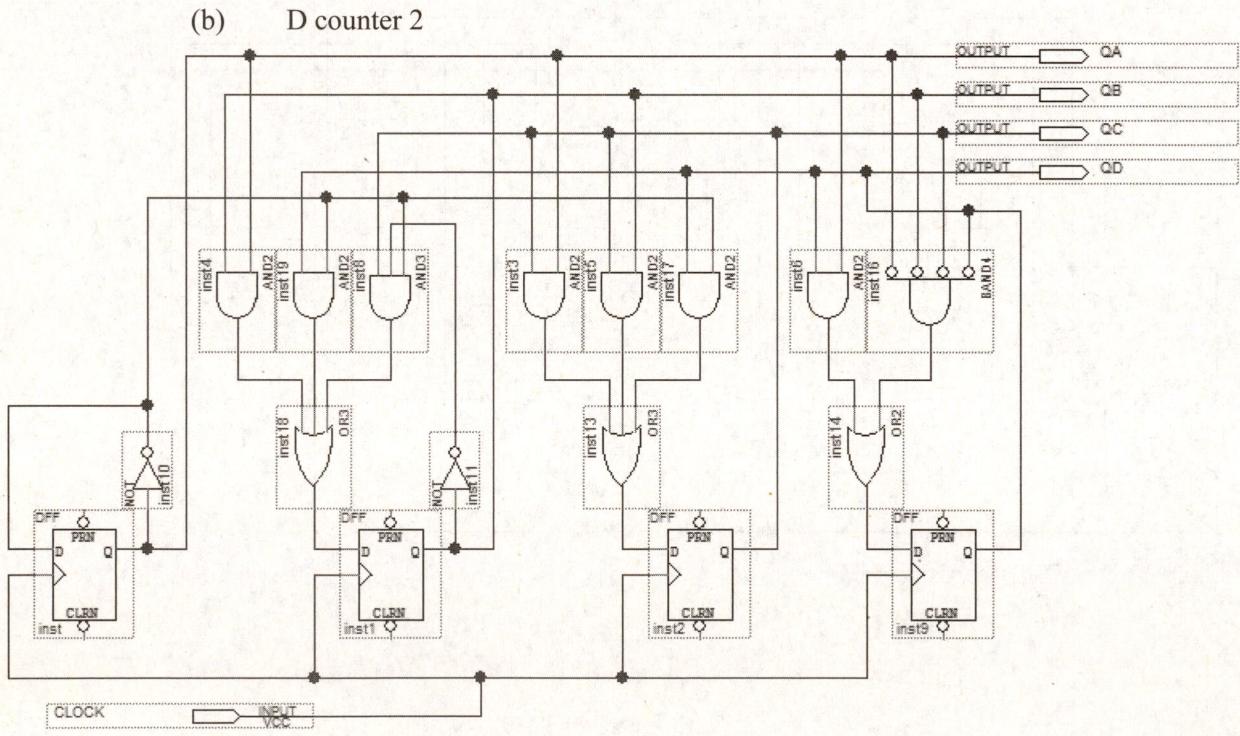

10.3 (a) JK counter 3

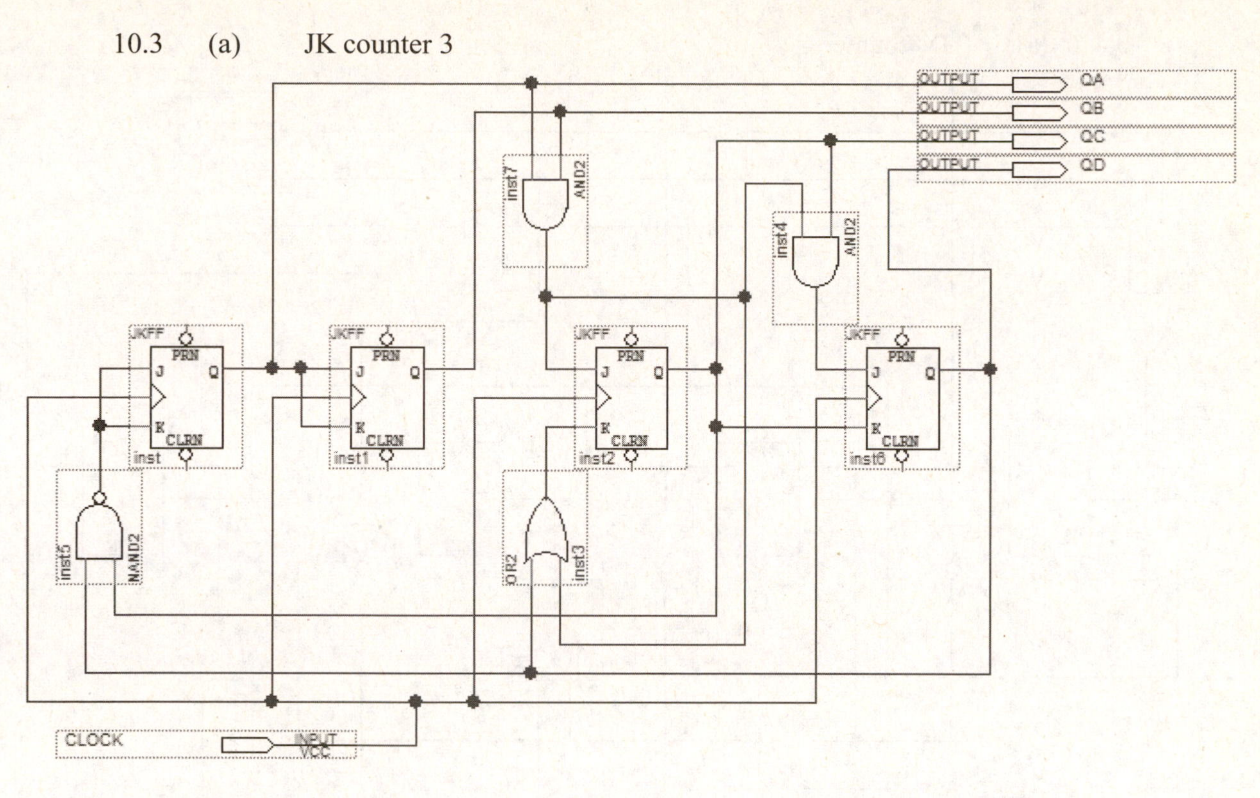

(b) D counter 3

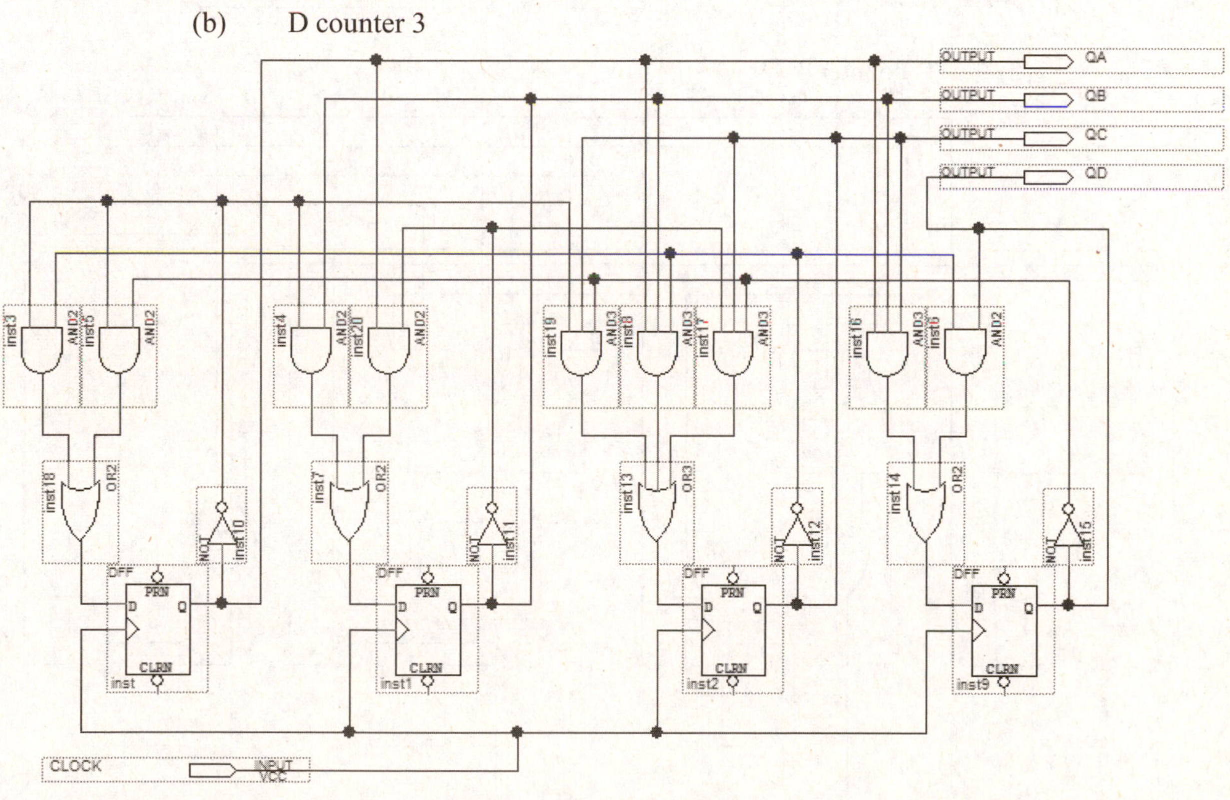

10.4 (a) JK counter 4

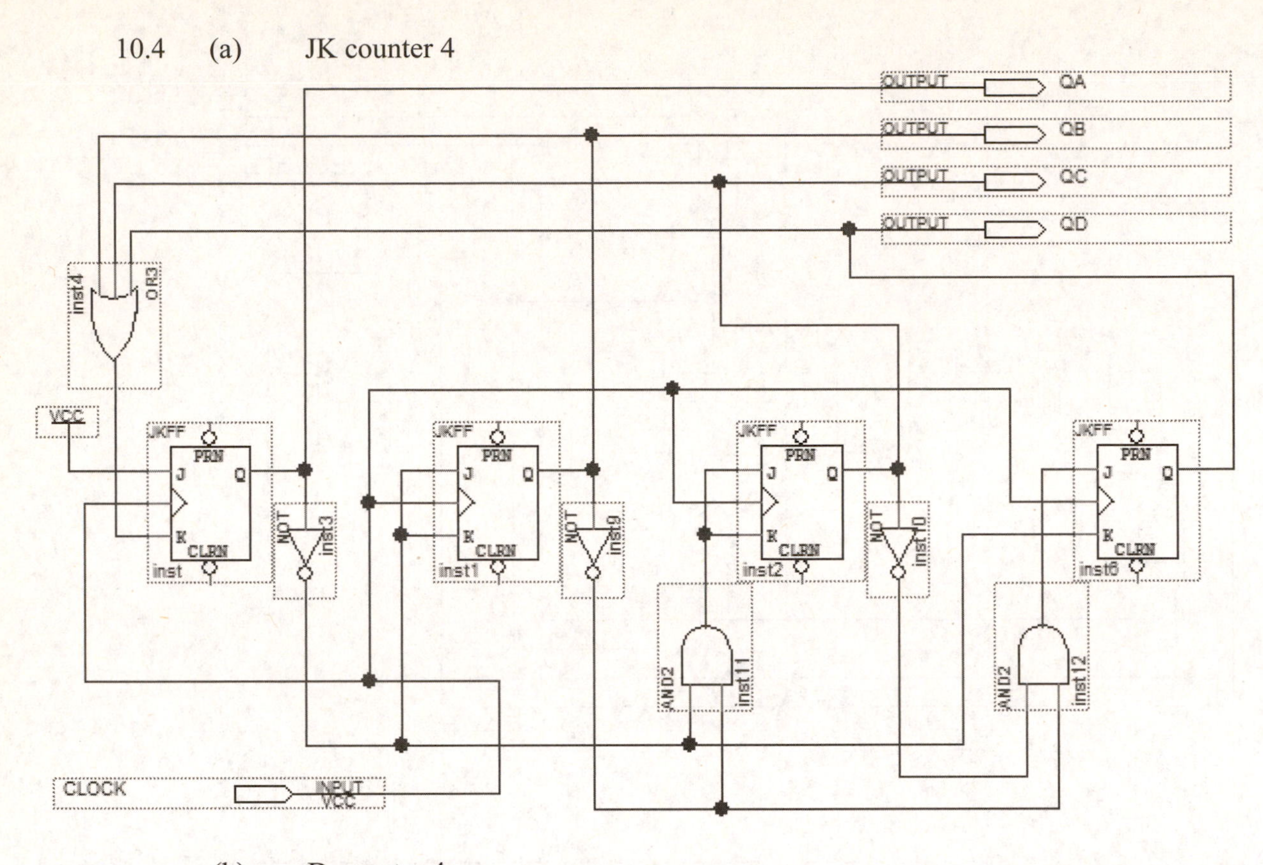

(b) D counter 4

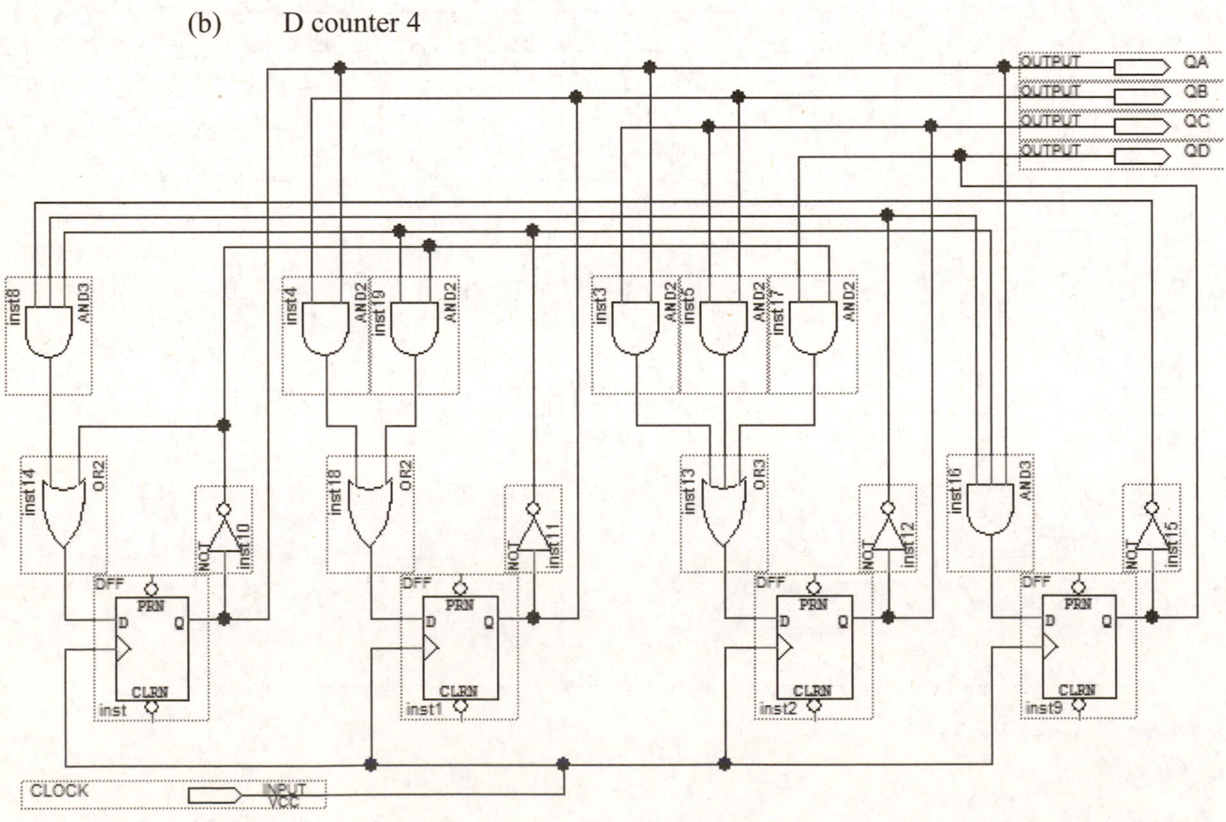

UNIT 11

CREATING AND TESTING COUNTER FUNCTIONS

Objectives

- To create counters with specified features using Quartus II megafunctions.
- To test the operation of counter functions.
- To simulate counter circuits.

Counter Circuit Features and Operation

Digital systems apply counters in many different configurations. Counter features that must be specified for an application include the number of flip-flops contained in the counter, the counter's modulus, synchronous or asynchronous counter clearing, synchronous or asynchronous counter loading, up/down count control, and various counter cascading implementations. A specific mixture of feature combinations will be needed for different design blocks in an application. Quartus's LPM_COUNTER megafunction makes it easy to create synchronous counters that include a desired set of counter features. The counter megafunction is found in the arithmetic library because an up count sequence is produced by adding one to the current counter state and a down count sequence is produced by subtracting one from the current counter state. See Quartus Notes available from the publisher's textbook web site at www.pearsonhighered.com/electronics.

Example 11-1

Create a mod-16 binary up-counter using a Quartus LPM_COUNTER megafunction. The 4-bit counter should also include the following functions: count enable, synchronous parallel load, and asynchronous clear.

The schematic symbol for the specified counter and MegaWizard settings are shown in Fig. 11-1. The counter outputs are named Q3 Q2 Q1 Q0 (with Q3 = MSB). The control inputs are labeled LOAD, EN, and CLR. The clock input is CLK. The parallel data inputs are labeled D3 D2 D1 D0.

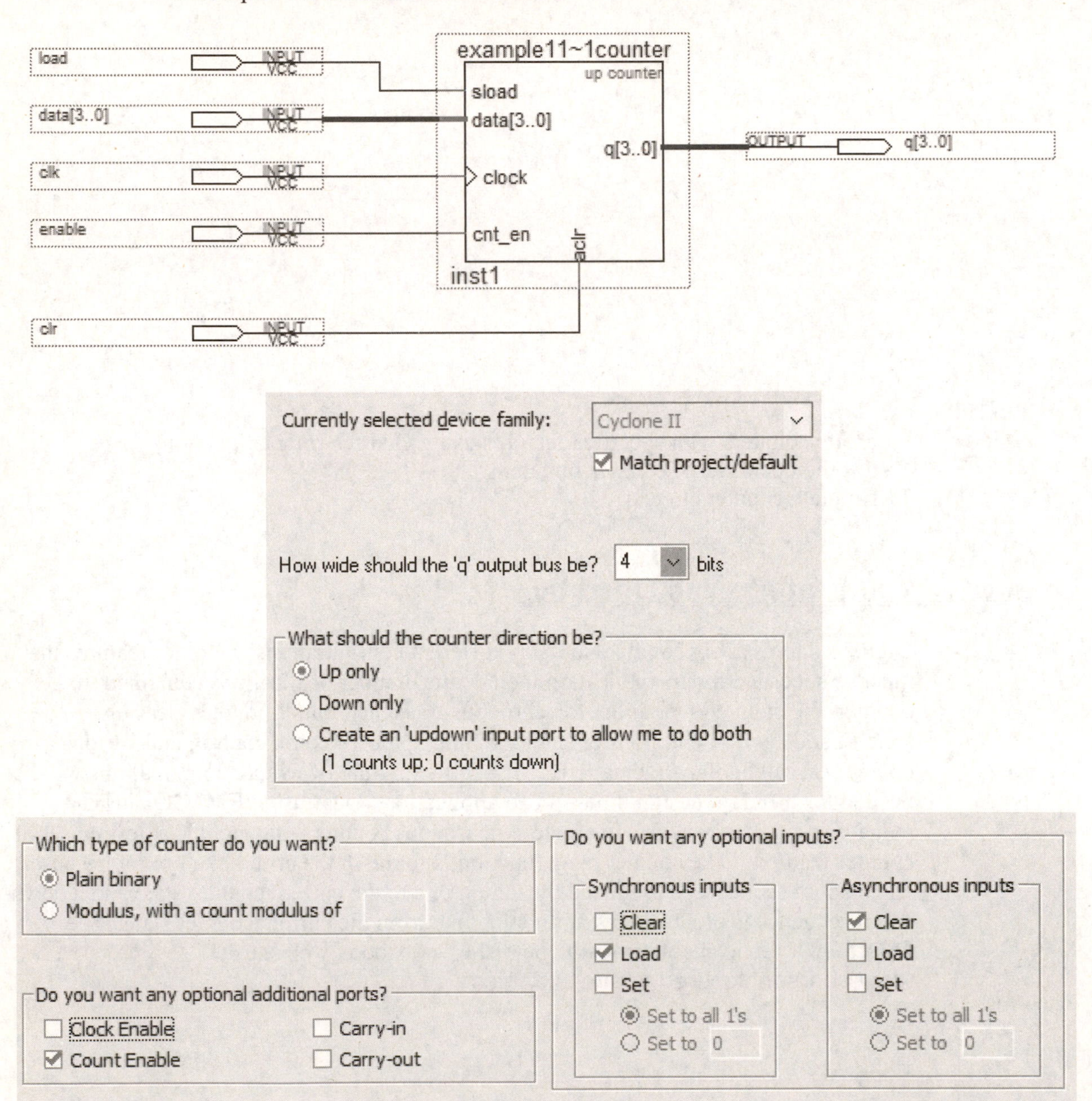

Fig. 11-1 Mod-16 binary up-counter using LPM_COUNTER and MegaWizard settings

Example 11-2

Create the counter circuit shown in Fig. 11-2 by modifying the circuit in Example 11-1. The LOAD input control switch is replaced with a 2-input AND gate whose inputs are Q3 and Q0 from the counter. Determine the counter's operation when EN = 1, CLR = 0, and D[3..0] = 0000.

When the LOAD input is HIGH, the counter will synchronously load the input applied to D[3..0]. The AND gate detects (decodes) when the counter has a HIGH output for both Q[3] and Q[0], which would first occur at counter state 1001. So, when the counter reaches 1001, 0000 will be loaded into the counter on the next clock. The modified circuit, therefore, produces a recycling count sequence of 0000 to 1001. This is a mod-10 or decade counter. See the functional simulation results in Fig. 11-3.

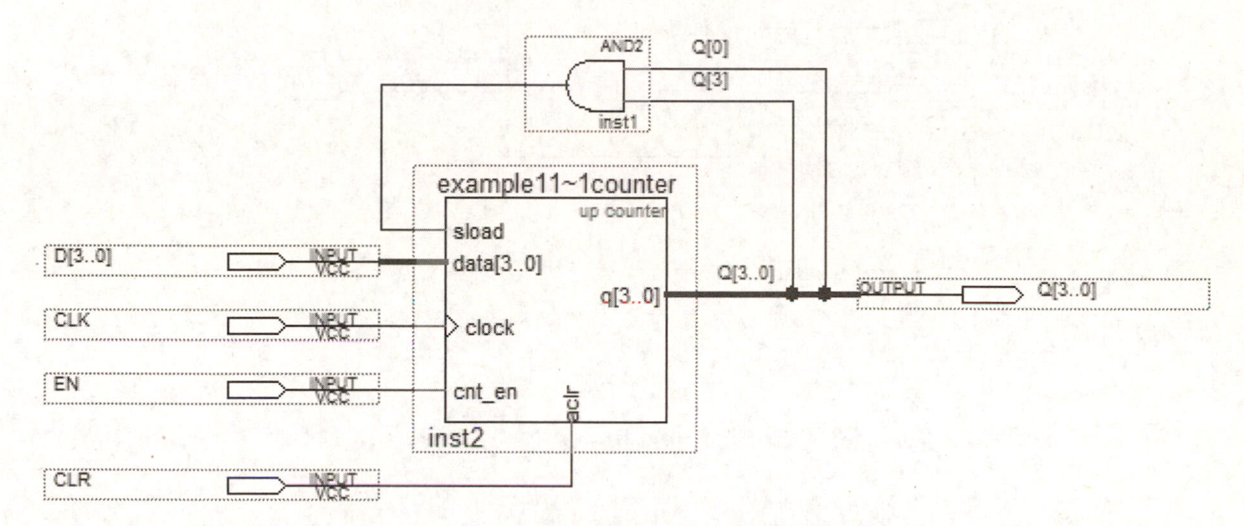

Fig. 11-2 Schematic for modified counter

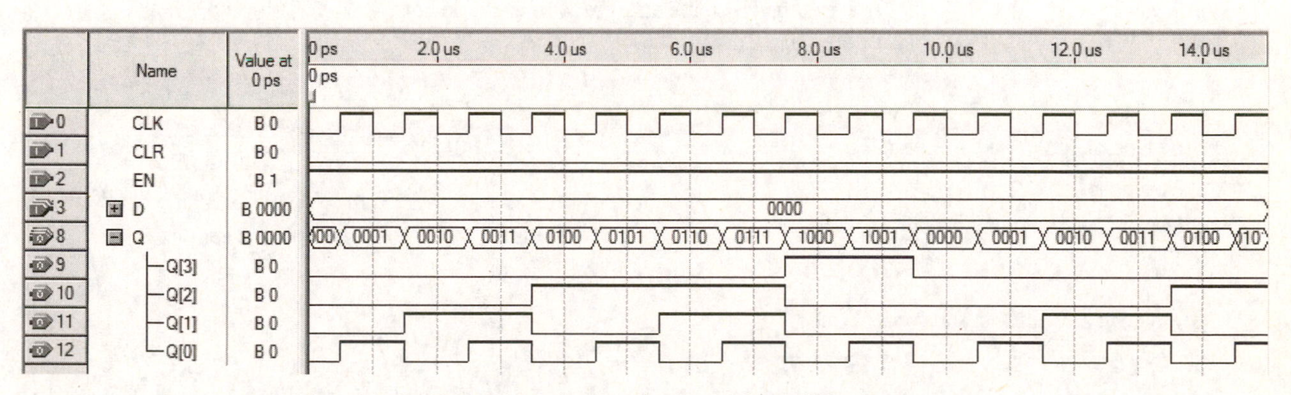

Fig. 11-3 Functional simulation of modified counter

Example 11-3

Determine the functional difference between the three types of enable controls (CLK_EN, CNT_EN, and CIN) available for LPM_COUNTERS.

A MOD-6, recycling up counter is used as a test circuit. The count sequence will be 0 through 5. The MOD-6 test circuit also includes a synchronous clear control (SCLR) and a terminal state decoder output called COUT. The block symbol for the test circuit and MegaWizard settings are shown in Fig. 11-4. All control inputs are active-HIGH.

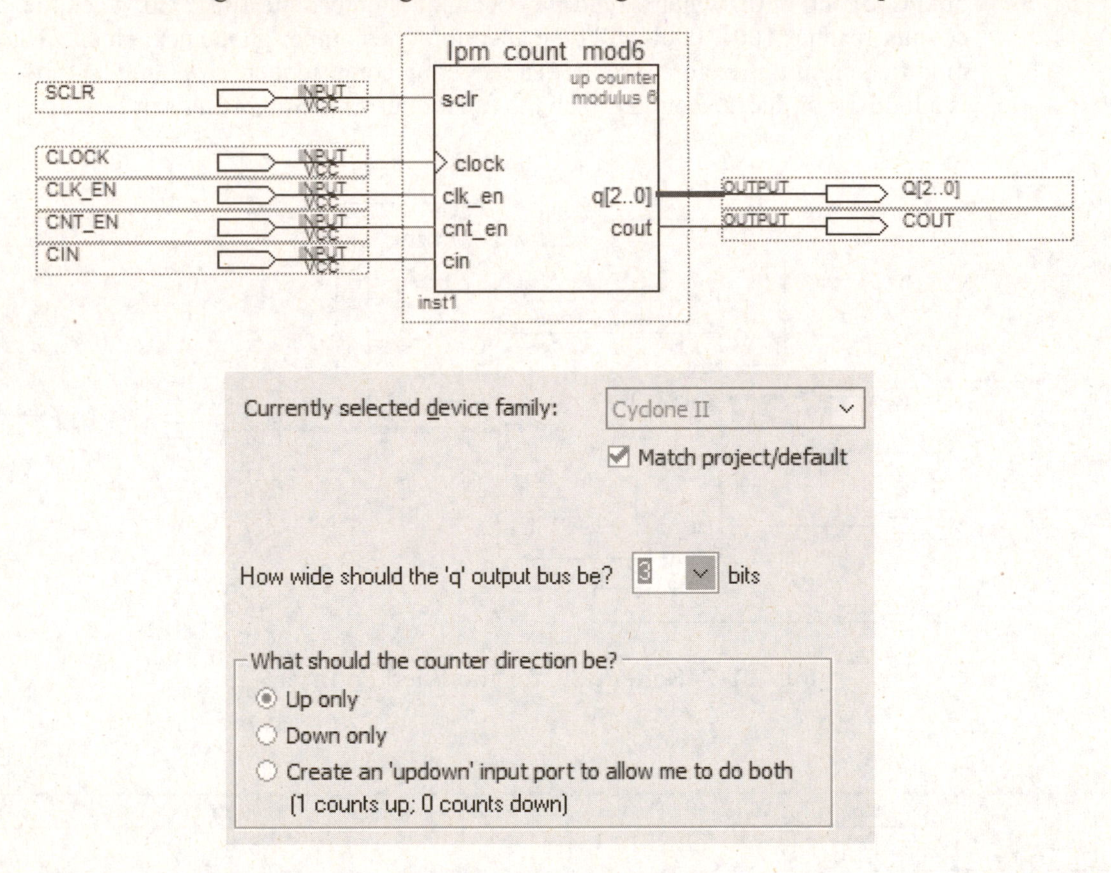

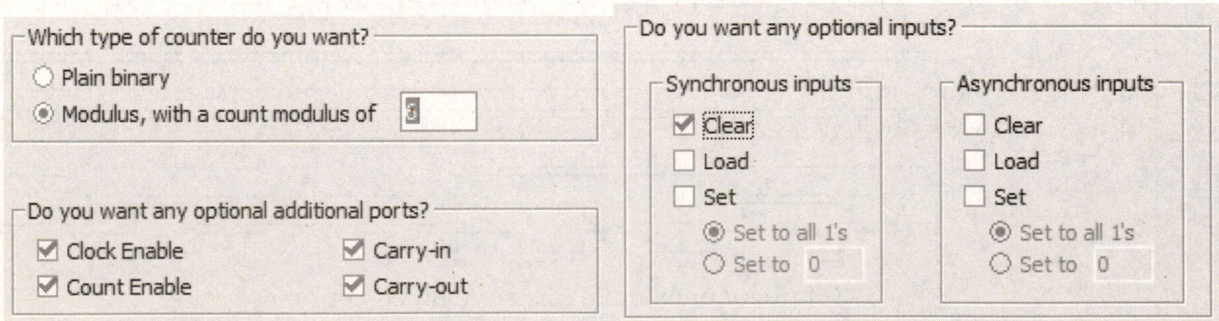

Fig. 11-4 MOD-6 counter (lpm_count_mod6) used as a test circuit

The results of a functional simulation for this test circuit are given in Fig. 11-5. The recycling count sequence can be seen in the time segment from 0 to 7μs including a HIGH for COUT when the counter reaches state 5. Each of the three enables will stop the count when they have a LOW input (time: 8-15μs). The counter is synchronously cleared when SCLR = 1 (time: 18.5μs). When the counter is at its terminal state and, one-at-a-time, each enable control has a LOW input applied during time range 24 to 29μs, we can see that the CIN enable also controls COUT. During the time range 30 to 44μs, the SCLR input is tested while each of the three enables individually is LOW. One of the enable controls, CLK_EN, will not only stop the count but it also disables other synchronous input controls. In summary, all three enable controls can be used to disable the count, but CIN also disables the terminal state decoder output (COUT) and CLK_EN also disables other synchronous control functions.

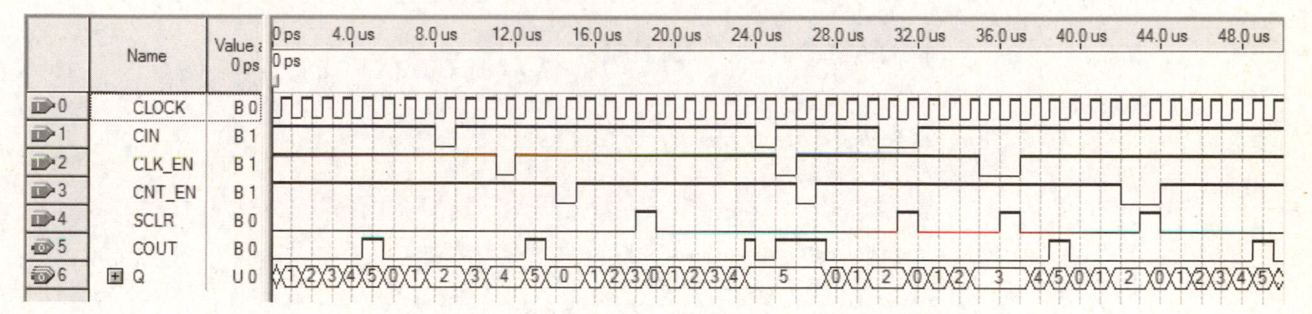

Fig. 11-5 Functional simulation results for Example 11-3

Example 11-4

Create a MOD-10 up/down counter using an LPM_COUNTER megafunction. The counter will count up when the up/down control, named UP_DN, is HIGH. The counter should have a count ENABLE control. Also add an active-HIGH output that detects the terminal state (TS) of the decade counter. The TS output should also be controlled by ENABLE. The terminal state for a decade up counter is 9, while the terminal state for the down counter is 0.

The schematic symbol and MegaWizard settings for the specified counter are shown in Fig. 11-6. The decade counter outputs are named Q3 Q2 Q1 Q0 (with Q3 = MSB). The updown and cin control inputs have been selected using the Quartus wizard for the LPM_COUNTER megafunction. The cin input stands for "carry-in," which is a count enable control. The difference between the count enable function and carry-in is that cin is designed to also automatically control the cout ("carry-out") output while the count enable only controls the count function. The function of cout is to detect the terminal state of the counter megafunction.

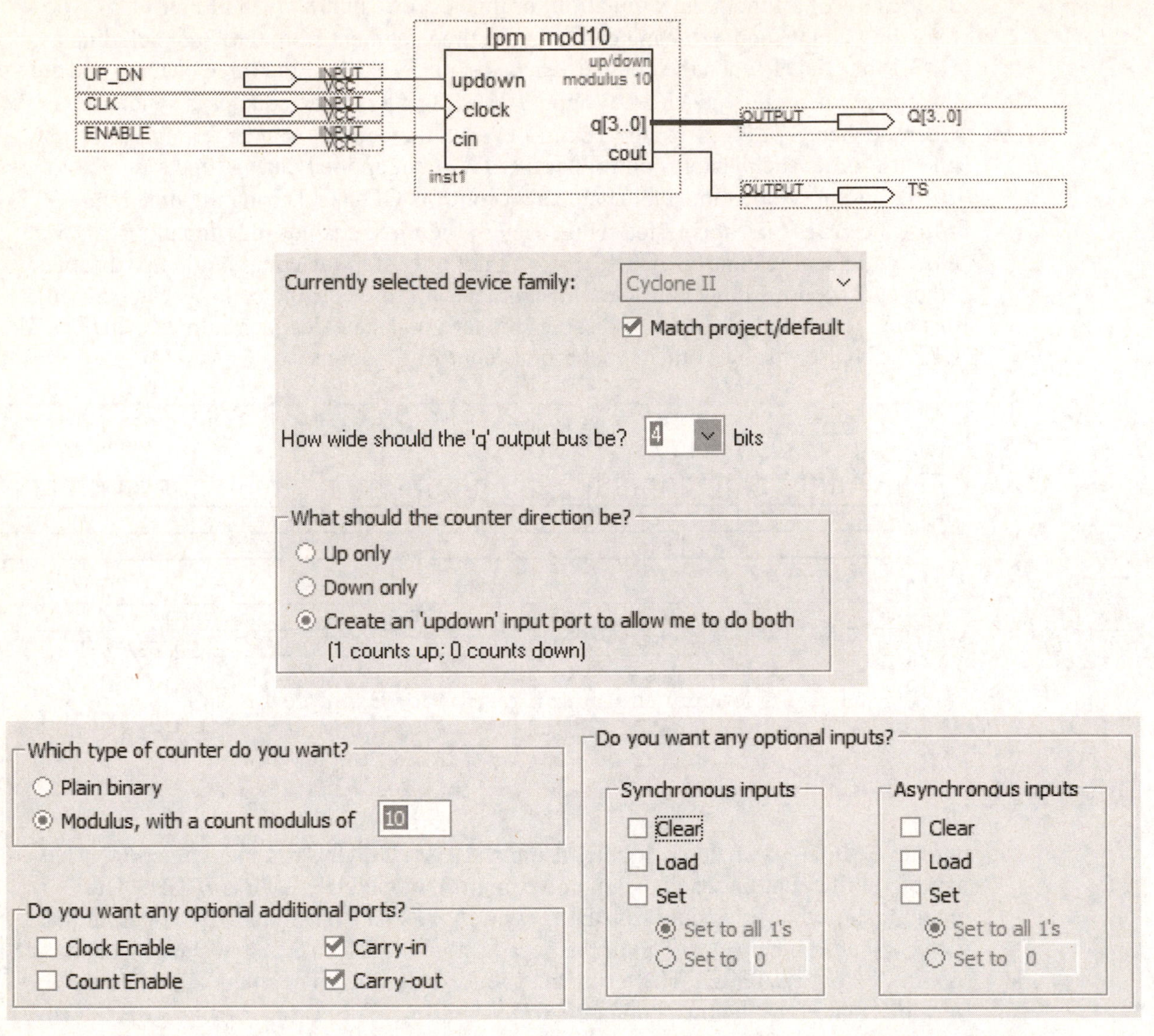

Fig. 11-6 Schematic for MOD-10 up/down counter using an LPM_COUNTER

Laboratory Projects

Construct and test each of the following logic circuits.

11.1 MOD-16 binary counter

Test the operation of the MOD-16 binary counter described in Example 11-1. Connect CLK to a manual clock input (debounced pushbutton). If you cannot observe (using a lamp or logic probe) the logic signal produced by the pushbutton, add another output port (see schematic below) so that you can monitor its operation on a lamp. Connect EN, CLR, and LOAD controls and data inputs D3 D2 D1 D0 to logic switches. Monitor the counter outputs Q3 Q2 Q1 Q0 on lamps.

Start your investigation with the three control inputs (LOAD, CLR, and EN) all LOW while you manually clock the counter. What happens? Next switch EN to a HIGH logic level and clock the counter. After a few clock pulses have been applied, switch EN LOW and continue clocking. What is the function of EN? Is EN an active-HIGH or active-LOW control? Continue clocking the counter until it recycles. Which output bit is the MSB? Which is the LSB? What is the count sequence? Does the counter count up or down? Why is this counter called a MOD-16 counter? Which clock edge triggers the counter to go to the next state?

Step the counter to a state other than 0000. Keep the EN input HIGH and switch CLR HIGH also. What happens? Do you need to press the CLK pushbutton for the CLR function? Is CLR synchronous or asynchronous? How do you know? Is the CLR function dependent upon the logic level applied to EN (i.e., do you need to keep EN HIGH for CLR to operate)? Is the CLR control input active-HIGH or active-LOW? Why do we need to step to a state other than 0000 to test the CLR function? Which control EN or CLR has a higher priority?

Next switch LOAD to a HIGH while EN and CLR are both LOW. Set the D[3..0] switches to a binary number and press the CLK pushbutton. Try several different binary numbers on D[3..0] and clock the counter. What happens? Is LOAD an active-HIGH or active-LOW control? Is LOAD synchronous or asynchronous? How do you know? What is the priority order (highest to lowest) for the control inputs (EN, CLR, & LOAD)?

Create a function table to describe the operation of this counter. Simulate the circuit. Demonstrate the operation of this counter to your lab instructor.

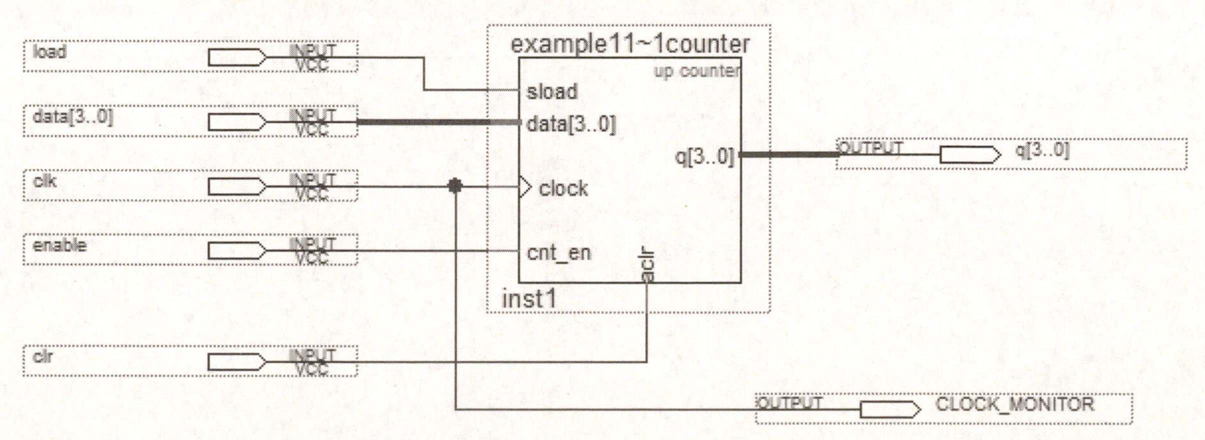

11.2 Binary counter signal frequencies

Apply a compatible clock signal that has a convenient frequency (such as 16 kHz) to the binary counter described in Example 11-1 while EN = 1 and LOAD = CLR = 0. Measure the actual signal frequencies of the clock input and the four counter outputs (Q0, Q1, Q2, and Q3) using either a frequency counter or an oscillocope. What is the frequency relationship between each of these signals?

11.3 Functional vs. timing simulation

Perform and compare two simulations of the Example 11-1 circuit – a functional simulation and a timing simulation of the same circuit with EN = 1 and LOAD = CLR = 0. Double-click "Edit Settings" task to change the Simulation Mode. Use a clock frequency of 50 MHz for your simulations. How are the simulations different? Why? What seems to be happening between some of the states in the timing simulation? Why? Why don't we see this on the lights?

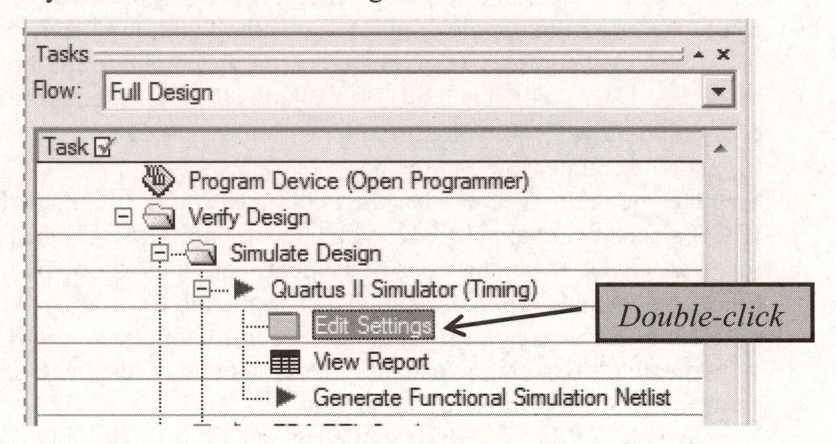

11.4 MOD-12 counters

(a) Use the lpm counter described in Example 11-1 to create a MOD-12 counter by controlling the CLR input. Remove the original input port connection for the CLR and, instead, connect to the CLR input a logic gate that detects ("decodes") the appropriate counter state to produce the MOD-12 count sequence. Which counter state should be detected? Construct and test this circuit solution with a debounced manual clock or a low-frequency compatible clock signal. What is the count sequence? What do you have to do to the other control inputs for the counter to count? What is meant by the term "transient" state? The timing diagram for this counter solution will have a "glitch." Where will it occur and why? Simulate (functional) this design with a 1-kHz clock.

(b) Create a second design solution by using the **same** decoder circuit from part (a) to control the LOAD input instead of the CLR. Remember to remove the original input port connection for the LOAD and reconnect the CLR to an input port. What input must now be on the D[3..0] inputs to produce a mod-12 count sequence? Construct and test the second circuit solution with a debounced manual clock or a low-frequency compatible clock signal. How are the two count sequences different? Why do the two MOD-12 counters have different sequences? What two things would you need to change in this counter design so that it has the same count sequence as the counter in part (a)? Simulate (functional) this design with a 1-kHz clock.

11.5 Self-stopping counter

Modify the MOD-16 counter in Example 11-1 so that it automatically stops counting when it reaches the counter's terminal (last) state. You will need to select a gate to detect (decode) the terminal state of the counter and then use the output from this decoder gate to control the counter's EN input. When the counter circuit hits the last state, it will disable itself and stop counting. What two ways can we "restart" the counter back at **0000**? How will these two restart solutions differ in their operation? Why does the counter start counting again after it has been restarted? Construct and test each of these designs with a debounced manual clock. Simulate (functional) your designs.

11.6 MOD-10 up/down counter

Construct the counter described in Example 11-4. Test the operation of the up/down counter using a debounced manual clock or a low-frequency compatible clock signal. Connect the control inputs (ENABLE and UP_DN) to logic switches. Connect the outputs (Q3 Q2 Q1 Q0 and TS) to lamps. Describe the operation of the counter. What does the ENABLE control do? What is the active level for ENABLE? What logic level on UP_DN will make the counter count down? When does the TS output go HIGH? What is the function of the TS output? How can we tell that the operation of TS is dependent upon the UP_DN input? Why does this counter have two "terminal" states? Is the TS output dependent on the ENABLE control? How can you prove this? Simulate (functional) this counter.

11.7 MOD-10 timing diagram

Use an oscilloscope to observe the input and output waveforms with ENABLE = 1 (6 signals: CLK, Q0, Q1, Q2, Q3, TS) for the MOD-10 counter described in Example 11-4. Apply a compatible clock signal that has a convenient frequency (such as 10 kHz). With a 4-channel scope you will only be able to view four waveforms at a time, so it will be necessary to swap the signals applied to the scope. **Always trigger the oscilloscope on the lowest frequency signal.** Carefully draw the waveforms to show the timing relationships between each of the circuit outputs and the clock signal when UP_DN = 0 and again when UP_DN = 1. What is the function of the TS output? Measure the actual clock input frequency. Measure the frequency of Q3 and the frequency of TS. What is the relationship between the frequencies of Q3, TS, and CLK? How does that compare to the counter's modulus?

11.8 Comparing counters

Compare the operation of the two LPM_COUNTER circuits shown in the following schematic. Make sure that you precisely select the features indicated in the schematic symbols for each counter using the MegaWizard. Specify a "set to 1" in the wizard for the upper counter in the schematic. Test each counter using a debounced manual clock or a low-frequency compatible clock signal. Connect all outputs (4 for each counter) to lamps. What is the count sequence for each counter? What is the modulus for each counter? Which circuit has a higher modulus? What is the terminal state for each counter? Are there any transient states? What is the function of cout? (Add an output port if you need to observe cout on a lamp.) What is the function of sset? Is sset a synchronous or an asynchronous control? Simulate (functional) the circuit.

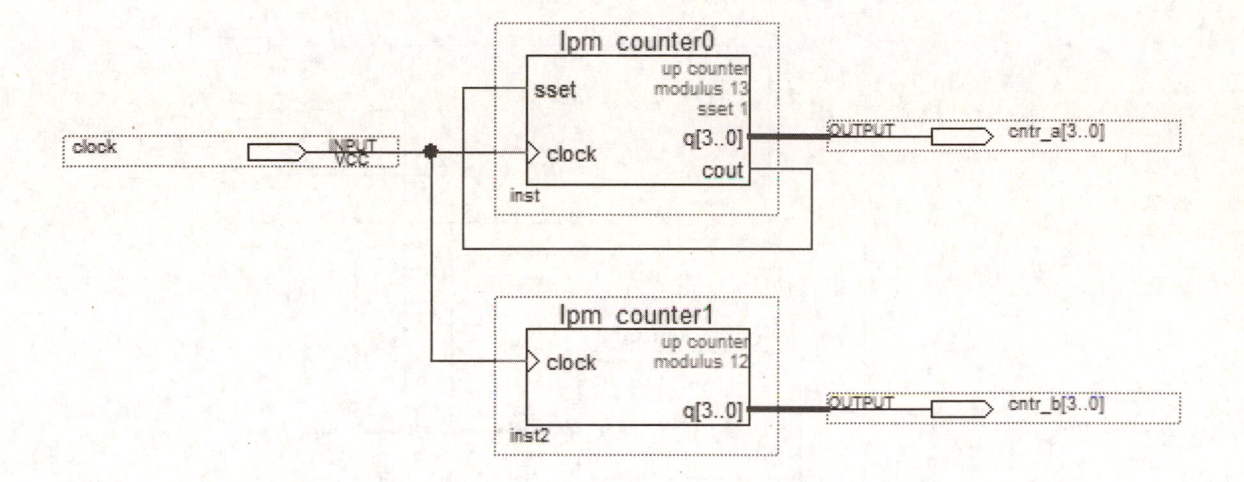

11.9 Counter waveforms

Construct and test the operation of the following counter circuit application. The LPM_CONSTANT has a value of **3** for the data input to the lower counter. Apply a compatible **10kHz** clock signal to CLK while monitoring CLK and the five outputs (F1 through F5) with an oscilloscope. With a 4-channel scope you will only be able to view four waveforms at a time, so it will be necessary to swap the signals applied to the scope. Keep CLK connected to the scope so that you can compare each of the output signals with it. **Always trigger the oscilloscope on the lowest frequency signal.** Carefully draw the waveforms to show the timing relationships between each of the circuit outputs and the clock signal. Measure the clock input and the output frequencies. What is the relationship between the input and output signal frequencies? How does that compare to the modulus of each counter? What is the duty cycle of each of the output waveforms? Make a table of signal frequencies and duty cycles. What else do you observe about the output signal pulses?

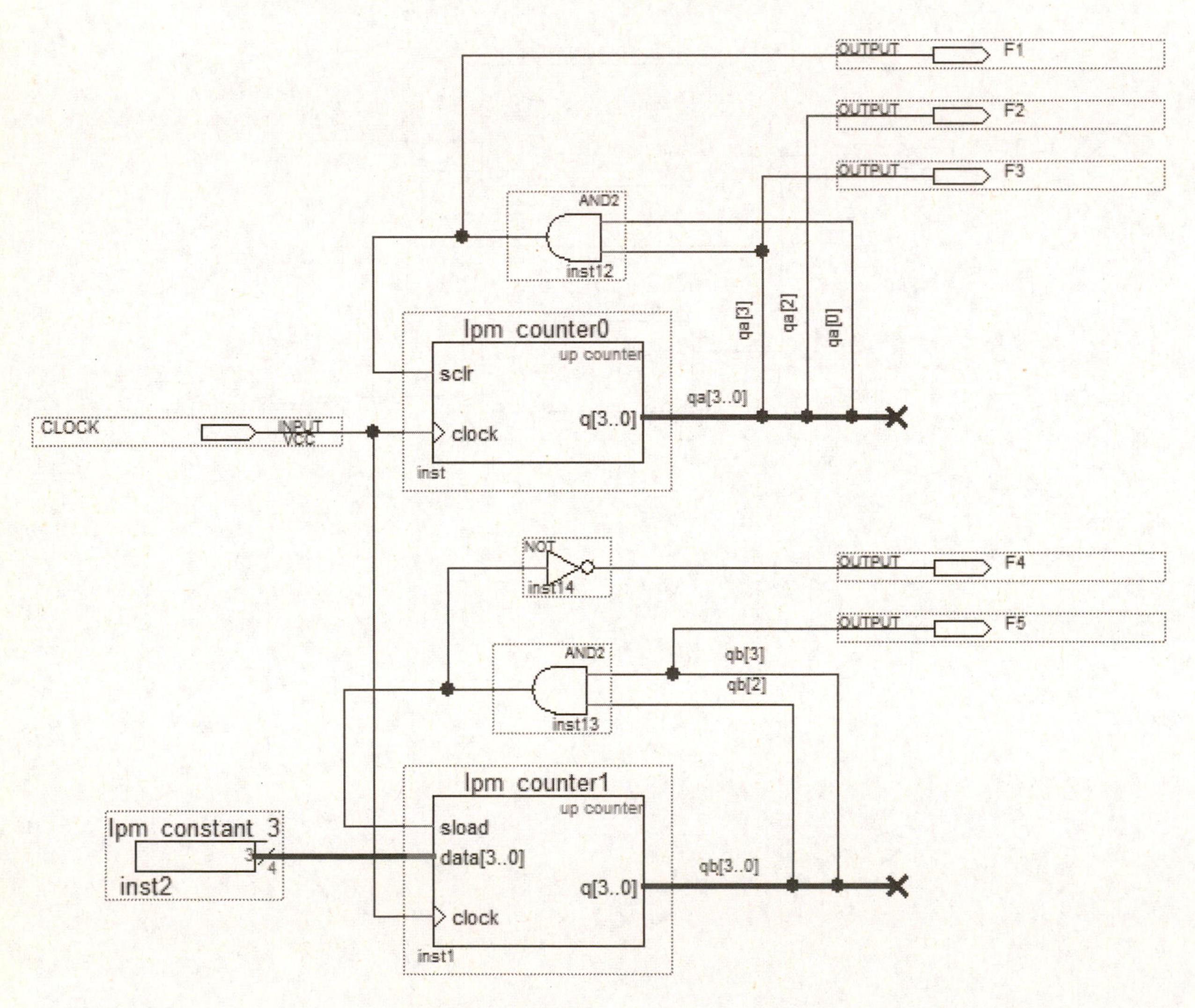

UNIT 12

APPLICATIONS USING MAXPLUS2 COUNTERS

Objectives

- To control the operation of standard counters.
- To produce specified count sequences using standard counters.
- To produce specified signal frequencies with frequency divider circuits using standard counters.
- To apply standard counters in design applications requiring the cascading of counters together.

Applications of Counters

Counters have a number of common applications, including the counting of items or objects, keeping track of time, frequency division, and controlling a sequence of activities. In frequency division, the input (applied to the clock) signal frequency is divided by a specified factor to produce the resultant lower output frequency of the divider. These circuits may be constructed using "maxplus2" standard counter functions in Quartus II, which are equivalent to standard MSI counter chips. See Quartus Notes available from the publisher's textbook web site at www.pearsonhighered.com/electronics.

Example 12-1

Design a MOD-8 counter that counts the sequence of 1 through 8 and then recycles using a standard binary counter. The counter also needs an active-HIGH enable named EN.

One solution shown in Fig. 12-1 uses a standard 74163 binary counter. Because the counter needs to recycle back to 1 after 8, it will be necessary to load the binary number 0001 into the counter at the proper time. The load function on the 74163 is synchronous; therefore the counter state 8 must be detected to control the active-LOW load (LDN) on the next clock pulse. This can be done simply with an inverter because the only time that QD goes HIGH in the specified count sequence is at state 8. The EN control is connected to the active-HIGH ENT enable on the 74163. This is one of the two enable controls on the 74163 – the other, labeled ENP, is tied to a HIGH. The active-LOW clear (CLR) is disabled by tying it HIGH also.

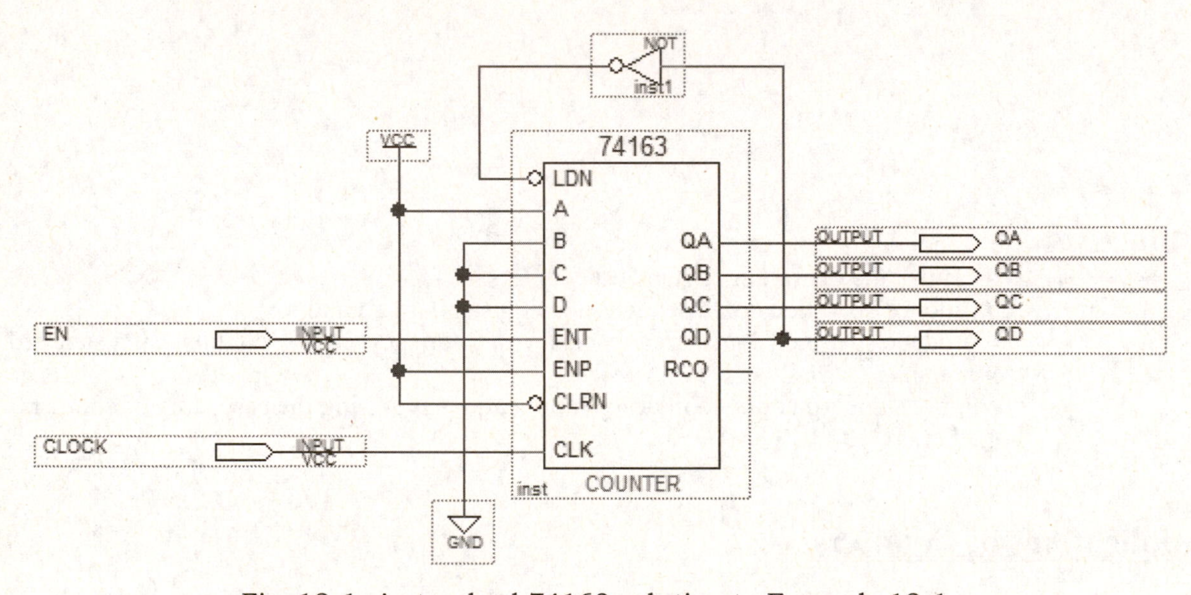

Fig. 12-1 A standard 74163 solution to Example 12-1

Another solution is shown in Fig. 12-2. This time we have used a standard 74191 binary counter. Because the load function on the 74191 is asynchronous, the next counter state 9 must be detected to control the active-LOW load (LDN). The state 9 will be a transient state and will not be counted in the modulus. It should be noted that this transient state will produce a glitch on the QA output, which may not be desirable. State 9 (for which QD and QA are simultaneously HIGH) can be decoded in this count sequence using a NAND gate. The count enable GN on a 74191 is active-LOW, so an inverter is used to connect EN to create an active-HIGH control for the counter circuit. The 74191 is an up/down counter that requires D/U to be tied LOW to count up.

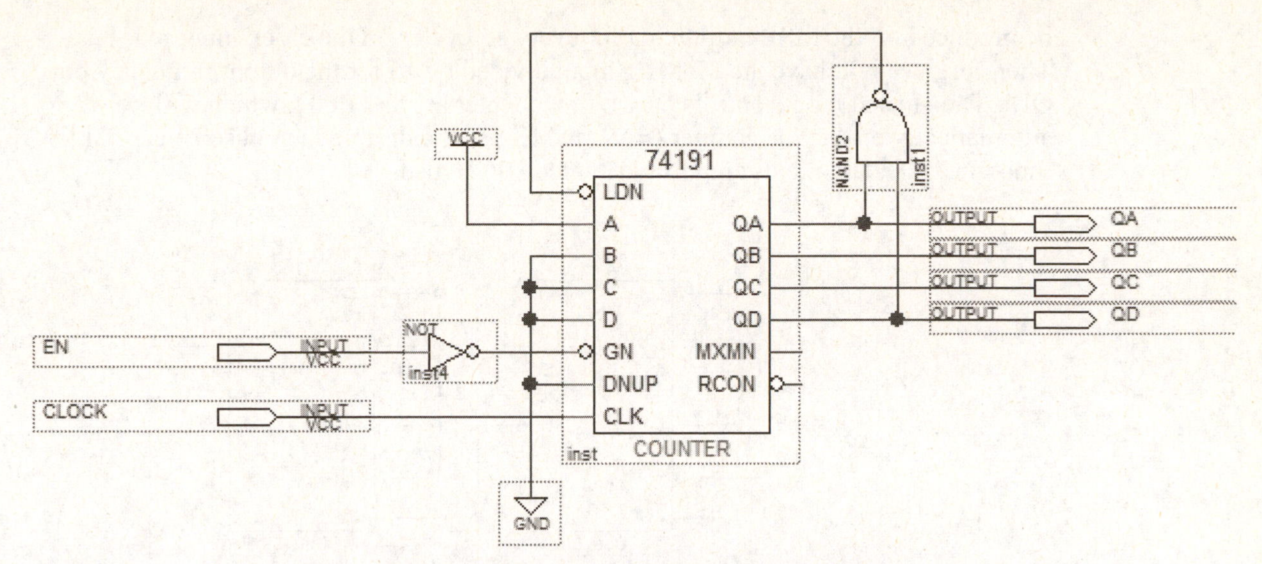

Fig. 12-2 A standard 74191 solution to Example 12-1

Either desired solution also can be easily implemented with the maxplus2 library in Quartus II. The simulation results are shown in Fig. 12-3. The simulation automatically starts at 0000_2. The count-up sequence recycles to 1 (due to the load) after state 1000_2 (8) so that the desired count sequence of 1 through 8 is produced. When EN = 0, the count will hold the current counter state.

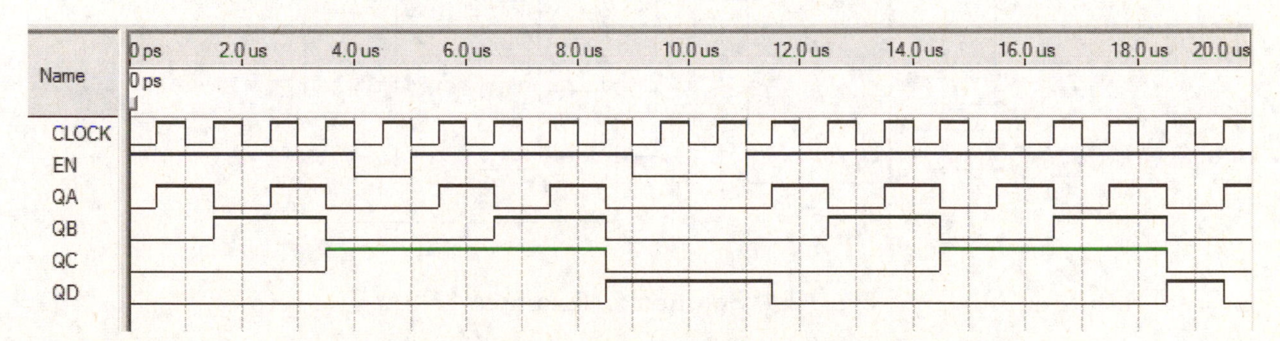

Fig. 12-3 Simulation results for Example 12-1 design

Example 12-2

Design a frequency divider circuit using the 74163 standard binary counter that will produce three different frequencies, 1.8 MHz, 150 kHz, and 100 kHz, from a 27-MHz clock signal.

A solution is shown in Fig. 12-4. The 27-MHz input signal must be divided by 15 to obtain 1.8 MHz, so we will need a MOD-15 counter. The MOD-15 counter is created with U1 by using the RCO (Ripple Carry Out) output to synchronously load in a 0001_2 (1). The function of the RCO is to detect the final (terminal) counter state, which would be 1111_2 (15) for a binary counter. The inverter is needed to match the active-HIGH output on RCO with the active-LOW input on LDN. One waveform cycle will be produced on the MSB output QD for every 15 cycles on the clock input pin. Therefore, we will have the 27-MHz input divided by 15 for the output frequency on QD. The unused input controls on the first counter are all HIGH, which will automatically enable the counter (ENT and ENP) and disable the unused clear (CLRN) function. The data inputs are arbitrarily tied HIGH also.

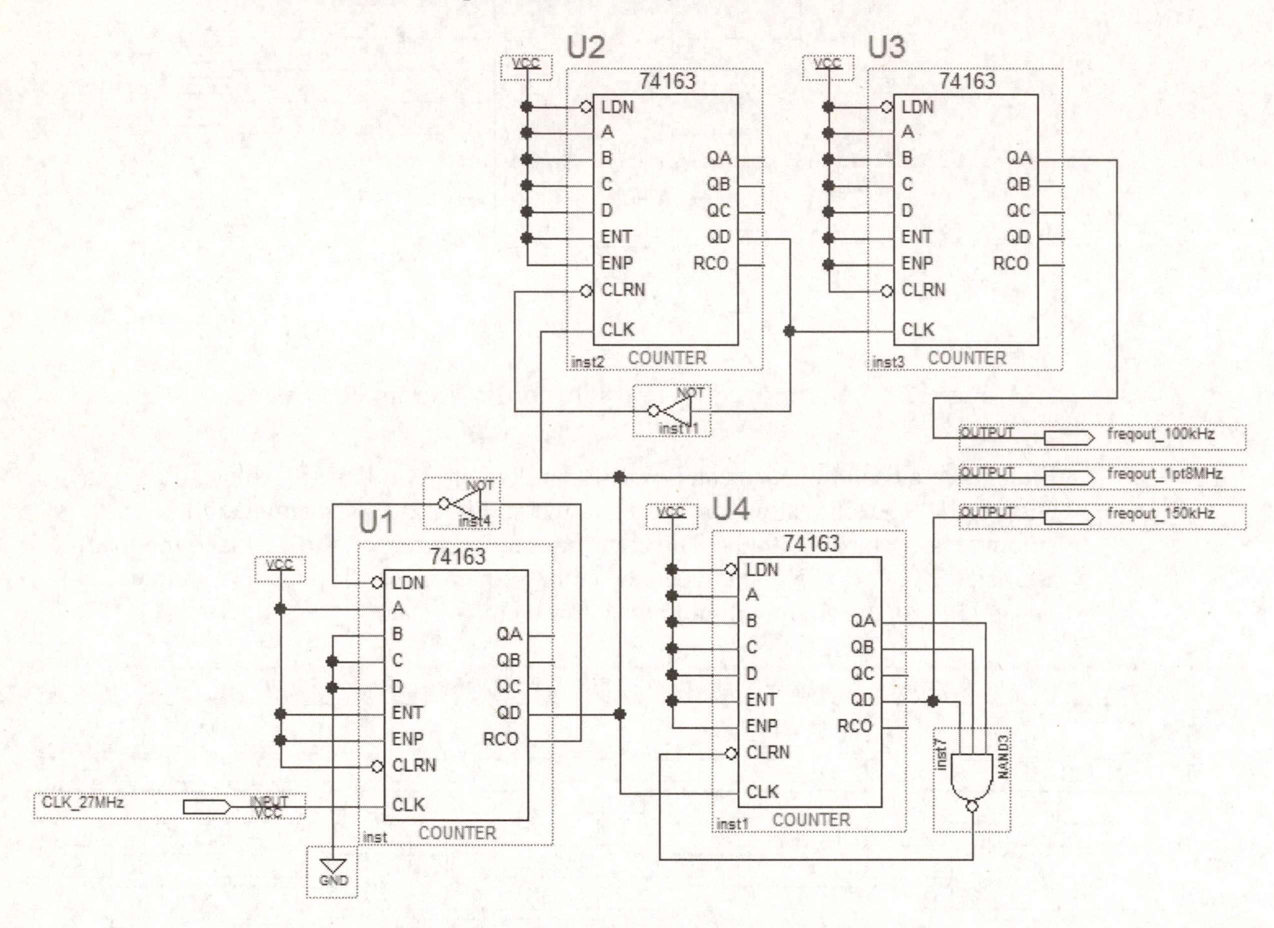

Fig. 12-4 Solution for Example 12-2

The 1.8-MHz signal must be divided by 18 to obtain the desired 100-kHz signal. This is done by creating two counters, a MOD-9 (U2) and a MOD-2 (U3). The active-LOW clear (CLRN) is synchronous in a 74163. Therefore, we will detect the counter state 1000_2 (8) with the inverter to clear the counter on the next clock pulse. The resulting 0 through 8 count sequence will provide a MOD-9 pattern. A 200-kHz signal is produced at the MSB QD pin on U2, which then is further divided by 2 with the MOD-16 binary count sequence of U3 resulting in a 100-kHz signal output at the LSB QA pin.

The 1.8-MHz signal must also be divided by 12 to obtain the final desired frequency of 150 kHz. A MOD-12 counter is created with U4. By detecting the state 1011_2 (11), it will synchronously clear the counter on the next clock pulse. U4 will produce a 0 through 11 count sequence to give us a MOD-12 pattern. Again the desired 150-kHz frequency is obtained from QD, the counter's MSB.

Example 12-3

Design a MOD-100, binary, down-counter circuit using a standard counter. The binary sequence should count from 99 to 0 and then recycle.

We will need a binary counter that can count down so a 74191 has been selected. This standard counter can count up or down. The largest state is 1100011_2 (99), so two 4-bit 74191 counters are needed to handle 7 bits. An example solution is given in Fig. 12-5. The 74191 counters are cascaded together using synchronous clocking and by tying the terminal state (0 for a down counter) detector output RCON from the least significant counter (outputs QD QC QB QA) to the enable (GN) input on the most significant counter (outputs QG QF QE). Both counters will count down if the DNUP control input is HIGH. The counter circuit is allowed to recycle to 11111111_2 after reaching the terminal state 0000000_2. The 11111111_2 state is decoded simply by detecting when the unused QD output goes HIGH. By connecting the decoding NOT gate output to the asynchronous, active-LOW LDN (load) inputs on both counters, the 11111111_2 will be a transient state. The 11111111_2 state will be replaced with 01100011_2 (99) to recycle the counter back to the "top of the count." The recycling count sequence is 99 down to 0 as desired.

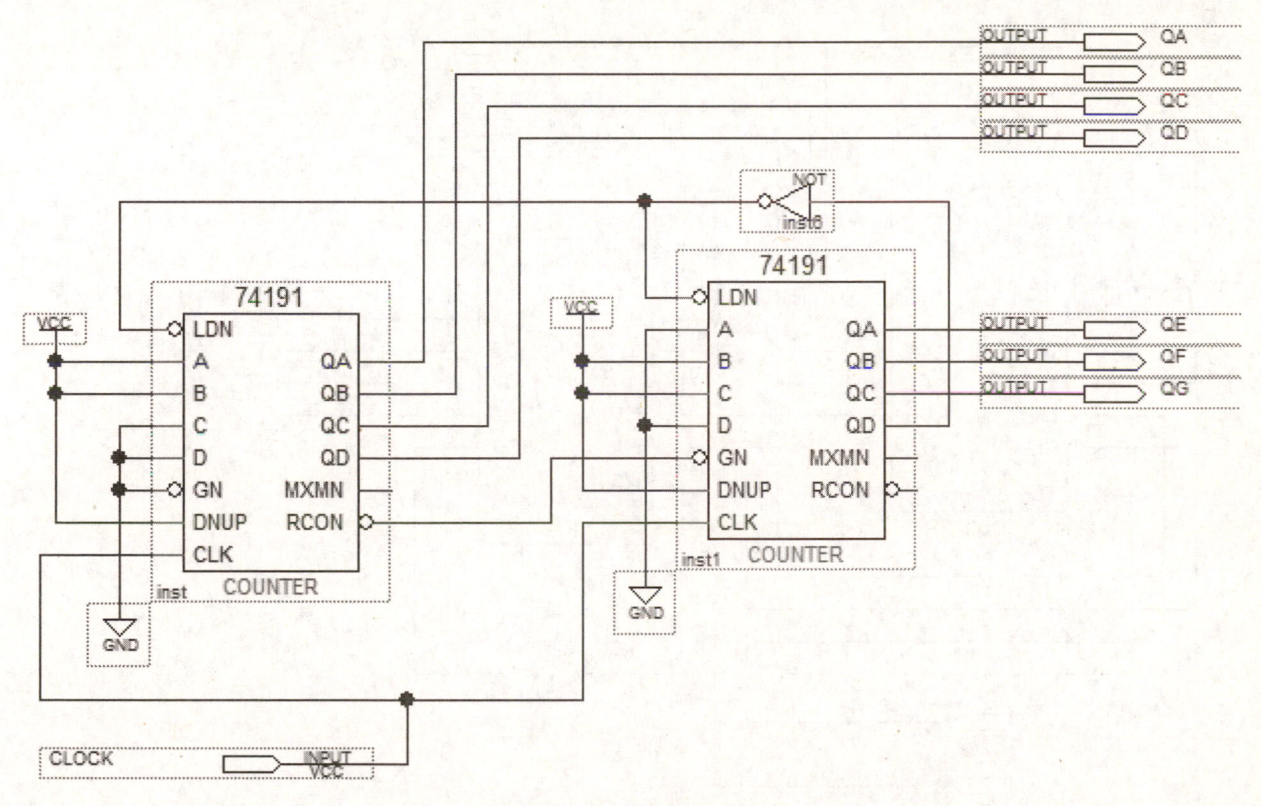

Fig. 12-5 Solution for Example 12-3

Example 12-4

Design a self-stopping **BCD** counter circuit using the 74160 decade counter. The count sequence should count up to 99 and then stop. The counter can be asynchronously reset with an active-LOW signal named RESTART. The counter also needs to have an active-LOW count enable named EN.

A solution is shown in Fig. 12-6. Two 74160 BCD counters are cascaded together by synchronously clocking both chips and connecting the RCO output of one 74160 to the ENT enable of the other. The 74160 on the left is the one's digit, and the one on the right is the ten's digit for the two-digit counter. Because the RCO output is a function of the chip's ENT input, both 74160s can be enabled/disabled by controlling the ENT pin on the least significant digit. The count enables are active-HIGH, so an inverter is needed for the EN control signal. The asynchronous clear pin for each chip is connected to the RESTART control signal. The active-LOW load function is disabled with a HIGH input. The one's RCO output will go HIGH when the counter reaches its terminal state 9. The ten's RCO output will go HIGH when both digits reach 9 (i.e., counter state 99 in BCD) because the ten's ENT is connected to the one's RCO. The final RCO then is used to disable both 74160s through the ENP count enable control.

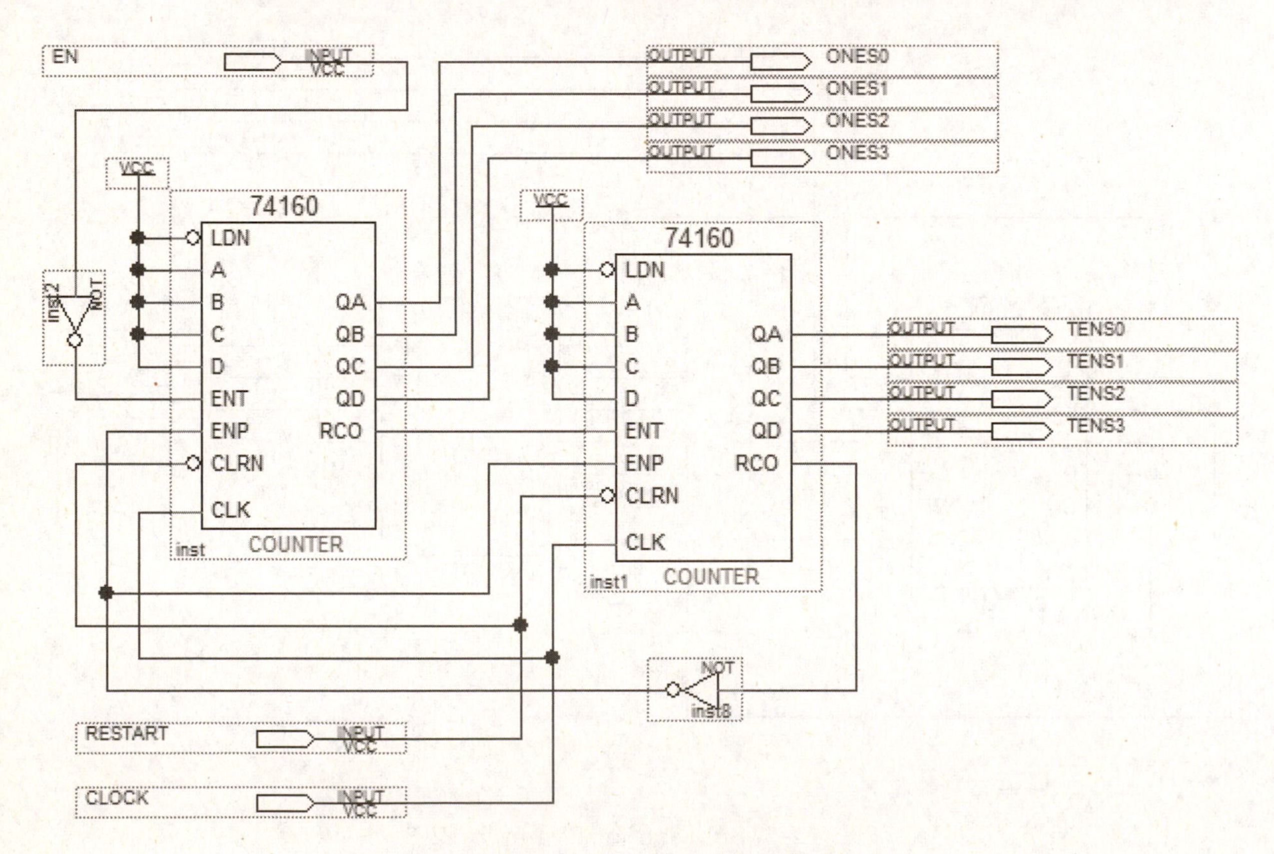

Fig. 12-6 Solution for Example 12-4

Example 12-5

Design a MOD-4096 binary counter circuit using the 74163 MOD-16 counters. The counter can be synchronously loaded with an active-LOW signal named ldn. The counter also needs to have an active-HIGH count enable named en. Decode the maximum count value (decimal 4095).

A solution is shown in Fig. 12-7. A 12-bit binary counter is needed to produce a MOD-4096 count sequence, which will require three 74163 maxplus2 functions to be cascaded together. The least significant bit in the MOD-4096 counter is labeled q0 and the most significant bit is q11. Synchronous clocking must be used because the counter is to be synchronously loaded. All three load control pins must be connected to ldn. The 12-bit input for the parallel data is labeled d[11..0]. The RCO output for each 74163 will go HIGH when the count reaches 1111_2 if the chip's ENT count enable input is HIGH. The counters are cascaded together by connecting the RCO output of the least significant counter (U1) to the ENT of the next counter (U2) and then, finally, the RCO from U2 to the ENT of the most significant counter (U3). Controlling the ENT on U1 will enable or disable the entire 12-bit counter. If ENT is LOW, the counter is disabled, and RCO will also be LOW. Therefore, the output signal labeled max will detect (decode) the maximum count value ($1111\ 1111\ 1111_2$) <u>only</u> when the counter is enabled. The data input and counter output signals are each grouped in buses. Note that the en input and max output are connected by labeling the respective signal lines.

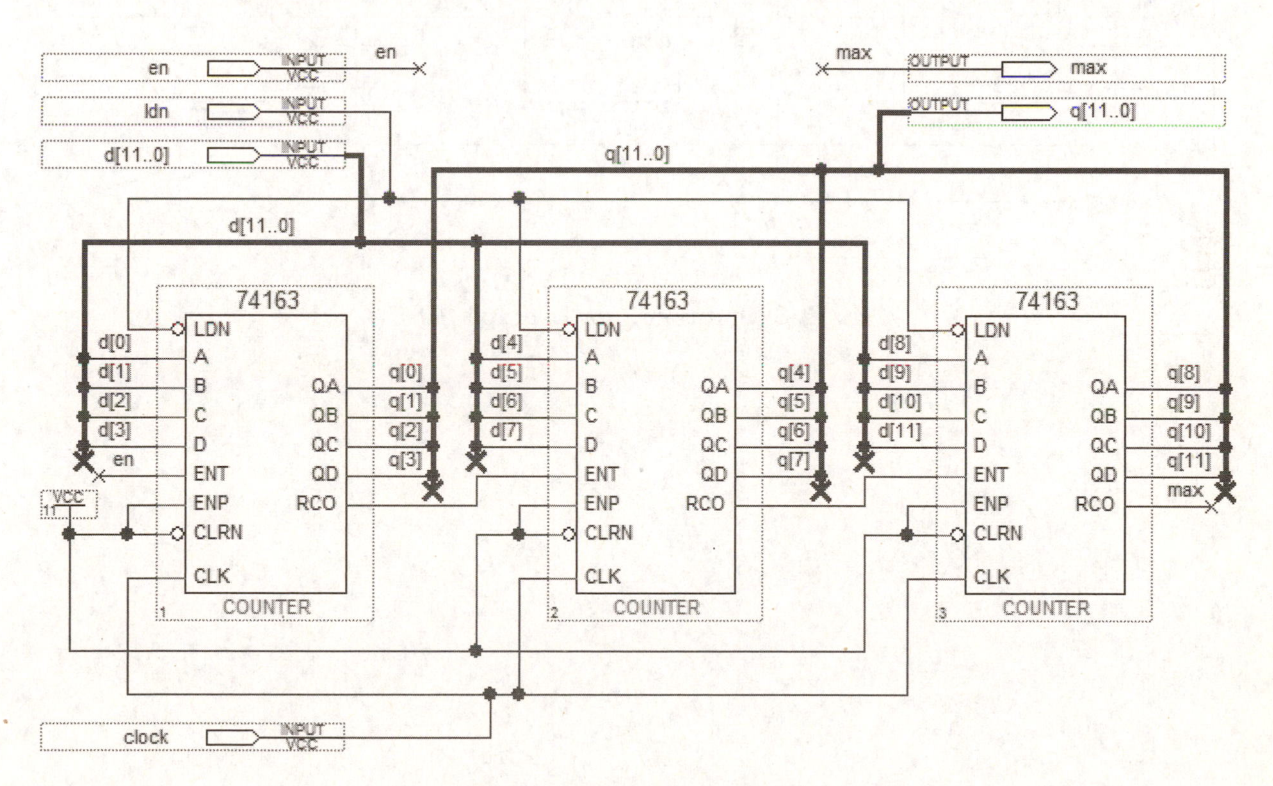

Fig. 12-7 Solution 1 for Example 12-5

Note that the problem statement did not specify that the decoder output should also be controlled by en, as it will be in Fig. 12-7. If that solution is undesirable, the design can be modified slightly (see Fig. 12-8). The ENT count enable on U1 is permanently enabled with the HIGH input, and the RCO to ENT cascading is done as in the previous solution. This will now also permanently enable the decoder output max. The count enable function instead can be controlled with the ENP enable on all three chips. The two count enables ENT and ENP on the 74160 to 74163 series of counters are slightly different in function. ENT also enables the RCO output, which decodes the maximum count value on the counter chip, but ENP does <u>not</u> control the RCO output.

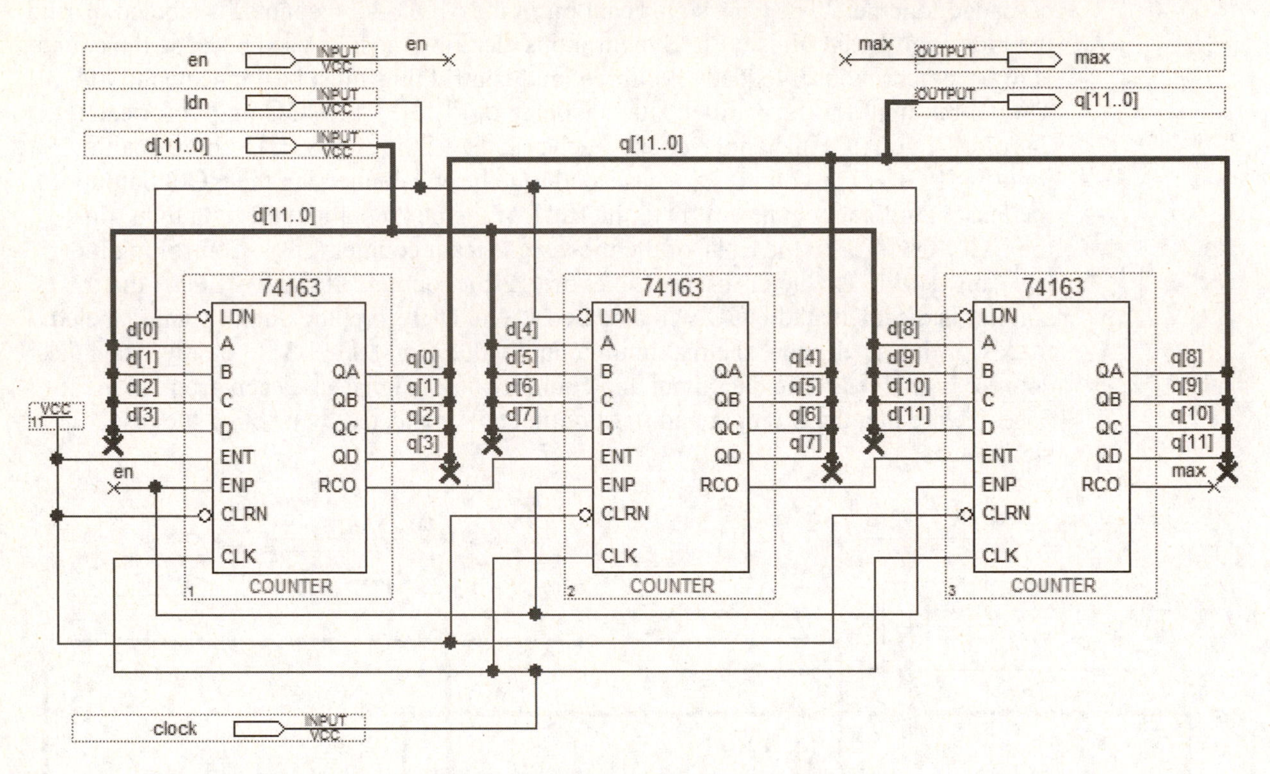

Fig. 12-8 Solution 2 for Example 12-5

Laboratory Projects

Construct and test the following applications of counter circuits using "maxplus2" standard counter functions in Quartus. Functionally simulate the circuits with Quartus.

12.1 MOD-13 count sequence

Design a MOD-13 counter using a 74163 counter that produces the sequence specified in the following state transition diagram. The counter should have an active-LOW enable named en. Note that two states have unconditional transitions (i.e., they are not controlled by en). How would you make all state transitions dependent on en?

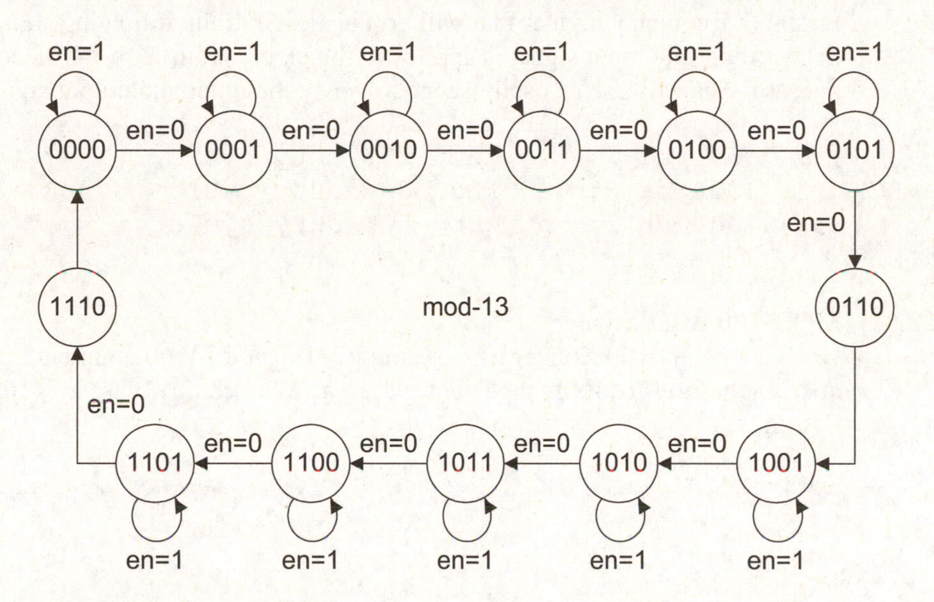

12.2 Frequency divider circuit

Design a logic circuit using 74163 counters to produce output frequencies of 25 kHz and 10 kHz. The input signal is a <u>compatible</u> signal with a frequency of 150 kHz. Use a frequency counter to verify proper operation of your circuit design.

12.3 MOD-200 binary counter

Design a MOD-200 binary counter using 74163 counters. The recycling count sequence should be from 1 to 200. The design should have an active-HIGH count enable control.

12.4 MOD-50, self-stopping down-counter

Design a MOD-50, self-stopping, down-counter using 74191 counters. The counter should count down from 50 to 1 and automatically stop. When the counter has reached the end of its count, an active-HIGH output signal named stop should be asserted. The count sequence will be asynchronously started with a signal named restart. The counter will load 50 when restart goes HIGH and then start counting down when restart is returned to a LOW again.

12.5 Waveform generator circuits

Design a frequency divider that will produce each of the following signals when a 100 kHz compatible input signal is applied to the clock input. Use 74163 counters and any necessary gates. Use an oscilloscope to verify the input and output signals.

```
Output 1:  10 kHz, 10% duty cycle
Output 2:  10 kHz, 20% duty cycle
Output 3:  10 kHz, 50% duty cycle
Output 4:  50 kHz, 50% duty cycle
```

12.6 MOD-100 BCD counter

Design a MOD-100 counter by cascading 74190 and 74160 counters together. Include an asynchronous reset in the counter design. Why can't we use a 74162 to create this design?

UNIT 13

APPLICATIONS USING LPM_COUNTER MEGAFUNCTIONS

Objectives

- To determine the specifications for lpm_counter megafunctions in design applications.
- To control the operation of lpm_counter megafunctions.
- To produce specified signal frequencies using frequency divider circuits.

Applications of Counters

Counters have a number of common applications, including the counting of items or objects, keeping track of time, frequency division, and controlling a sequence of activities. In frequency division, the input (applied to the clock) signal frequency is divided by a specified factor to produce the resultant lower output frequency of the divider. These circuits can be easily implemented using the flexible lpm_counter megafunction in Quartus II. See Quartus Notes available from the publisher's textbook web site at www.pearsonhighered.com/electronics.

Example 13-1

Create a recycling counter that will count the sequence 1 to 20. The counter also needs to have an enable control.

An LPM_COUNTER megafunction can be easily used to create the desired counter (see Fig. 13-1). The modulus for this counter is specified in the MegaWizard to be 21 even though we only need a MOD-20. The up-counter will automatically start at state 0 and count through 20. We will eliminate state 0 by synchronously loading 00001 after the terminal state of 20 occurs. The output port carry-out (cout) is enabled for the mod-21 counter. This output will automatically detect the counter's terminal state and will be used to enable (with sload) the loading of data[4..0] on the next clock. The parallel data input has the constant value 00001 applied (using the LPM_CONSTANT megafunction). The enable input (ENABLE) is placed on the clk_en control port of lpm_counter0. The clk_en port is used to enable the counter because it will also control the synchronous load function that recycles the count to 1 upon reaching the terminal state, unlike the cnt_en that only controls the count function. The functional simulation results are shown in Fig. 13-2.

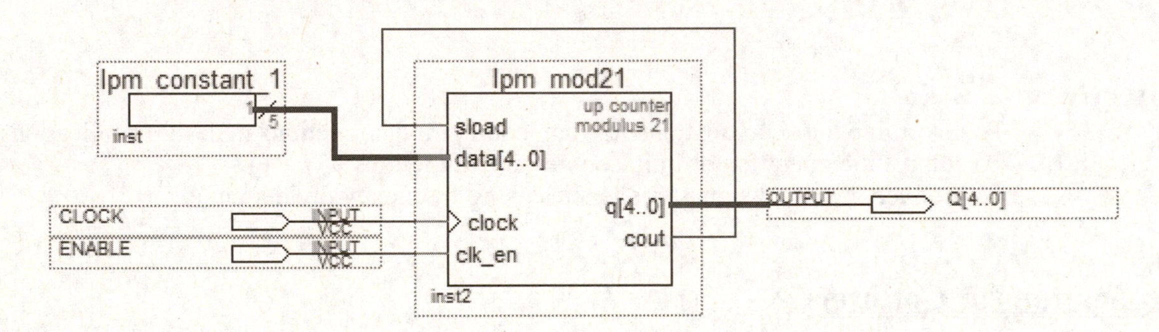

Fig. 13-1 Block diagram for Example 13-1 (a recycling 1 to 20 counter)

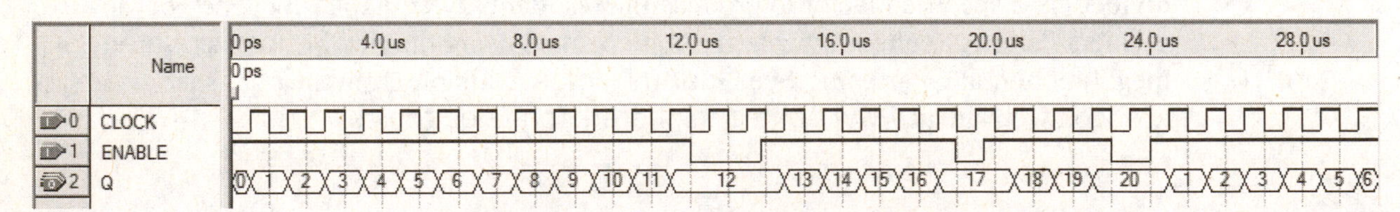

Fig. 13-2 Functional simulation of Example 13-1

Example 13-2

Create a self-stopping counter that will count the sequence 1 to 20. The counter also needs to have a synchronous restart control.

The same LPM_COUNTER megafunction that was created for Example 13-1 can be used, with a couple of wiring changes, to produce the desired counter (see Fig. 13-3). The modulus for this counter is again specified in the MegaWizard to be 21 even though we only need a MOD-20. The up-counter will automatically start at state 0 and count through 20. The carry-out (cout) port will automatically detect the counter's terminal state (20) and will be used to disable the counter. We will restart the counter by synchronously loading the constant value 00001 via the data[4..0] port when the sload control is asserted by RESTART. This application requires us to use LPM_COUNTER's cnt_en port to disable the counter instead of using either clk_en or cin because clk_en would also disable sload and cin controls the carry-out function. The functional simulation results are shown in Fig. 13-4.

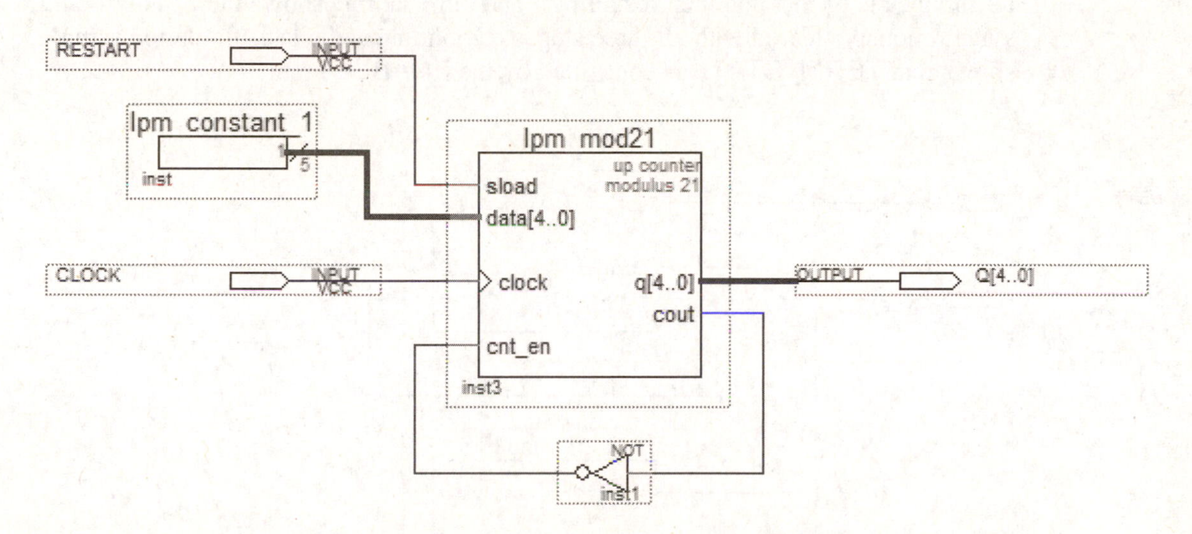

Fig. 13-3 Block diagram for Example 13-2 (a self-stopping 1 to 20 counter)

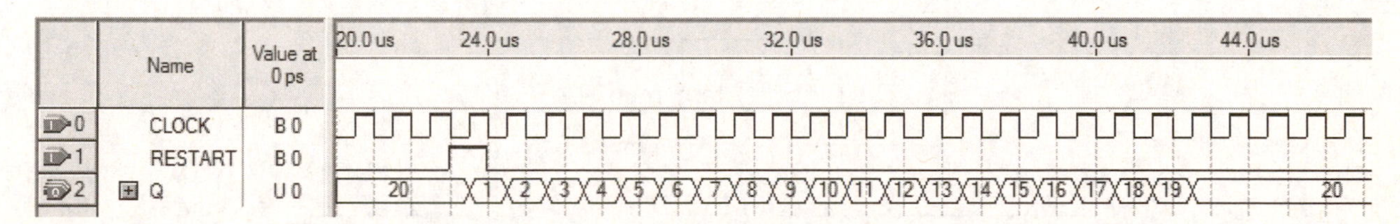

Fig. 13-4 Functional simulation of Example 13-2

Example 13-3

Create a recycling MOD-1000 BCD counter. Include count enable and synchronous clear controls. Detect the terminal state for the counter.

A MOD-10 counter is created using LPM_COUNTER. This same BCD counter is repeated 3 times and cascaded together to produce a MOD-1000 BCD counter (see Fig. 13-5). The carry-in (cin) and carry-out (cout) ports make the cascading very easy because cout will only detect the terminal state (9) of a decade counter when it is enabled by cin. The entire MOD-1000 counter will be enabled or disabled by controlling the cin on the least significant digit (LSD). The enable control will feed through all decade stages. The TERM_STATE output will detect state 999 when the counter is enabled with a HIGH logic level on ENABLE. The synchronous CLEAR control is connected to each of the three BCD counter stages. Sample functional simulation results are shown in Fig. 13-6. The first simulation snapshot shows the counter being disabled. The second snapshot illustrates the counter reaching the terminal state of 999 and then recycling. The third sample shows the counter being synchronously cleared with all three stages responding. The last simulation sample shows that TERM_STATE is controlled by the ENABLE input.

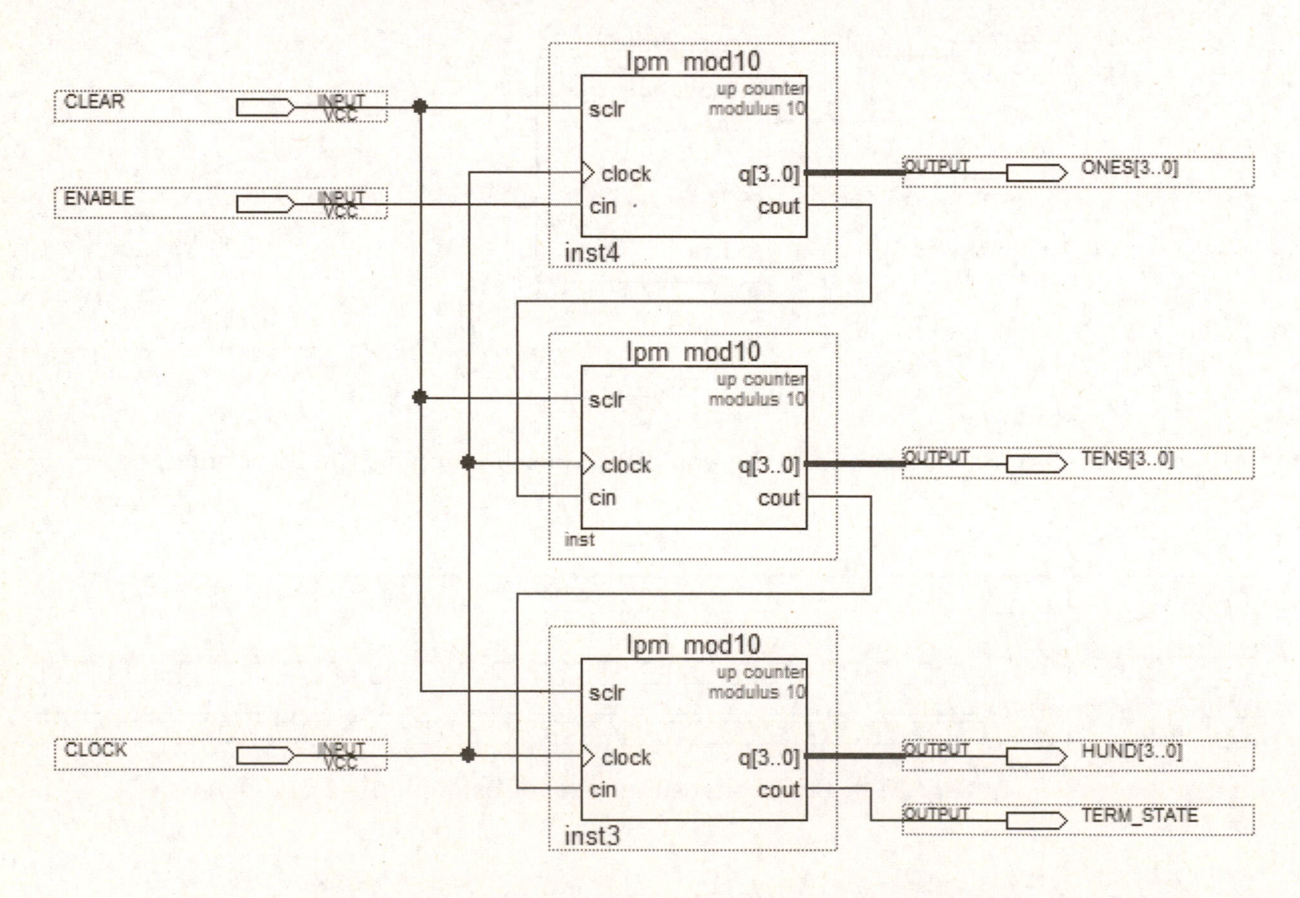

Fig. 13-5 Block diagram for Example 13-3 (a MOD-1000 BCD counter)

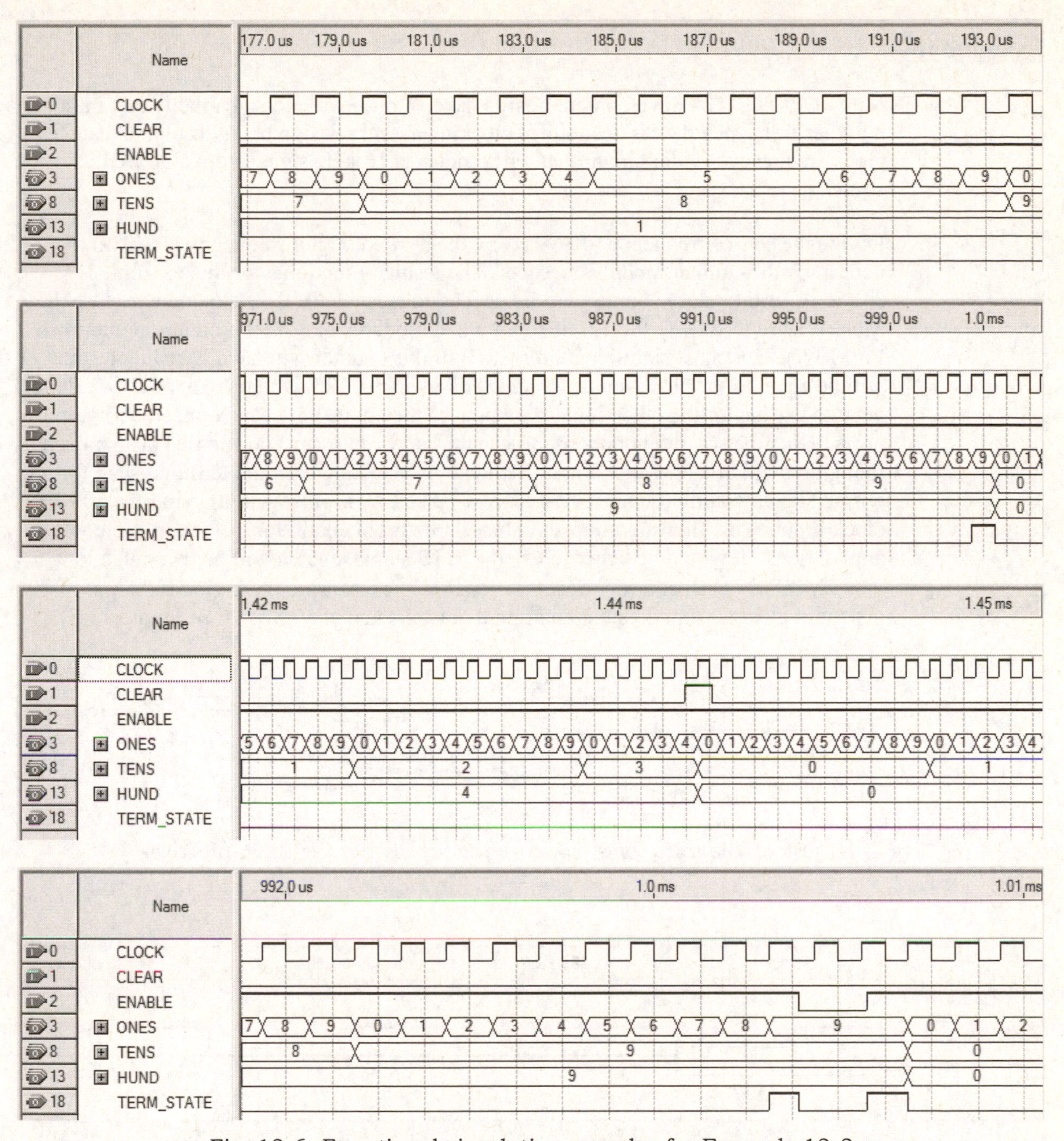

Fig. 13-6 Functional simulation samples for Example 13-3

Example 13-4

Many CPLD/FPGA development boards have an on-board clock available that might be a higher frequency than is desired for clocking another design being tested. Create a clock frequency divider circuit that will produce a 16-kHz signal from a 50 MHz clock source.

Divide the source frequency (50 MHz) by the desired output frequency (16 kHz) to determine the counter modulus needed. The resultant modulus for this example is 3125. A 12-bit binary counter will be needed to create a MOD-3125 counter. Use the MegaWizard to specify the 12-bit counter and modulus of 3125 for an up-counter (see Fig. 13-7). The most significant bit (MSB) of this counter will have a frequency that will be equal to the clock input (50 MHz) divided by 3125, which is 16 kHz. We do not need to use any of the other counter outputs so they have not been connected to output ports. Note that we did not choose to use the carry-out (cout) port in this design even though it would be the same output frequency. Detecting only the terminal state for our output signal would produce a very narrow pulse (20 ns pulse width) using the 50 MHz clock input. This short duration signal might be problematic for a clock in some applications. It would be better to use the MSB signal since it will have a much longer pulse width. A correct design result is verified by measuring the period (62.5 μs) of clkout in a circuit functional simulation that is clocked at 50 MHz (see Figure 13-8).

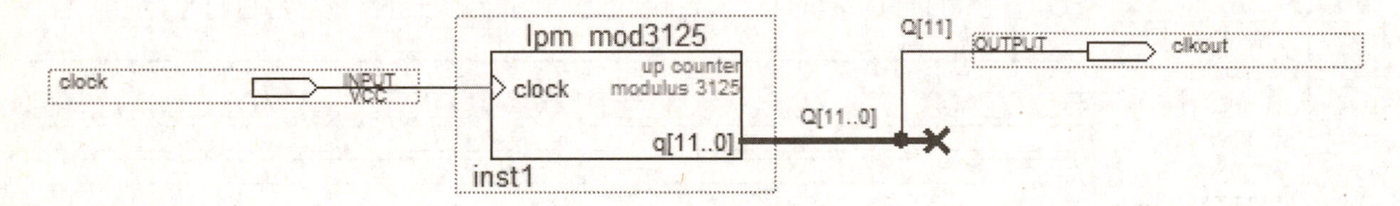

Fig. 13-7 Block diagram for a clock frequency divider for Example 13-4

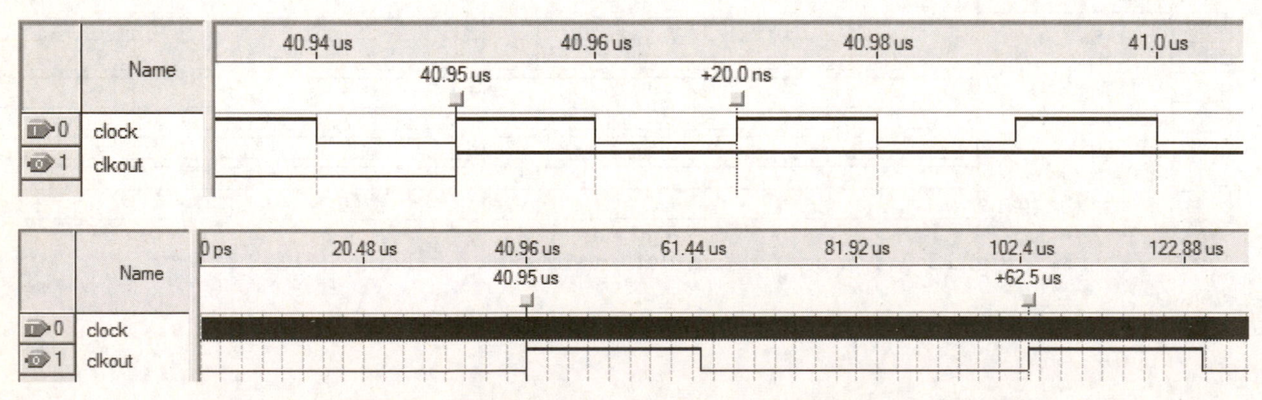

Fig. 13-8 Measuring periods of clock and clkout using time bars

Laboratory Projects

Design each of the following synchronous counters using LPM_COUNTER megafunctions. Simulate each circuit design. Construct and test the operation of the counters using a low-frequency compatible clock signal unless otherwise specified.

13.1 MOD-256 binary down-counter
Design a recycling, MOD-256 binary down-counter. The counter should have an active-LOW synchronous load controlled by LDN.

13.2 MOD-200 binary counter
Design a recycling, MOD-200 binary counter. The count sequence should be from 1 to 200. The counter should have an active-LOW enable named EN.

13.3 Self-stopping binary counter
Design a binary counter that will count from 0 to 100 and then automatically stop. The counter is asynchronously restarted by an input named GO. GO can only restart the counter when it has reached its terminal state.

13.4 Automatic up/down counter
Design a MOD-32 up/down counter that will automatically switch count directions when it reaches the terminal count. Use a T flip-flop to control the count direction with the T enable input controlled by the counter's carry-out port. Trigger the counter on the rising clock edge and the T flip-flop on the falling clock edge.

13.5 MOD-100, up/down, BCD counter
Design a MOD-100, up/down, BCD counter. The counter should have a count enable, a higher-priority synchronous load function and an output (also controlled by the count enable) that detects the counter's terminal state.

13.6 MOD-50 BCD counter
Design a MOD-50 BCD counter with a recycling count from 0 to 49. The counter should have an active-LOW count enable, an active-LOW synchronous clear, and an active-LOW output that detects the counter's terminal state.

13.7 MOD-100, self-stopping, BCD down-counter
Design a MOD-100, self-stopping, BCD down-counter. The counter should count down from 99 to 0 and then automatically stop. When the counter has reached the end of its count, an active-HIGH output signal named stop should be asserted. The counter is asynchronously reloaded with 99 when the input named restart goes HIGH. The counter will count down when restart is brought LOW again.

13.8 Clock frequency divider

Design a frequency divider circuit that will output a 25 kHz signal when driven by a 50-MHz clock.

13.9 Frequency divider

Design a frequency divider that will output 4 pulse frequencies: 150 kHz, 100 kHz, 37.5 kHz, and 12.5 kHz. A 27-MHz clock will be used to drive the frequency divider.

13.10 One-shot counter

Design a counter circuit that will produce a single output pulse of a specified duration after it has been triggered. The pulse width will be determined by the clock frequency applied to the counter and a 4-bit input variable. Use a 4-bit, binary, down-counter that can be synchronously loaded with the desired length of time (pulse width = data[3..0] × clock period). The synchronous load control is the trigger input. The down-counter needs to stop counting when it reaches the terminal state 0. When the one-shot circuit is triggered, the counter will be loaded with the data that sets the pulse duration time. The output pulse can be terminated before it has completed its specified time interval by an asynchronous reset control on the counter. The pulse output signal will be HIGH when the counter is at any state except for the terminal state of the counter. Test this circuit using a 1-kHz clock signal for the counter and measure the pulse width with an oscilloscope. What is the range for the output pulse width that can be produced with this circuit? Measure the output pulse width produced by various time[3..0] settings using an oscilloscope and the following test circuit to repeatedly trigger the one-shot.

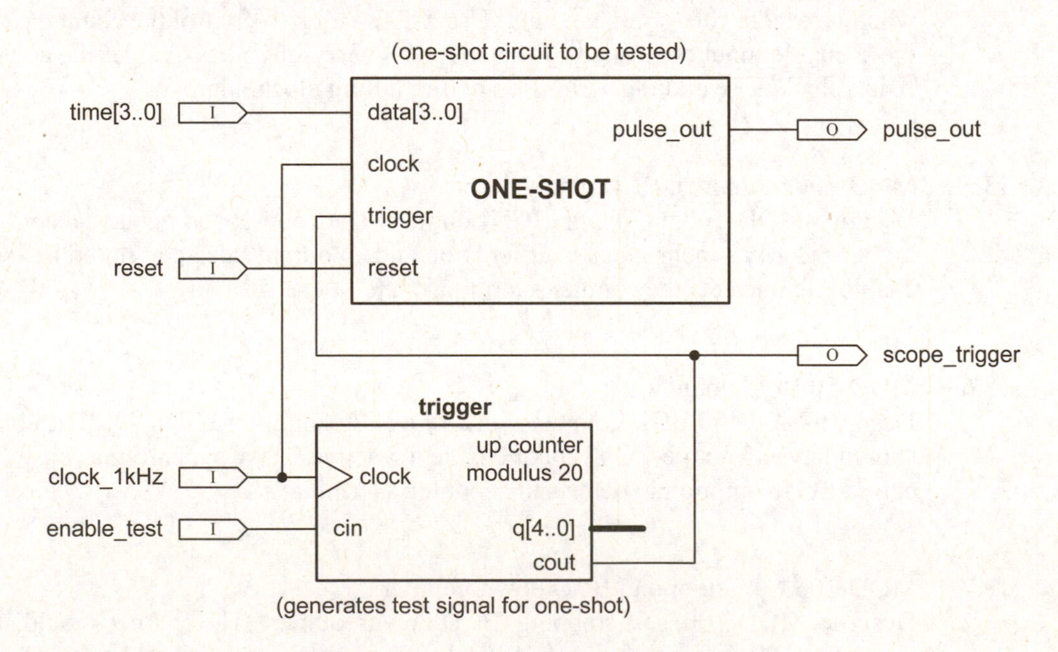

UNIT 14A

COUNTER DESIGNS USING AHDL

Objective

- To design sequential logic circuits with CPLDs or FPGAs using the Altera Hardware Description Language (AHDL).

Sequential Circuits Using AHDL

The design and implementation of sequential circuits using the Altera Hardware Description Language is carried out in much the same fashion as for combinational circuits. The flip-flops contained in the Altera PLDs can be treated as SR, JK, D, or T-type flip-flops in which the control inputs are produced by the programmable logic in the PLD. Due to ease of use, D flip-flops are most commonly utilized in hardware description languages. AHDL is a HIGH-level language that provides many options for defining sequential circuits, including IF/THEN and CASE statements and TABLEs. See Quartus Notes (www.pearsonhighered.com/electronics).

Fig. 14-1 shows an AHDL design description that creates a single D flip-flop. The SUBDESIGN section declares the input and output ports for d_ff.tdf. A single D flip-flop (the primitive name is DFF in Quartus II) is declared in the VARIABLE section. This is called a register declaration in AHDL. The flip-flop, whose instance name is q_out, will be connected to the subdesign output port because this name was also declared to be an output port in the SUBDESIGN section. All primitives have input ports and output ports. The connections to the primitives necessary for your design will be described in the Logic section (between the BEGIN and END) of the AHDL file. DFF primitives have clock and d inputs and a q output. These ports are named .clk, .d, and .q, respectively. These port names are appended to the instance name (q_out) given in the register declaration. All primitives have only one output, so you can also use just the instance name without the .q added onto the instance name. Flip-flop primitives have optional active-LOW preset (.prn) and clear (.clrn) input ports. If they are not needed in a design, they are simply omitted from the Logic section. In this case, the primitive's input ports .clk, .prn, .clrn, and .d are connected to the subdesign's input ports named clock, preset, clear, and data, respectively. This is simply a wiring list for the desired flip-flop. Results for the simulation of d_ff.tdf are given in Fig. 14-2.

```
SUBDESIGN d_ff
(
data, clock, preset, clear                :INPUT;
q_out                                     :OUTPUT;
)

VARIABLE
q_out         :DFF;
      -- create single D flip-flop named q_out
BEGIN
q_out.clk = clock;        -- connect clock port
q_out.prn = preset;       -- active-low preset port
q_out.clrn = clear;       -- active-low clear port
q_out.d = data;           -- D input port
END;
```

Fig. 14-1 AHDL description creating a single D flip-flop

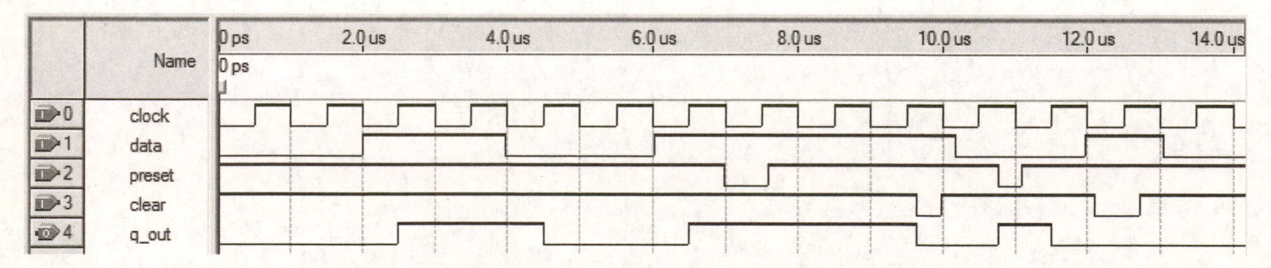

Fig. 14-2 Simulation results for the AHDL D flip-flop

Example 14-1

Design a MOD-16, binary up/down counter using AHDL. The count direction is controlled by dir (dir = 1 for count up).

An AHDL text file solution named updncntr.tdf is shown in Fig. 14-3. Four flip-flops will be needed for a MOD-16 counter. Accordingly, four D flip-flops (DFF) named count[3..0] are declared in the VARIABLE section. The D flip-flops form essentially a "buried" register inside the updncntr block. A register was created because it is very easy to produce the desired binary count sequence by either incrementing or decrementing the current state contained in the register by one. The Logic section (after the keyword BEGIN) describes the functionality of this block. The first statement *count[ ].clk = clock* identifies the name of the signal (clock) that is connected to the clock port (.clk) on the set of flip-flops (named count[ ]). The IF/THEN statement tests the input dir. If dir is true (or HIGH), then the inputs (.d port) to the set of D flip-flops (count[ ]) will be one more than the current output (.q port) from the register. Otherwise (dir = 0), the new input should be one less than the current output from the register. The result of incrementing or decrementing the register's contents, of course, is a count-up or a count-down sequence, respectively. The register is buried, which means that it is not connected directly to the SUBDESIGN's output port. The statement *q[ ] = count[ ]* will connect the register's outputs (the .q port is understood) to the output port for updncntr. The results of simulating the AHDL design file are shown in Fig. 14-4.

```
SUBDESIGN  updncntr
(
      clock, dir          :INPUT;
      q[3..0]             :OUTPUT;
)
VARIABLE
      count[3..0]         :DFF;     -- create 4-bit register
BEGIN
      count[].clk = clock;      -- connect clock to register

      IF  dir  THEN             -- dir=1 to count up
            count[].d = count[].q + 1;
      ELSE                      -- dir=0 to count down
            count[].d = count[].q - 1;
      END IF;

      q[] = count[].q;        -- connects buried reg to port
END;
```

Fig. 14-3 AHDL file for the binary up/down counter design for Example 14-1

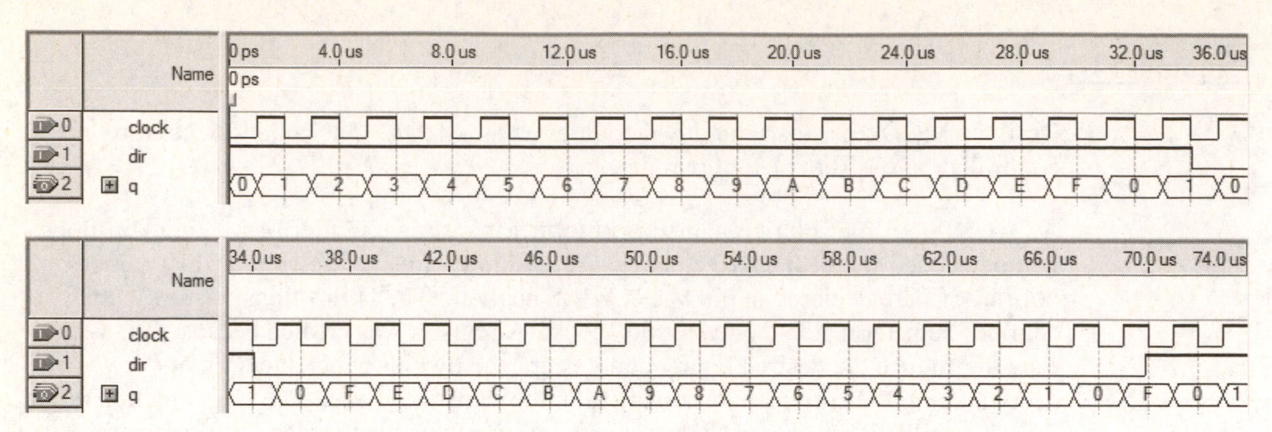

Fig. 14-4 Simulation of the AHDL file for Example 14-1

Example 14-2

Design a decade counter using AHDL. The MOD-10 counter has a count enable, parallel load, and asynchronous clear controls as shown in Table 14-1. The counter should also produce a ripple carry output signal (rco) that goes HIGH during the last state in the count sequence when the counter is enabled.

clear	load	enable	clock	function
1	X	X	X	CLEAR
0	0	0	↓	HOLD
0	1	X	↓	LOAD
0	0	1	↓	COUNT UP

Table 14-1 Example 14-2 functions

An AHDL text file solution named mod10.tdf is shown in Fig. 14-5. Four flip-flops will be needed for the decade counter. Accordingly, four D flip-flops (DFF) named counter[3..0] are declared in the VARIABLE section. In this example, the outputs from the D flip-flops are also connected to the output port for the block. The register is not "buried" in this example. In the Logic section, the output rco is declared to be normally LOW with the DEFAULTS statement. The clock port for the set of D flip-flops (named counter[ ]) is declared with the *counter[ ].clk = !clock* statement. The NOT symbol (!) preceding the signal name is due to the need for a negative-edge-triggered counter (see Table 14-1) when the internal flip-flops are always positive-edge-triggered. All flip-flops have an optional asynchronous clear capability. The statement *counter[ ].clrn = !clear* is used to declare that the clear port (which is always active-LOW) on the set of D flip-flops should be controlled by the NOT clear signal, because Table 14-1 indicates an active-HIGH clear control is desired for the decade counter.

The IF/THEN statement is actually a nested IF/THEN in this example. IF statements automatically establish a priority of action by the order in which they are written. The first tested condition that evaluates to being true (HIGH) will determine the outcome of the IF/THEN statement. The outer IF statement first tests to determine if load is active.

If so, then the statement *counter[ ].d = d[ ]* will be applied. This statement will parallel load the 4-bit input d[3..0] into the register when clocked. If load is not true, then the next priority is to test enable. If enable is true, then a secondary test will be made with the nested IF. The register counter[ ] is tested to see if the contents equals 9. If so, then the value to be input at the next clock should be 0000_2. This will recycle the counter back to zero after reaching the terminal count of 9 for our decade counter. Also, the rco output should go HIGH at this state. If the count has not yet reached 9, then the counter should be incremented on the next clock (*counter[ ].d = counter[ ].q + 1*). Finally, if both load and enable tests should fail, then the counter should perform the remaining function of holding the current count. For a D flip-flop to hold the same data when it is clocked, the output must be fed back into the input, which can be accomplished with the statement, *counter[ ].d = counter[ ].q*. Simulation of the design is shown in Fig. 14-6.

```
SUBDESIGN  mod10
(
      clock, load, enable, clear, d[3..0]            :INPUT;
      counter[3..0], rco                             :OUTPUT;
)
VARIABLE
      counter[3..0]                                  :DFF;
BEGIN
      DEFAULTS
            rco = GND;         -- rco is normally low output
      END DEFAULTS;

      counter[].clk  = !clock;       -- clock on NGT
      counter[].clrn = !clear;       -- active-high clear

      IF  load  THEN    -- synchronous load highest priority
            counter[].d = d[];       -- parallel data inputs
      ELSIF  enable  THEN            -- test for count enable
            IF counter[].q == 9  THEN      -- at max count?
                  counter[].d = B"0000";   -- then recycle &
                  rco = VCC;               -- assert rco
            ELSE                     -- count up
                  counter[].d = counter[].q + 1;
            END IF;
      ELSE                           -- hold count if disabled
            counter[].d = counter[].q;
      END IF;
END;
```

Fig. 14-5 AHDL design file for the decade counter of Example 14-2

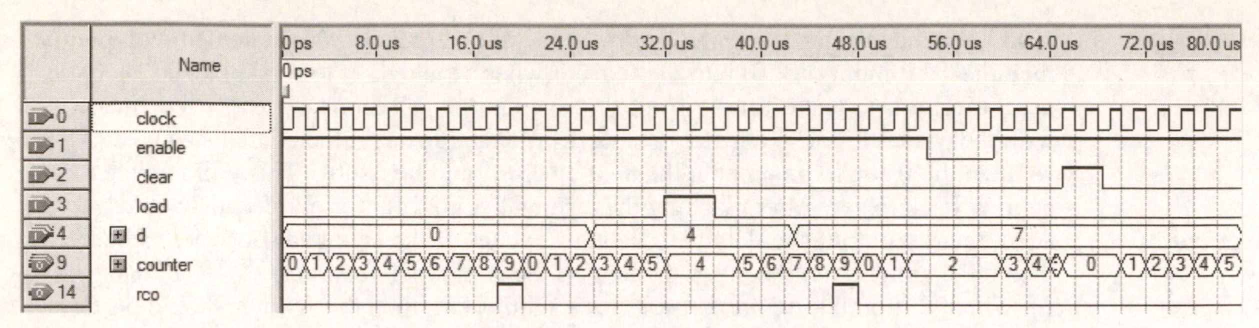

Fig. 14-6 Simulation results for Example 14-2

Example 14-3

Design a counter using AHDL that produces the 4-bit, irregular, recycling count sequence given in the timing diagram of Fig. 14-7. Also include an asynchronous reset to zero control for this counter and a count enable.

Fig. 14-7 Timing diagram for the sequential design of Example 14-3

An AHDL solution is shown in Fig. 14-8. The VARIABLE section has a state machine named seq defined with the keyword MACHINE. A state machine is basically a sequential circuit. In the definition of a state machine, the phrase “OF BITS” is optional. This phrase is used if it is necessary to name the individual output bits of the sequential circuit. In this example, the outputs are named qd, qc, qb, qa to match the names given in the timing diagram above. The phrase “WITH STATES” must be given for a state machine. This phrase identifies how many states are present in the sequential circuit and the names used to represent those states. In this example, the states have been named arbitrarily s0, s1, s2, ..., s8. Since there are nine states in this list and we have defined a 4-bit state machine, which could have 16 combinations, there are evidently some bit combinations that we are not concerned with. Another option in defining state machines is the specific bit pattern associated with each state. Some applications may not care what the states look like, and just care that the states occur in a specified order. In this example, the timing diagram gave us the specific pattern for each state in the sequence so that information is given for each state named by the WITH STATES phrase. Note the punctuation used in defining the state machine.

AHDL

```
SUBDESIGN  irreg_cnt
(
      ck1, reset, count                    :INPUT;
      qa, qb, qc, qd                       :OUTPUT;
)
VARIABLE
      seq          :MACHINE OF BITS (qd, qc, qb, qa)
                      WITH STATES (
                         s0 = B"0000",
                         s1 = B"1000",
                         s2 = B"0101",
                         s3 = B"1010",
                         s4 = B"0001",
                         s5 = B"1001",
                         s6 = B"0110",
                         s7 = B"1011",
                         s8 = B"0011");
%
state machine seq has 4 output bits that are connected
to output ports, bit patterns are defined for each state

counter modulus is defined by sequence of state names
defined in logic section below
%

BEGIN
      seq.clk = ck1;              -- PGT on ck1 to clock seq
      seq.reset = reset;          -- asynchronous reset port
      seq.ena = count;            -- active-high enable port

      TABLE
      % present-state                 next-state %
            seq          =>              seq;
            s0           =>              s1;
            s1           =>              s2;
            s2           =>              s3;
            s3           =>              s4;
            s4           =>              s5;
            s5           =>              s6;
            s6           =>              s7;
            s7           =>              s8;
            s8           =>              s0;
      END TABLE;
END;
```

Fig. 14-8 AHDL file for the irregular counter in Example 14-3

In the Logic section, we again see the clock port for the state machine assigned. A state machine has an optional, active-HIGH, asynchronous reset port available. The reset port name for a state machine is .reset. This example specified an asynchronous reset control that is accomplished with the statement *seq.reset = reset.* A state machine also has an optional active-HIGH enable port available. The design specification for a count enable is achieved with the statement *seq.ena = count.*

A TABLE is used to define the sequence for this state machine. It gives the "present state/next state" pattern for the state machine as defined in the timing diagram of Fig. 14-7. For any present state listed on the left side of the table, the next state to be produced after being clocked is given on the right side. The state machine recycles with the line *s8 => s0.* It is important to note that AHDL is a "concurrent language" and, as such, the line-by-line order in the TABLE does not matter; only the next state in each individual line matters.

Simulation results are shown in Fig. 14-9. The desired recycling, irregular count sequence is produced by this state machine. The simulation also verifies the asynchronous reset and count enable.

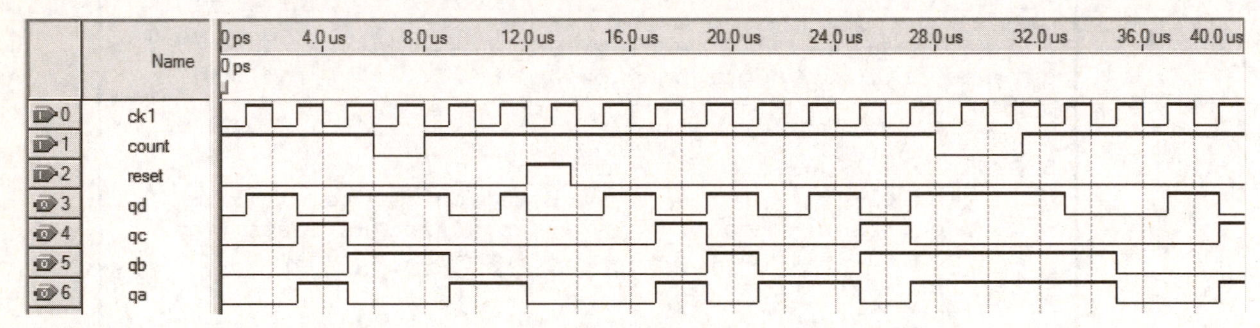

Fig. 14-9 Simulation results for Example 14-3

An alternate AHDL solution for Example 14-3 is shown in Fig. 14-10. A CASE statement is used in the Logic section to produce the desired irregular sequence for this design. Each state is identified with a WHEN clause. The present state will then determine the appropriate behavioral assignment for the state machine to produce the required next state.

```
SUBDESIGN  irreg_cnt2
(
      ck1, reset, count                 :INPUT;
      qa, qb, qc, qd                    :OUTPUT;
)
VARIABLE
      seq           :MACHINE OF BITS (qd, qc, qb, qa)
                       WITH STATES (
                          s0 = B"0000",
                          s1 = B"1000",
                          s2 = B"0101",
                          s3 = B"1010",
                          s4 = B"0001",
                          s5 = B"1001",
                          s6 = B"0110",
                          s7 = B"1011",
                          s8 = B"0011");
%
state machine seq has 4 output bits that are connected
to output ports, bit patterns are defined for each state

counter modulus is defined by sequence of state names
defined in logic section below
%

BEGIN
      seq.clk = ck1;              -- PGT on ck1 to clock seq
      seq.reset = reset;          -- asynchronous reset port
      seq.ena = count;            -- active-high enable port

      -- desired sequence defined by CASE statement
      CASE  seq  IS
            WHEN  s0  =>  seq = s1;
            WHEN  s1  =>  seq = s2;
            WHEN  s2  =>  seq = s3;
            WHEN  s3  =>  seq = s4;
            WHEN  s4  =>  seq = s5;
            WHEN  s5  =>  seq = s6;
            WHEN  s6  =>  seq = s7;
            WHEN  s7  =>  seq = s8;
            WHEN  s8  =>  seq = s0;
      END CASE;
END;
```

Fig. 14-10 Alternate AHDL file for the irregular counter in Example 14-3

Example 14-4

Design a bidirectional, full-step controller for a stepper motor using AHDL. The controller will produce a clockwise sequence pattern when cw = 1 and a counterclockwise sequence pattern when cw = 0. The stepper motor sequence is shown in Fig. 14-11.

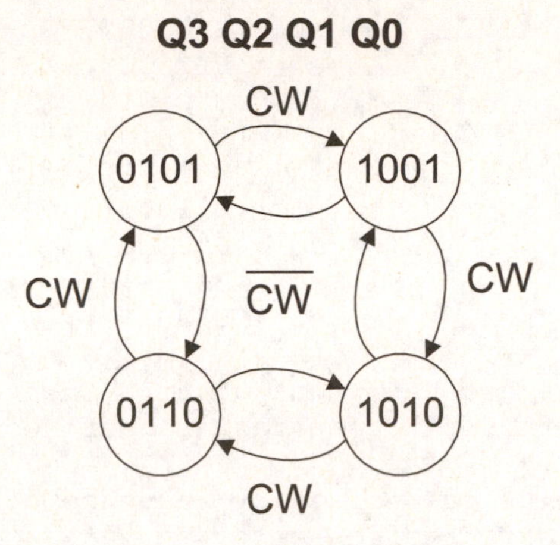

Fig. 14-11 State transition diagram for the stepper controller of Example 14-4

An example solution using AHDL is shown in Fig. 14-12. The SUBDESIGN name is given in single quotes because it includes the non-alphanumeric hyphen character. The application's bit combinations are specified in the state machine that is defined in the VARIABLE section. The state 0000_2 (initial) is included because the chip will automatically reset all flip-flops at power-up, and it will be necessary to define how to get out of that state and into the specified sequence to drive the stepper motor. The state machine's sequence is defined in the Logic section using a TABLE. The table includes the input control variable cw along with the present state for the machine called stepper on the input side of the table. The present state and the logic level for cw will determine the correct next state that should be produced by the state machine, except for the initial state produced at power-up. At power-up, we do not care what the next state is. We just want the machine to get into the specified count sequence. The next state 1010_2 was chosen arbitrarily from the list of valid states for the stepper motor controller.

Simulation results for this design are shown in Fig. 14-13. The count sequence for each direction is correct for the recycling state machine.

```
SUBDESIGN  'full-step'  -- single-quotes due to hyphen
(
        step, cw      :INPUT;
        q[3..0]       :OUTPUT;
)
VARIABLE
        stepper       :MACHINE OF BITS (q[3..0])
                          WITH STATES (
                              initial = B"0000",
                              s1 = B"1010",
                              s2 = B"0110",
                              s3 = B"0101",
                              s4 = B"1001");
        -- power-on state (0000) defined to start up stepper
BEGIN
        stepper.clk = step;               -- clock input

        -- present-state next-state table defines sequence
        TABLE
            stepper,   cw    =>    stepper;
            initial,   X     =>    s1;
            s1,        1     =>    s2;
            s1,        0     =>    s4;
            s2,        1     =>    s3;
            s2,        0     =>    s1;
            s3,        1     =>    s4;
            s3,        0     =>    s2;
            s4,        1     =>    s1;
            s4,        0     =>    s3;
        END TABLE;
END;
```

Fig. 14-12 AHDL file for the stepper motor controller

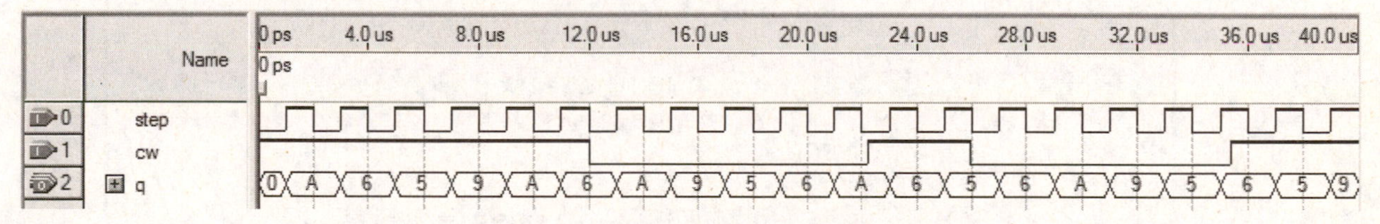

Fig. 14-13 Simulation results for Example 14-4

An alternative AHDL design for Example 14-4 is shown in Fig. 14-14. A CASE statement is used to test for each defined state of the state machine with the WHEN clauses. The power-up state (initial) will always be followed by s1 after a ↑-edge on the clock input (step). Each of the other next state assignments is dependent on the direction control (cw). The state assignment will produce the clockwise count sequence if cw = 1, and the state assignment after ELSE will produce the counterclockwise sequence if cw = 0.

```
SUBDESIGN  'full-step2'    -- single-quotes due to hyphen
(
      step, cw      :INPUT;
      q[3..0]       :OUTPUT;
)
VARIABLE
      stepper       :MACHINE OF BITS (q[3..0])
                       WITH STATES (
                          initial = B"0000",
                          s1 = B"1010",
                          s2 = B"0110",
                          s3 = B"0101",
                          s4 = B"1001");
      -- power-on state (0000) defined to start up stepper
BEGIN
      stepper.clk = step;            -- clock input

      -- stepper sequence controlled with CASE & IF
      CASE  stepper  IS    -- CASE determines present-state
            WHEN  initial  =>   stepper = s1;
            WHEN  s1    =>
                  IF  cw  THEN  stepper = s2;    -- go CW
                  ELSE  stepper = s4;            -- go CCW
                  END IF;
            WHEN  s2    =>
                  IF  cw  THEN  stepper = s3;
                  ELSE  stepper = s1;
                  END IF;
            WHEN  s3    =>
                  IF  cw  THEN  stepper = s4;
                  ELSE  stepper = s2;
                  END IF;
            WHEN  s4    =>
                  IF  cw  THEN  stepper = s1;
                  ELSE  stepper = s3;
                  END IF;
      END CASE;
END;
```

Fig. 14-14 Alternate AHDL file for the stepper motor controller

Example 14-5

Design a sequential circuit that will wait in an idle state for a trigger input (start) to occur and then output a single pulse four clock cycles later. The sequential circuit will then wait for another trigger signal. The state transition diagram for this delay circuit is shown in Fig. 14-15. The state machine should also be self-correcting.

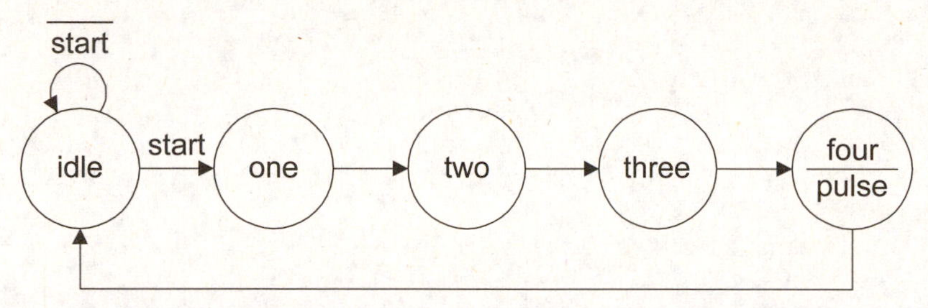

Fig. 14-15 State transition diagram for Example 14-5

An AHDL solution is shown in Fig. 14-16. A 3-bit state machine named mach is defined in the VARIABLE section. The bit patterns have not been defined for this state machine since we do not really care what they look like. The compiler will assign bit patterns for each state. This state machine will not be connected to output ports on the SUBDESIGN. The state machine will be "buried." A CASE statement is used to define the sequence of states for our state machine. When the state machine is in the idle state, it will be waiting for the start signal to go HIGH (and the ↑ clock) in order to proceed to the next state (one). After starting the sequence, the state machine will step through each state on each ↑ clock. Upon reaching state four, the pulse expression will be true until the next ↑ clock, when the state machine returns to idle. Notice that the only output needed is the signal named pulse.

The results of the simulation for Example 14-5 are shown in Fig. 14-17. The buried state machine (notice its "buried handle" labeled with an "s") has been included in the simulation to verify its correct operation. **See Quartus Notes for the setup procedure.** Only the state names are given by the simulator because the bit values were not user-defined. Note that the second start signal did not initiate the machine sequence because it occurred after the ↑ clock.

```
SUBDESIGN  delay
(
      clock          :INPUT;
      start          :INPUT;
      pulse          :OUTPUT;
)
VARIABLE
      mach           :MACHINE              -- "buried" machine
                        WITH STATES (
                           idle,
                           one,
                           two,
                           three,
                           four);          -- machine states
BEGIN
      mach.clk = clock;

      CASE  mach  IS
            WHEN idle =>         -- mach waits here
                  IF start THEN  mach = one;     -- go
                  ELSE           mach = idle;    -- wait
                  END IF;
            WHEN one =>          -- advances when clocked
                  mach = two;
            WHEN two =>          -- advances when clocked
                  mach = three;
            WHEN three =>        -- advances when clocked
                  mach = four;
            WHEN four =>         -- recycle to idle
                  mach = idle;
      END CASE;

      pulse = mach == four;    -- detect when at state four
END;
```

Fig. 14-16 AHDL file for Example 14-5

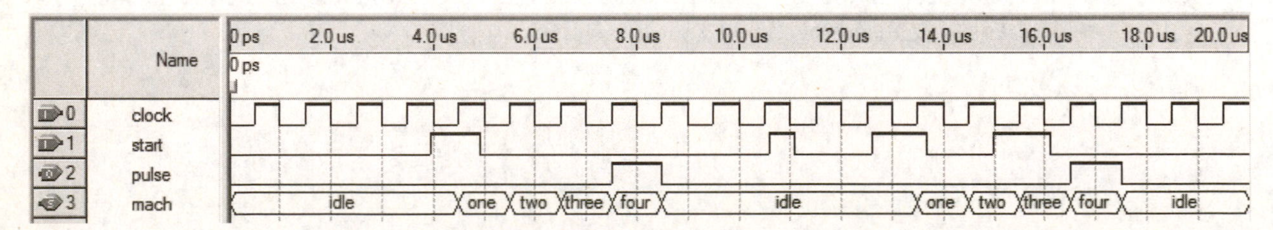

Fig. 14-17 Simulation results for Example 14-5

Example 14-6

Design the state machine (using AHDL) that is defined by the state transition diagram given in Fig. 14-18. The state machine has three inputs (trig and delay[1..0]) that control the sequence of states. The logic circuit will produce the three output signals named outs[3..1] according to the current state for the machine. The states are named, but the bit patterns for each state are not defined in this design.

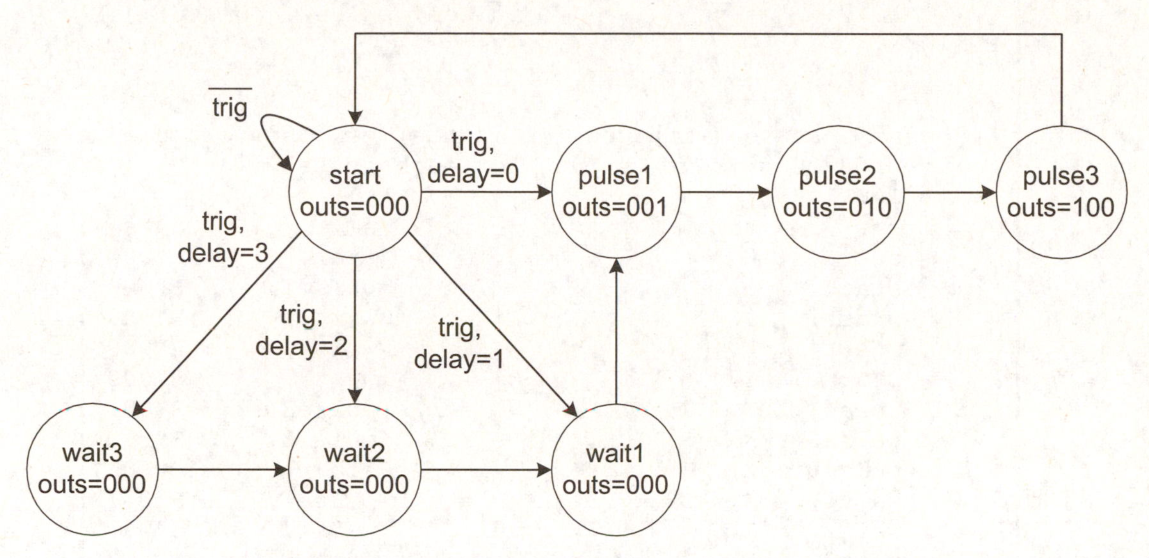

Fig. 14-18 State transition diagram for Example 14-6

An example AHDL solution named state_machine is shown in Fig. 14-19. The state machine (sm) is buried because its outputs are not connected to the output ports for the SUBDESIGN. The states for the state machine sm are defined by name only (start, wait1, wait2, wait3, pulse1, pulse2, pulse3). The optional bit patterns for the MACHINE definition were left for the compiler to determine because we do not care what they actually look like. Assigning bit names is also optional in AHDL. A TABLE is used in the Logic section to describe the operation of the state machine and the outputs named outs[3..1]. The input side of the table lists the current states and input conditions that will determine the next state for the state machine. If an input signal has no effect on a particular state, it has an X, which indicates that we "don't care" about that variable. The output side of the table lists the resulting next states for each present state/input combination. The right side of the table also gives the outputs to be produced for each present state. Be careful! These are combinational outputs (outs[3..1]) that match up with the appropriate present state, not the next state!

```
SUBDESIGN  state_machine
(
      clock, trig, delay[1..0]        :INPUT;
      outs[3..1]                      :OUTPUT;
)
VARIABLE
      sm    :MACHINE             -- buried state machine sm
               WITH STATES (start, wait1, wait2, wait3,
                              pulse1, pulse2, pulse3);
BEGIN
      sm.clk = clock;

%  present-state (with control inputs) next-state table
   to produce current outputs for each state  %
      TABLE
      sm,       trig,   delay[]    =>    sm,          outs[];
      start,    0,      X          =>    start,       B"000";
      start,    1,      0          =>    pulse1,      B"000";
      start,    1,      1          =>    wait1,       B"000";
      start,    1,      2          =>    wait2,       B"000";
      start,    1,      3          =>    wait3,       B"000";
      pulse1,   X,      X          =>    pulse2,      B"001";
      pulse2,   X,      X          =>    pulse3,      B"010";
      pulse3,   X,      X          =>    start,       B"100";
      wait1,    X,      X          =>    pulse1,      B"000";
      wait2,    X,      X          =>    wait1,       B"000";
      wait3,    X,      X          =>    wait2,       B"000";
      END TABLE;
END;
```

Fig. 14-19 AHDL file for Example 14-6

Simulation results for Example 14-6 are shown in Fig. 14-20. The buried state machine sm has been included in the simulation to make it easier to verify the circuit's operation.

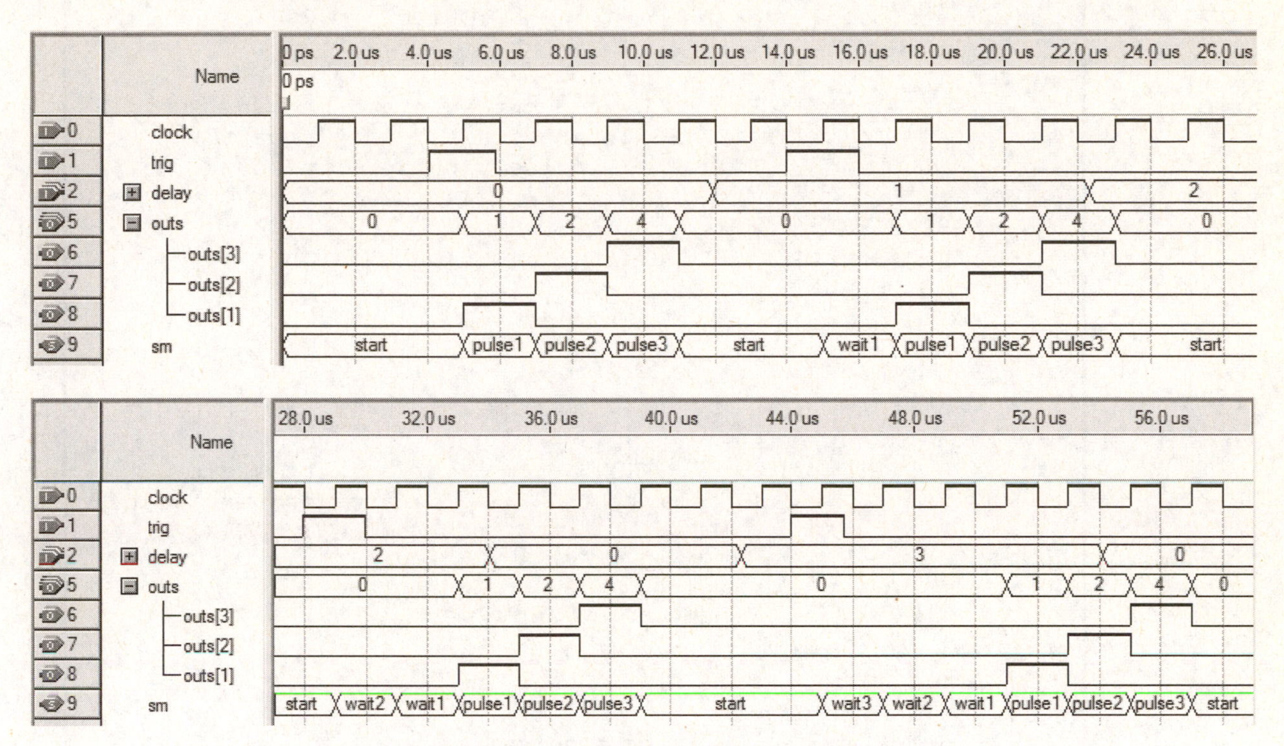

Fig. 14-20 Simulation of Example 14-6

An alternative AHDL technique for defining the state machine for Example 14-6 is shown in Fig. 14-21. A CASE statement is used to identify the states and describe the appropriate behavior(s). Multiple behavioral assignments are separated by semicolons. A DEFAULTS statement is used to declare the three output bits to be normally LOW. Active-HIGH signals will automatically default to LOW levels, so it is not necessary to use a DEFAULTS statement like we did in this example solution, but it makes the design's operation clearer to the reader. Only the output conditions that are different then need to be assigned. Next state assignments are also given for each CASE clause. Most state transitions are unconditional in this design, requiring only a ↑ clock. To get out of the state named start, however, will also require that the input signal trig be active (HIGH). The next state after start is also dependent upon the input control named delay[1..0], which is determined by the CASE statement testing for the appropriate input value and assigning the correct state machine behavior.

```
SUBDESIGN  state_machine2
(
      clock, trig, delay[1..0]          :INPUT;
      outs[3..1]                        :OUTPUT;
)
VARIABLE
      sm    :MACHINE              -- buried state machine sm
               WITH STATES (start, wait1, wait2, wait3,
                              pulse1, pulse2, pulse3);
BEGIN
      DEFAULTS
            outs[] = B"000";            -- outs default to 0
      END DEFAULTS;

      sm.clk = clock;

      CASE  sm  IS                      -- identify sm states
       WHEN  start  =>
            IF trig       THEN          -- trig exits start
                 CASE delay[] IS        -- determines next state
                      WHEN 0    =>      sm = pulse1;
                      WHEN 1    =>      sm = wait1;
                      WHEN 2    =>      sm = wait2;
                      WHEN 3    =>      sm = wait3;
                 END CASE;
            ELSE                        sm = start;
            END IF;
       WHEN wait1 =>      sm = pulse1;
       WHEN wait2 =>      sm = wait1;
       WHEN wait3 =>      sm = wait2;
       WHEN pulse1 =>     sm = pulse2;     outs[] = B"001";
       WHEN pulse2 =>     sm = pulse3;     outs[] = B"010";
       WHEN pulse3 =>     sm = start;      outs[] = B"100";
      END CASE;
END;
```

Fig. 14-21 Alternate AHDL file for Example 14-6

Example 14-7

Design a MOD-1000, recycling BCD counter using the mod10.tdf design in Example 14-2. The counter will have the functions given in Table 14-2. The AHDL file for the mod10 counter is repeated in Fig. 14-22.

clear	load	enable	clock	function
1	X	X	X	CLEAR
0	0	0	↓	HOLD
0	1	X	↓	LOAD
0	0	1	↓	COUNT UP

Table 14-2 Example 14-7 functions

```
SUBDESIGN  mod10
(     clock, load, enable, clear, d[3..0]          :INPUT;
      counter[3..0], rco                           :OUTPUT;     )
VARIABLE
      counter[3..0]                                :DFF;
BEGIN
      DEFAULTS
            rco = GND;         -- rco is normally low output
      END DEFAULTS;
      counter[].clk  = !clock;       -- clock on NGT
      counter[].clrn = !clear;       -- active-high clear
      IF  load  THEN    -- synchronous load highest priority
            counter[].d = d[];       -- parallel data inputs
      ELSIF  enable  THEN            -- test for count enable
            IF counter[].q == 9  THEN      -- at max count?
                  counter[].d = B"0000";   -- then recycle &
                  rco = VCC;               -- assert rco
            ELSE                           -- count up
                  counter[].d = counter[].q + 1;
            END IF;
      ELSE                           -- hold count if disabled
            counter[].d = counter[].q;
      END IF;
END;
```

Fig. 14-22 AHDL design file for the decade counter from Example 14-2

A top-level schematic/block diagram file (Fig. 14-23) can be created using the default symbol for the mod10.tdf design. Three copies of the mod10 symbol will be needed, one for the ones digit, one for the tens digit, and one for the hundreds digit. The three mod10 stages are cascaded together by connecting the rco output (which detects the terminal state 9 when the counter is enabled) to the next digit's enable control. The enable input on the ones digit will control all of the mod10 counters. The enable input on the tens digit will be HIGH only if the ones digit is a 9 and the MOD-1000 counter is enabled. Likewise, the hundreds digit will be enabled only if the tens digit and the ones digit are both 9 (and the counter is enabled). The load and clear controls are connected to all three counter stages.

Sample simulation results for the MOD-1000 counter are shown in Fig. 14-24.

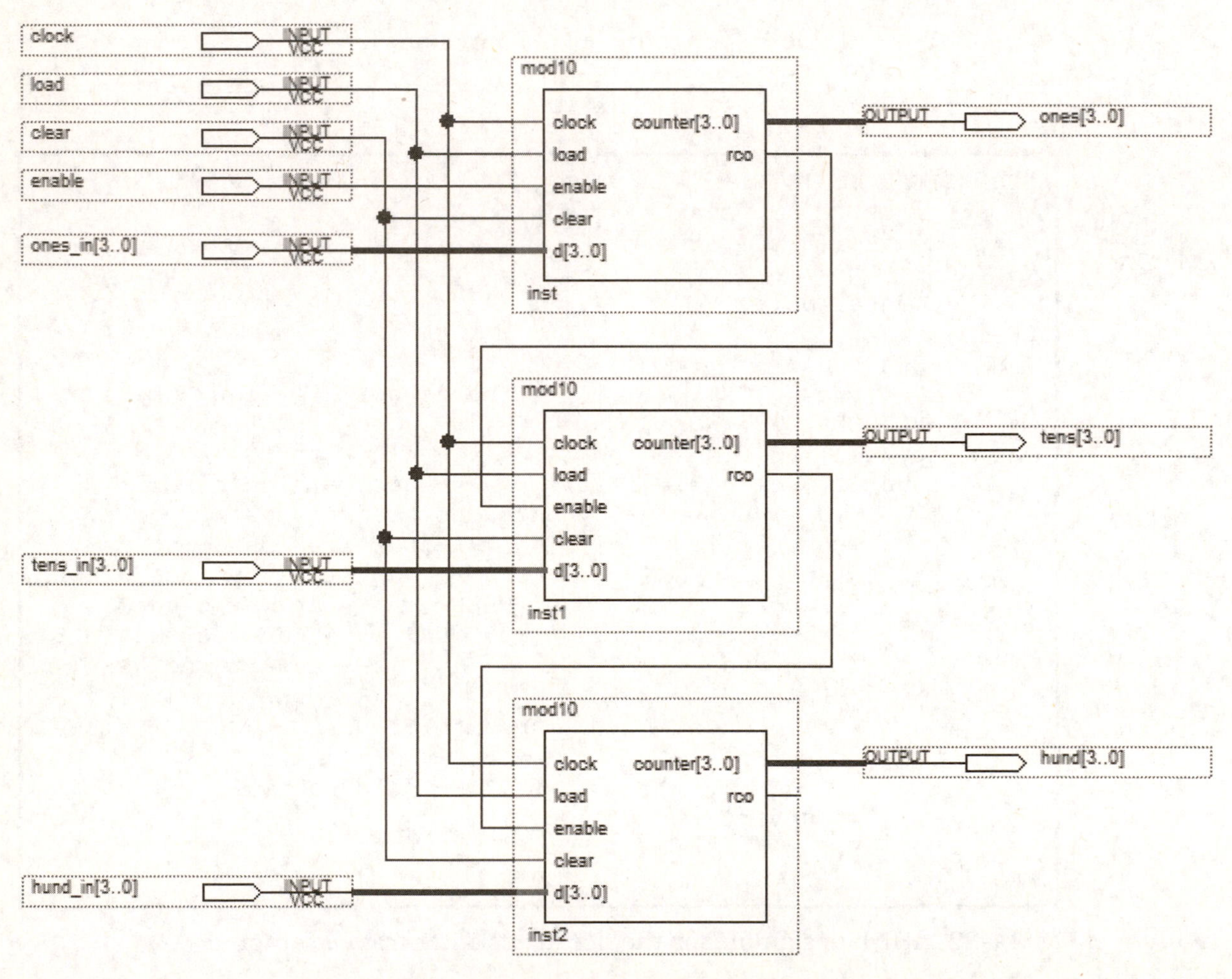

Fig. 14-23 Top-level design file for Example 14-7

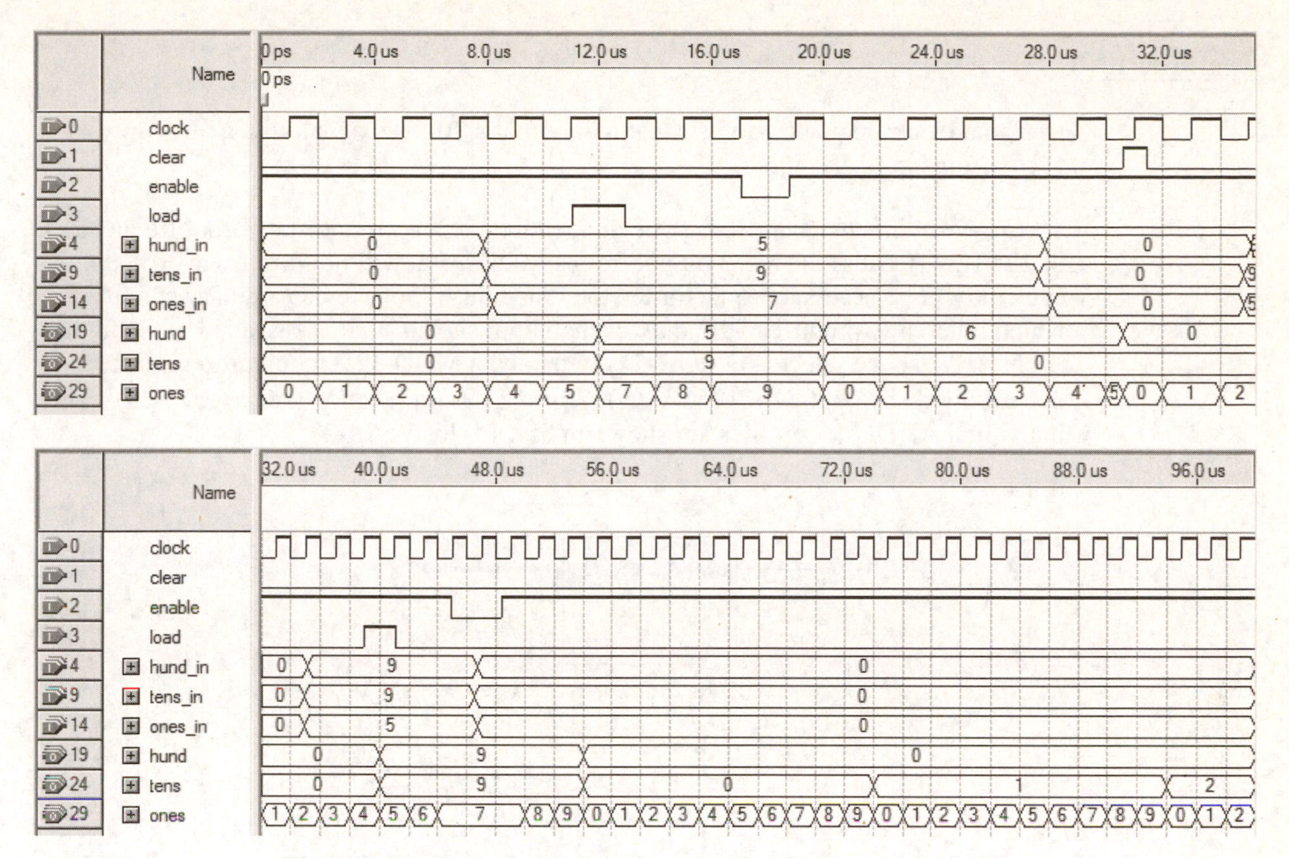

Fig. 14-24 Sample simulation results for Example 14-7

Example 14-8

Design a clock frequency divider that will output a 20-kHz signal with a 50% duty cycle using a 50-MHz input.

The 50-MHz input signal needs to be divided by 2500 to produce an output frequency of 20 kHz. Because the MSB output from a MOD-2500 will not produce a 50% duty cycle, it will be necessary to factor the total modulus (2500) to use a binary counter, which inherently produces 50% duty cycle output signals, that is cascaded with another counter. We have selected a MOD-2 counter and a MOD-1250 counter to produce an overall divide by 2500 ($2 \times 1250 = 2500$) for our design solution (see Fig. 14-25). The individual AHDL design files are shown in Fig. 14-26 and 14-27.

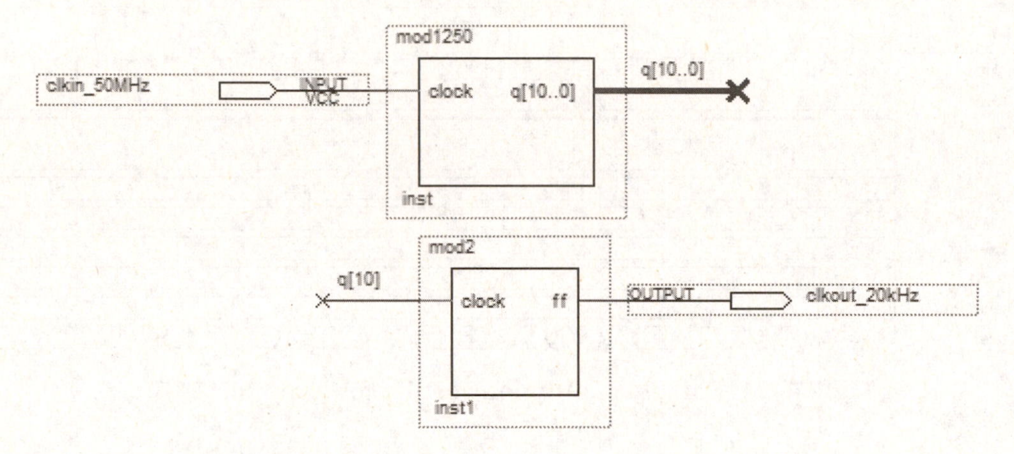

Fig. 14-25 Top-level design file for Example 14-8

```
SUBDESIGN  mod1250
(
      clock        :INPUT;
      q[10..0]     :OUTPUT;
)
VARIABLE
      q[10..0]     :DFF;             -- 11-bit counter needed
BEGIN
      q[].clk = clock;

      IF q[] == 1249    THEN         -- detect terminal state
            q[].d = 0;               -- and recycle
      ELSE
            q[].d = q[].q + 1;       -- otherwise, count up
      END IF;
END;
```

Fig. 14-26 AHDL design for MOD-1250 counter block

```
SUBDESIGN  mod2
(
        clock                   :INPUT;
        ff                      :OUTPUT;
)
VARIABLE
        ff                      :DFF;         -- only need one flip-flop
BEGIN
        ff.clk = clock;

        ff.d = !ff.q;           -- feed in opposite state when clocked
END;
```

Fig. 14-27 AHDL design for MOD-2 counter block

The simulation results are shown in Fig. 14-28. The period measurement of the output signal using the time bars in the simulator is 50.0 μs. This is the correct time value for a 20-kHz waveform. See Quartus Notes for the setup procedure.

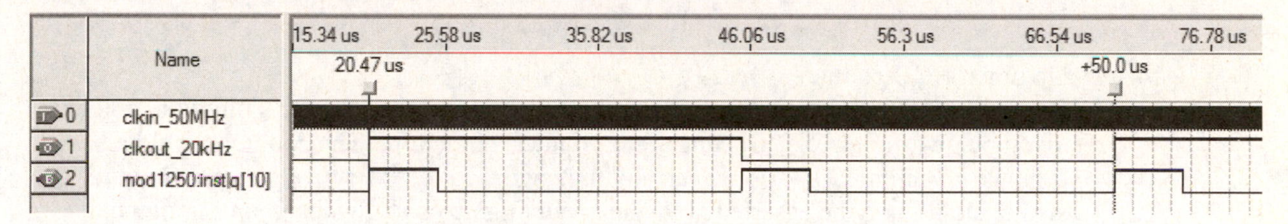

Fig. 14-28 Sample simulation results for Example 14-8

Laboratory Projects

Design the following counter applications using AHDL. Simulate and test the circuit designs.

14A.1 Binary down counter

Design a recycling, 4-bit binary down counter. Counting is controlled by an active-LOW enable named en. The counter can also be synchronously parallel loaded when an input named ld is HIGH. The following function table describes the counter's operation (note the function priorities). Label the parallel inputs data[3..0] and the counter outputs q[3..0]. The circuit also should produce an output signal called last, which goes HIGH whenever the counter reaches its terminal state (0000).

ld	en	operation
1	X	Load
0	0	Count
0	1	Hold

14A.2 Up/down BCD counter

Design a MOD-10, recycling, BCD up/down counter. The counter's function table is given below. Inputs c[1..0] control the counter's function. The counter should also produce an active-LOW ripple carry output signal called carryn. The carry output signal should only be asserted at 9 when counting up or at 0 when counting down.

c1	c0	operation
0	0	Reset
0	1	Count Down
1	0	Count Up
1	1	Hold Count

14A.3 MOD-60 BCD counter

Design a MOD-60, recycling BCD counter. The counter should have an active-LOW, synchronous reset (resn) signal. The count sequence will be 0 through 59_{10} in BCD. Create a hierarchical design with separate *tdf* files for the tens digit and the ones digit. Use a *bdf* file for the top-level file that interconnects the two *tdf* files.

14A.4 MOD-100 binary counter

Design a MOD-100, recycling binary counter. The counter is triggered with ↓ clocks and should have an active-HIGH count enable called enable. The counter will also output a single active-HIGH pulse (called pulse) during each MOD-100 count sequence. The pulse occurs at the end of the count sequence during either the last 5 or 10 counter states. The pulse width will be controlled by an input signal called w, as shown in the table below. Essentially, w selects the duty cycle of the divide-by-100 output signal. Compare functional and timing simulation results. Why does the timing simulation show glitches?

W	PULSE width	duty cycle
0	5 clock periods	5%
1	10 clock periods	10%

14A.5 Stepper motor sequence controller

Design a half-step sequencer to control a stepper motor. The sequencer should produce the appropriate sequence of states to drive the motor either clockwise (CW) or counterclockwise (CCW) without using a binary counter (i.e., produce the half-step count sequence directly). The stepper direction is controlled by an input signal named cw. The stepper enable is called go. The function table is given below. Make sure the sequencer is self-starting. Hint: The enable input to a state machine is the .ena port.

GO	CW	function
0	X	HALT
1	0	CCW
1	1	CW

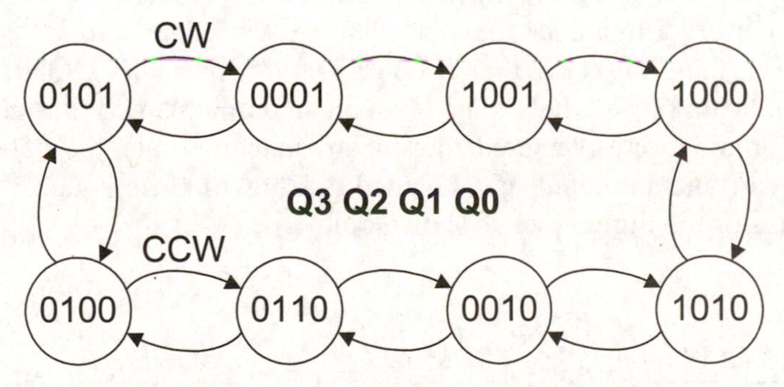

Half-step sequence for stepper motor control

14A.6 Variable frequency divider

Design a variable frequency divider. The frequency divider should divide the input frequency by one of four different factors. The divide-by factor is controlled by two mode controls, as described by the following function table. The mode controls are used to change the modulus of the counter used for the frequency division. The output signal is the most significant bit of the necessary counter for each divider.

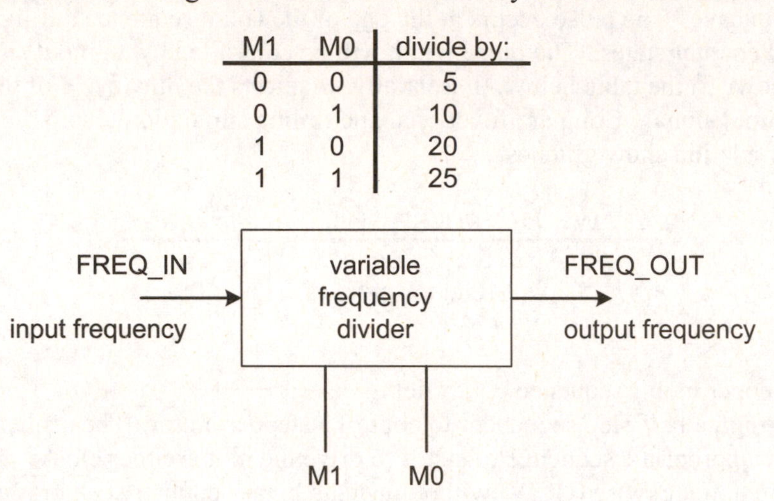

M1	M0	divide by:
0	0	5
0	1	10
1	0	20
1	1	25

14A.7 Digital lock

Design a digital lock circuit. The lock will have a 4-bit data input and an enter signal. Use a state machine with 5 states to keep track of the sequencing through the combination. **The enter signal is a manual clock signal for the state machine.** A sequence of four 4-bit numbers will make up the lock combination (that you will select for your design). Each 4-bit input number will be applied (via switches), and then the manual enter signal (↑clock) will be pulsed. The machine states must be sequenced in the proper order to unlock the lock. The START state is followed by the states STATE1, then STATE2, then STATE3, and then finally DONE, which "unlocks" the lock (indicated by a HIGH logic level on an output named unlock). If an incorrect input combination is "entered" during any machine state, the state machine will return to START and an output signal named ready will be activated. The state transition diagram for the digital lock is shown below.

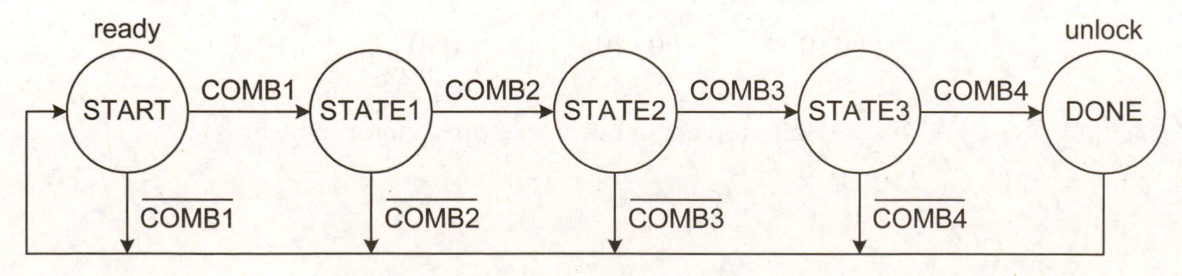

Returns to START if incorrect combination is applied

14A.8 Programmable frequency divider

Design an 8-bit programmable frequency divider. The input signal frequency (freq_in) will be divided by a variable 8-bit binary input value (B7 through B0). The output signal (freq_out) that is produced should be a single pulse each time the divider circuit counts through the modulus represented by B7..0. Hint: Detect the counter state 00000001_2 for the output frequency signal, synchronously load in the frequency division factor (b[7..0]) when the counter reaches this state (00000001_2), count down, and repeat.

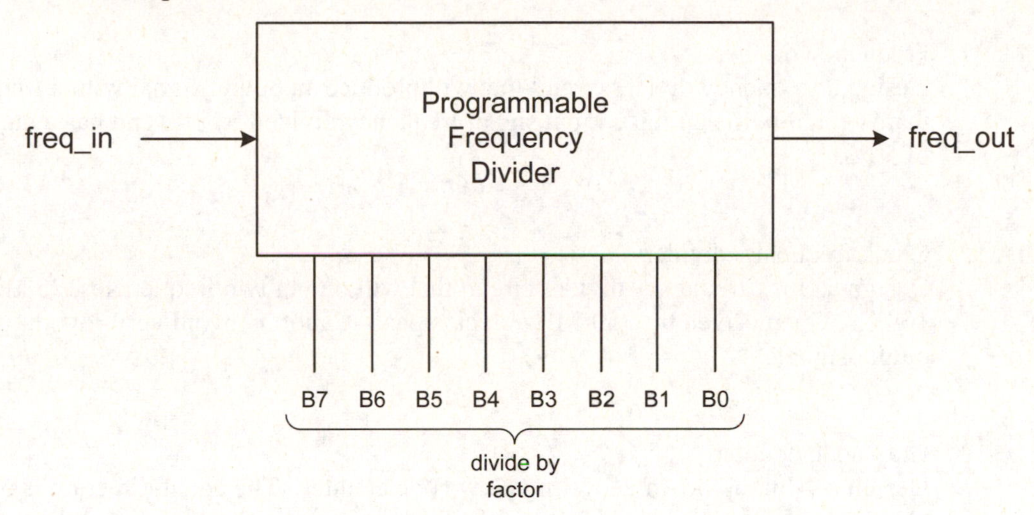

14A.9 State machine

Design the following state machine. Make the design self-correcting by forcing any other state to go to 001_2 on the next ↑ clock.

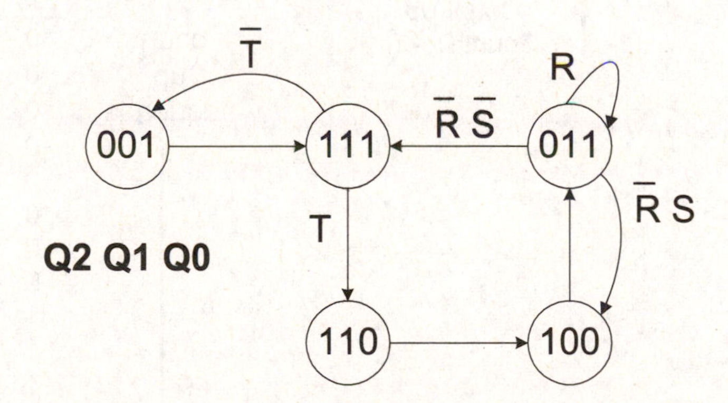

14A.10 Variable pulse delay

Design a state machine that will output a single time-delayed pulse after receiving a triggering signal named start. The state machine can only be triggered to start the timing delay for the output pulse when it is in an idle state. The length of the delay is variable and controlled by an input named delay. The output pulse produced has a duration of one clock cycle and will begin on the fifth rising edge of clock when delay = 0 or on the tenth rising edge of clock when delay = 1. At the end of the variable timing cycle for the state machine, the state machine will return to its idle state and wait for another trigger input on start.

14A.11 Frequency divider

Design a frequency divider circuit that will produce an output signal with a frequency that is equal to the circuit's input signal frequency divided by 100 and has a duty cycle of 50%.

14A.12 Clock frequency divider

Design a clock frequency divider circuit that will output two frequencies, 25 kHz and 10 kHz, when driven by a 50-MHz clock signal. Caution: avoid very-low duty cycle output signals!

14A.13 Gray code counter

Design a 4-bit, up/down, recycling Gray code counter. The count direction is controlled by a signal called dir, as indicated in the following function table. Label the counter outputs q[3..0]. The circuit also should produce an output signal called index, which goes LOW whenever the counter state is 0000.

dir	operation
0	Count Up
1	Count Down

Gray code sequence (count up sequence: top to bottom ↓; count down sequence: bottom to top ↑)

q3	q2	q1	q0
0	0	0	0
0	0	0	1
0	0	1	1
0	0	1	0
0	1	1	0
0	1	1	1
0	1	0	1
0	1	0	0
1	1	0	0
1	1	0	1
1	1	1	1
1	1	1	0
1	0	1	0
1	0	1	1
1	0	0	1
1	0	0	0

UNIT 14V

COUNTER DESIGNS USING VHDL

Objective

- To design sequential logic circuits with CPLDs or FPGAs using VHDL.

Sequential Circuits Using VHDL

The design and implementation of sequential circuits using VHDL is carried out in much the same fashion as for combinational circuits. The flip-flops contained in Altera's CPLDs or FPGAs can be treated as SR, JK, D, or T-type flip-flops. The control inputs for the flip-flops are produced by the programmable logic in the PLD. Due to ease of use, D flip-flops are most commonly utilized in hardware description languages. The high-level language VHDL provides many options for defining sequential circuits, including IF/THEN and CASE statements. See Quartus Notes (www.pearsonhighered.com/electronics).

Shown in Fig. 14-1 is a VHDL design description that creates a single D flip-flop. The ENTITY design unit declares the input and output ports for d_ff.vhd to be a BIT data type. The behavioral description for a single D flip-flop is described in the ARCHITECTURE design unit. This is accomplished within a PROCESS in VHDL because the behavior of a flip-flop is sequential. Sequential behavior can only be defined in a PROCESS. It is important to remember that the PROCESS statement itself is concurrent; just the statements inside the PROCESS are sequential. The sensitivity list for this PROCESS includes the signals clock, preset, and clear. If any of these signals should change, then the PROCESS will be invoked to determine the resultant output (q_out). The asynchronous inputs clear and preset are active-LOW and have priority in the D flip-flop's operation. They are, therefore, tested first in the IF statement, and the corresponding signal assignment behavior is described after each

THEN. If neither of the asynchronous inputs invoked the PROCESS, then it must have been the clock signal that changed. The given Boolean expression (*clock'EVENT AND clock = '1'*) tests for a positive edge on the clock signal. The 'EVENT portion of the expression is a predefined property called an attribute for a named signal. The 'EVENT attribute is used to test if a transition has occurred on the associated signal, clock in this example. There are two possible transitions that may occur on any signal, positive-going and negative-going. We, of course, must define which of these is the clock edge that triggers our flip-flop. The other portion of the Boolean expression (*AND clock = '1'*) defines the desired edge. The *clock = '1'* will select the positive edge. If we had instead used the expression *(clock'EVENT AND clock = '0')*, then the flip-flop would be triggered on the negative edge. So, if the flip-flop receives the correct clock edge and the asynchronous inputs clear and preset are both inactive (HIGH), then the flip-flop will have the input called data stored. It is important to note that in using a behavioral description for a flip-flop or latch in VHDL, the actual device is implied by an incomplete description within the PROCESS. The memory characteristic for these devices is implied because the IF statement does not explicitly describe how this device should act if none of the tested Boolean expressions is true. There is no final ELSE with the corresponding signal assignment. Results for the simulation of d_ff.vhd are given in Fig. 14-2.

VHDL

```
ENTITY d_ff IS
PORT (
data, clock, preset, clear          :IN BIT;
q_out                               :OUT BIT
);
END d_ff;

ARCHITECTURE simple OF d_ff IS
BEGIN
      PROCESS (clock, preset, clear)
            -- invoke process if any listed signal changes
      BEGIN
                  -- priority order: clear, preset, clock
            IF    (clear = '0')     THEN  q_out <= '0';
            ELSIF (preset = '0')    THEN  q_out <= '1';
            ELSIF (clock'EVENT AND clock = '1')
                                    THEN  q_out <= data;
            END IF;
                  -- memory is implied, no ELSE is listed
      END PROCESS;
END simple;
```

Fig. 14-1 VHDL description creating a single D flip-flop

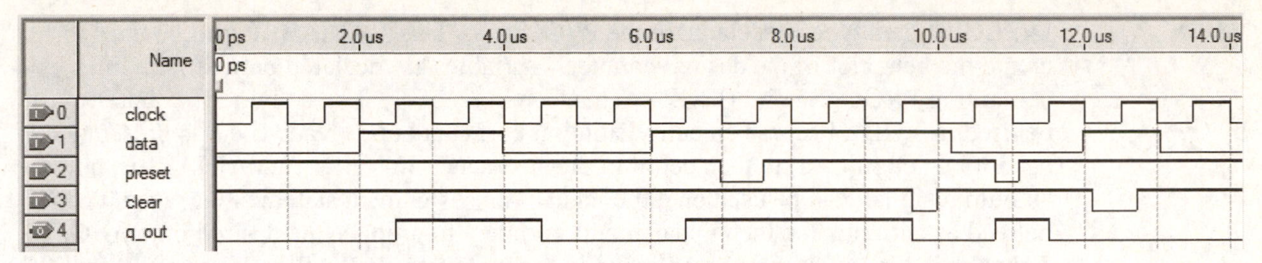

Fig. 14-2 Simulation results for the VHDL D flip-flop

Example 14-1

Design a MOD-16, binary up/down counter using VHDL. The count direction is controlled by dir (dir = 1 for count up).

```
ENTITY updncntr IS
PORT (
        clock, dir            :IN BIT;
        q                     :OUT INTEGER RANGE 0 TO 15
                              -- output q will need 4 bits
);
END updncntr;

ARCHITECTURE a OF updncntr IS
BEGIN
        PROCESS (clock)              -- look for clock change
        VARIABLE count        :INTEGER RANGE 0 TO 15;
                -- create a variable for counter value

        BEGIN
                IF (clock'EVENT AND clock = '1')    THEN
                        IF  (dir = '1')   THEN    -- count up
                                count := count + 1;
                        ELSE                      -- count down
                                count := count - 1;
                        END IF;
                END IF;       -- count holds with no clock
        q <= count;           -- count value to output port
        END PROCESS;
END a;
```

Fig. 14-3 VHDL file for the binary up/down counter design for Example 14-1

A VHDL text file solution named updncntr.vhd is shown in Fig. 14-3. In order to allow the output port named q to be treated as a numerical value that can be incremented or decremented, it is assigned an INTEGER data type in the ENTITY declaration. By declaring a RANGE of 0 to 15 for this output port, the compiler will determine that four flip-flops will be needed. The actual count sequence and, therefore, the counter's modulus still must be defined in the PROCESS statement that is contained in the ARCHITECTURE design unit. An internal VARIABLE named count with an

INTEGER data type is declared in the PROCESS. This VARIABLE will be used to create the behavior of the desired counter. Variables are declared between the lines containing PROCESS and BEGIN. This technique of creating a "buried" counter must be used in VHDL because an entity's output port cannot be "read" (i.e., its present state cannot be determined) to then count to the next state. In VHDL, a VARIABLE, unlike an output signal, can be used on either side of an assignment statement because it can be read as an input and then updated appropriately to produce the desired output. Only a change in the clock signal (contained in the sensitivity list) will invoke the PROCESS and thereby produce the next state. The first IF statement tests for a logic level change, with the 'EVENT attribute AND also a HIGH for the clock signal. This notation in VHDL is used to describe a positive-going edge for the named signal. If clock instead has a negative-going transition (NGT), the IF statement fails the test and the counter will hold its current state. If the signal clock experienced a positive-going transition (PGT), then the following nested IF statement will be evaluated to determine the condition of the input signal dir. If dir is HIGH, then the VARIABLE count will be incremented by adding 1 to its present state in the variable assignment statement *count := count + 1*. Otherwise (i.e., dir = 0), the VARIABLE count will be decremented with the variable assignment statement *count := count – 1*. The result of incrementing or decrementing the register's contents is a count-up or a count-down sequence, respectively. The 4-bit VARIABLE count will automatically recycle on the next ↑ clock when it has reached the terminal state. The signal assignment statement, *q <= count*, will connect the VARIABLE count to the output port for updncntr. This signal assignment statement must be placed within the PROCESS because variables are only defined locally. The results of simulating the VHDL design file are shown in Fig. 14-4.

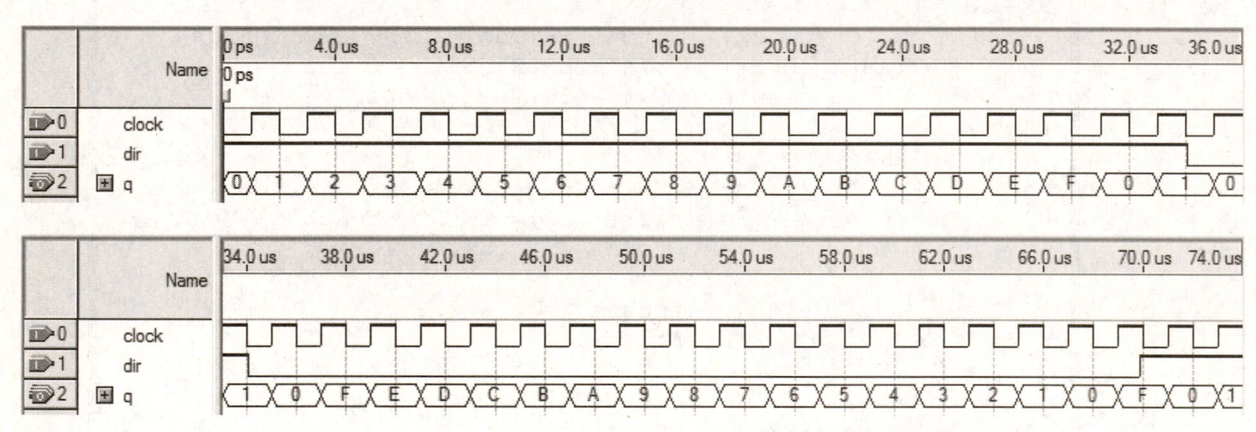

Fig. 14-4 Simulation of the VHDL file for Example 14-1

Example 14-2

Design a decade counter using VHDL. The MOD-10 counter has a count enable, parallel load, and asynchronous clear controls as shown in Table 14-1. The counter should also produce a ripple carry output (rco) signal that goes HIGH during the last state in the count sequence when the counter is enabled.

clear	load	enable	clock	Function
1	X	X	X	CLEAR
0	0	0	↓	HOLD
0	1	X	↓	LOAD
0	0	1	↓	COUNT UP

Table 14-1 Example 14-2 functions

A VHDL solution named mod10.vhd is shown in Fig. 14-5. The ENTITY declaration includes the output signal q specified as an INTEGER data type whose values will range from 0 to 15. The compiler uses this information to determine that four bits will be needed for the output port. Actually, we will only have values up to 9 for this BCD counter design but four bits will still be necessary. The range of values declared in the ENTITY is just used to determine the number of bits that will be necessary. The count sequence and therefore the counter's modulus are determined by the description contained in the ARCHITECTURE design unit. A VHDL PROCESS statement contains the sequential statements that describe the behavior of the mod-10 counter. The sensitivity list only needs to include the clock, asynchronous clear, and enable signals. A VARIABLE named counter with a data type of INTEGER is declared within the PROCESS. Variables can exist in a PROCESS (or PROCEDURE or FUNCTION). The VARIABLE counter is used to keep track of and update the value in the counter. In VHDL, the output ports cannot be "read"; they can only be "written (output) to." So we would not know what the current count value is. A variable, on the other hand, can be "read" and updated to the next state. When the PROCESS is invoked, the entity's output will be updated with the signal assignment statement *q <= counter*. Variables are declared between PROCESS and BEGIN. The first IF statement within the PROCESS will test the counter to determine if it is at the terminal state 9 when the counter is enabled. If this condition is true, rco will output a HIGH; otherwise rco will be LOW. The nested IF statements are used to define the behavior of the mod-10 counter. According to the function table (Table 14-1), the asynchronous clear should have priority over the other counter functions, so it is evaluated first. Notice that the assignment statement for variables uses a different symbol (:=) than is used for signals. If the test for clear (active-HIGH input) fails, then the ELSIF will next test for a negative edge (*clock'EVENT AND clock = '0'*) on clock. IF statements automatically establish a priority of action by the order in which they are written. The first tested condition that evaluates to being true (HIGH) will determine the outcome of the IF/THEN statement. If a negative edge occurred on clock, then the next level of IF statements will test to see if the counter is to be loaded; in which case, counter should get the value of d (*counter := d*). Note that the load function has priority (because it is tested first) over the count function. If the counter is enabled but load is LOW and a negative edge is applied to clock, then the count value should be incremented (*counter := counter + 1*), unless we have already reached the terminal state of 9 in our decade counter. The innermost IF statement will cause the counter to recycle to 0 after reaching 9. If the counter is clocked but disabled, then the count should stay the same. In VHDL, this behavior of not changing the state is implied by not describing any other action. The actual memory characteristic of a register or counter is implied, so you do not need to describe it with the phrase *ELSE counter := counter* in the second-level IF statement. Simulation of the design is shown in Fig. 14-6.

```
ENTITY mod10 IS
PORT (
      clock                    :IN BIT;
      load, enable, clear      :IN BIT;
      d                        :IN INTEGER RANGE 0 TO 15;
      q                        :OUT INTEGER RANGE 0 TO 15;
      rco                      :OUT BIT
);
END mod10;

ARCHITECTURE bcd OF mod10 IS
BEGIN
      PROCESS (clock, clear, enable)
                  -- invoke process if signal changes
            VARIABLE  counter       :INTEGER RANGE 0 TO 15;
      BEGIN
            IF (clear = '1') THEN   counter := 0;
                  -- asynchronous clear has priority
            ELSIF (clock'EVENT AND clock = '0') THEN
                  IF (load = '1')  THEN   counter := d;
                        -- synchronous load
                  ELSIF  (enable = '1')  THEN
                        IF (counter = 9)  THEN  -- recycle
                              counter := 0;
                        ELSE        counter := counter + 1;
                        END IF;
                  -- hold count behavior is implied
                  END IF;
            END IF;

                  -- rco detects terminal count when enabled
            IF ((counter = 9) AND (enable = '1')) THEN
                        rco <= '1';
            ELSE        rco <= '0';
            END IF;

            q <= counter;       -- output counter to ports
      END PROCESS;
END bcd;
```

Fig. 14-5 VHDL design file for the decade counter of Example 14-2

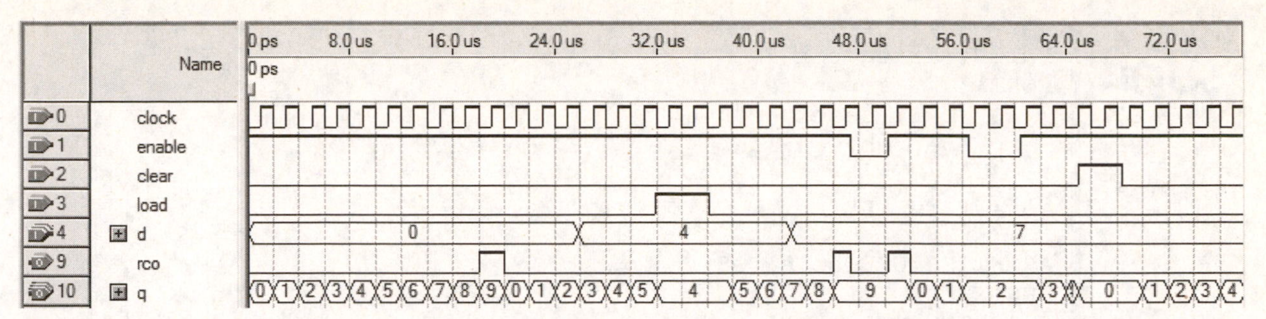

Fig. 14-6 Simulation results for Example 14-2

Example 14-3

Design a counter using VHDL that produces the 4-bit, irregular, recycling count sequence given in the timing diagram of Fig. 14-7. Also include an asynchronous reset to zero control for this counter and a count enable.

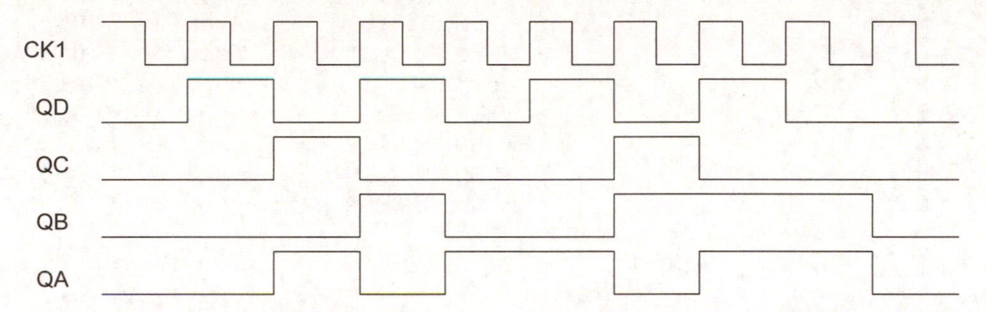

Fig. 14-7 Timing diagram for the sequential design of Example 14-3

A VHDL solution is shown in Fig. 14-8. The sensitivity list for this PROCESS contains the signals ck1 and reset. The VARIABLE seq, a BIT_VECTOR defined within the PROCESS, is used to define the irregular count sequence for this 4-bit counter. The reset signal is asynchronous and therefore has priority over the counting function, so it must be tested in the IF statement first. If the counter is being reset, then the variable assignment statement will make seq = 0000. If the test for the reset function fails, we will then test to see if the counter is disabled (*count = '0'*), which will keep the counter at the same state (*seq := seq*). Finally, if the counter is not being reset but it is enabled, then we will need to check to see if the correct positive-going clock edge has been applied (*ck1'EVENT AND ck1 = '1'*). A CASE statement is used to define the irregular count sequence for this counter. Each of the used states for this counter is identified with a WHEN clause, and the appropriate next state is assigned to the variable seq. All possible combinations of the expression to be evaluated (seq) must be covered in the WHEN statements. The WHEN OTHERS choice will take care of the unused states for this counter. The appropriate bit in the VARIABLE seq is assigned to the corresponding output port by the set of signal assignment statements. Simulation results are shown in Fig. 14-9. The desired recycling, irregular count sequence is produced by this state machine. The asynchronous reset and count enable are also verified by the simulation.

```
ENTITY irreg_cnt IS
PORT (
      ck1, reset, count              :IN BIT;
      qa, qb, qc, qd                 :OUT BIT
);
END irreg_cnt;

ARCHITECTURE soln1 OF irreg_cnt IS
BEGIN
      PROCESS (ck1, reset)      -- process sensitivity list
            VARIABLE seq        :BIT_VECTOR (3 DOWNTO 0);
      BEGIN
            IF (reset = '1')          THEN  seq := "0000";
            ELSIF (count = '0')       THEN  seq := seq;
            ELSIF (ck1'EVENT AND ck1 = '1')       THEN
                  CASE seq IS           -- define sequence
                        WHEN "0000" =>     seq := "1000";
                        WHEN "1000" =>     seq := "0101";
                        WHEN "0101" =>     seq := "1010";
                        WHEN "1010" =>     seq := "0001";
                        WHEN "0001" =>     seq := "1001";
                        WHEN "1001" =>     seq := "0110";
                        WHEN "0110" =>     seq := "1011";
                        WHEN "1011" =>     seq := "0011";
                        WHEN "0011" =>     seq := "0000";
                        WHEN OTHERS =>     seq := "0000";
                  END CASE;
            END IF;
            qa <= seq(0);               -- output each bit
            qb <= seq(1);
            qc <= seq(2);
            qd <= seq(3);
      END PROCESS;
END soln1;
```

Fig. 14-8 VHDL file for the irregular counter of Example 14-3

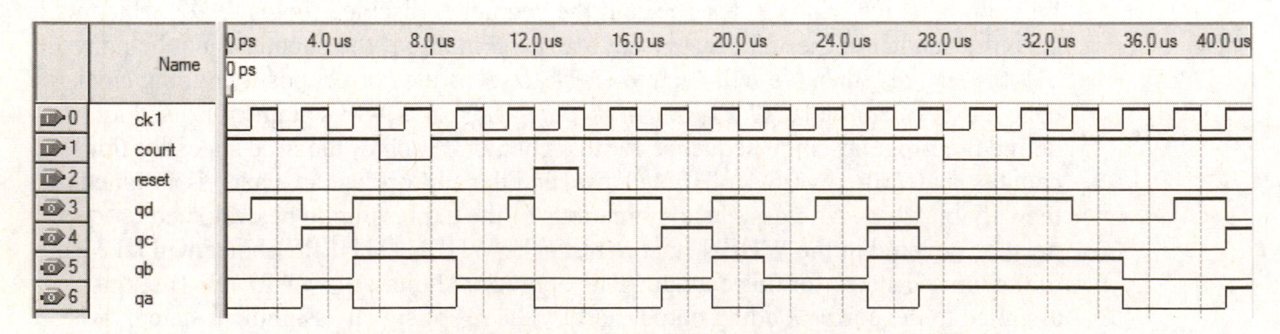

Fig. 14-9 Simulation results for Example 14-3

An alternate VHDL solution for Example 14-3 is shown in Fig. 14-10. In this solution, a SIGNAL named output is created as a BIT_VECTOR. Like the first solution, a VARIABLE named seq is also created. But this time, seq is given the data type INTEGER. As an INTEGER, seq can then be incremented to produce a mod-9 counter. The CASE statement determines the state of the dummy counter seq, which will then select the appropriate bit pattern for output. Each of the nine used states is identified with a WHEN clause. This is essentially creating a code converter so that the proper state sequence is produced.

```
ENTITY irreg_cnt IS
PORT (
      ck1, reset, count          :IN BIT;
      qa, qb, qc, qd             :OUT BIT
);
END irreg_cnt;

ARCHITECTURE soln2 OF irreg_cnt IS
      SIGNAL output              :BIT_VECTOR (3 DOWNTO 0);
BEGIN
      PROCESS (ck1, reset)       -- process sensitivity list
            VARIABLE seq         :INTEGER RANGE 0 TO 15;
      BEGIN
            IF (reset = '1')           THEN  seq := 0;
            ELSIF (count = '0')        THEN  seq := seq;
            ELSIF (ck1'EVENT AND ck1 = '1')        THEN
                  IF (seq = 8)         THEN  seq := 0;
                  ELSE                       seq := seq + 1;
                  END IF;
            END IF;
            CASE seq IS
                  WHEN 0 =>            output <= "0000";
                  WHEN 1 =>            output <= "1000";
                  WHEN 2 =>            output <= "0101";
                  WHEN 3 =>            output <= "1010";
                  WHEN 4 =>            output <= "0001";
                  WHEN 5 =>            output <= "1001";
                  WHEN 6 =>            output <= "0110";
                  WHEN 7 =>            output <= "1011";
                  WHEN 8 =>            output <= "0011";
                  WHEN OTHERS =>       output <= "0000";
            END CASE;
            qa <= output(0);           -- output each bit
            qb <= output(1);
            qc <= output(2);
            qd <= output(3);
      END PROCESS;
END soln2;
```

Fig. 14-10 Alternate VHDL file for the irregular counter of Example 14-3

Example 14-4

Design a bidirectional, full-step controller for a stepper motor using VHDL. The controller will produce a clockwise sequence pattern when cw = 1 and a counterclockwise sequence pattern when cw = 0. The stepper motor sequence is shown in Fig. 14-11.

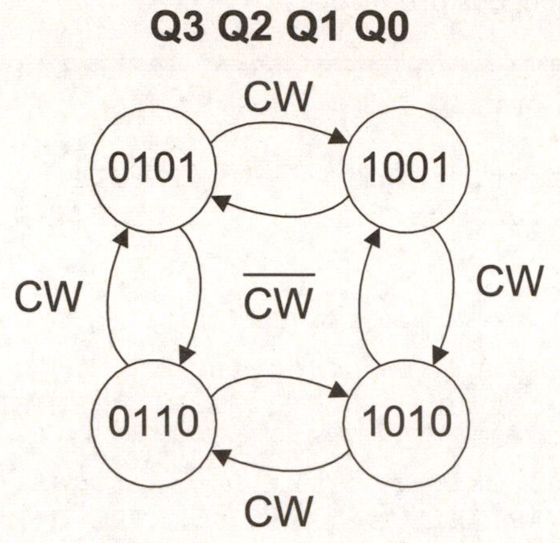

Fig. 14-11 State transition diagram for the stepper controller of Example 14-4

An example solution using VHDL is shown in Fig. 14-12. The input BIT named step is the clock input to the stepper controller. A VARIABLE named stepper is declared to be a 4-bit BIT_VECTOR data type in the PROCESS statement. The present state and the logic level for cw will determine the correct next state that should be produced by the state machine. The IF/ELSE statement is used to test cw and determine the count direction for the stepper controller. Each of the two count directions is specified with a separate CASE statement. For each value of cw, the correct variable assignment is determined based on the current value of stepper. The chip will automatically reset all flip-flops at power-up and it will be necessary to define how to get out of that state and into the specified sequence to drive the stepper motor. At power-up, we do not care what the next state is. We just want the counter to get into the specified count sequence. The next state 1010_2 was chosen arbitrarily from the list of valid states for the stepper motor controller. The WHEN OTHERS in the CASE statement will automatically take care of state 0000_2, as well as any other undefined state. The value for stepper is assigned to the output port q. Simulation results for this design are shown in Fig. 14-13. The count sequence for each direction is correct for the recycling state machine.

```
ENTITY full_step IS
PORT (
      step, cw            :IN BIT;
      q                   :OUT BIT_VECTOR (3 DOWNTO 0)
);
END full_step;

ARCHITECTURE solution1 OF full_step IS
BEGIN
    PROCESS (step)            -- step is clock input
        VARIABLE  stepper     :BIT_VECTOR (3 DOWNTO 0);
    BEGIN
        IF (step'EVENT AND step = '1') THEN
            IF (cw = '1')    THEN
                CASE stepper IS          -- CW sequence
                    WHEN "1010" =>   stepper := "0110" ;
                    WHEN "0110" =>   stepper := "0101" ;
                    WHEN "0101" =>   stepper := "1001" ;
                    WHEN "1001" =>   stepper := "1010" ;
                    WHEN OTHERS =>   stepper := "1010" ;
                END CASE;
            ELSE
                CASE stepper IS          -- CCW sequence
                    WHEN "1010" =>   stepper := "1001" ;
                    WHEN "1001" =>   stepper := "0101" ;
                    WHEN "0101" =>   stepper := "0110" ;
                    WHEN "0110" =>   stepper := "1010" ;
                    WHEN OTHERS =>   stepper := "1010" ;
                END CASE;
            END IF;
        END IF;
    q <= stepper;        -- output to port
    END PROCESS;
END solution1;
```

Fig. 14-12 VHDL file for the stepper motor controller

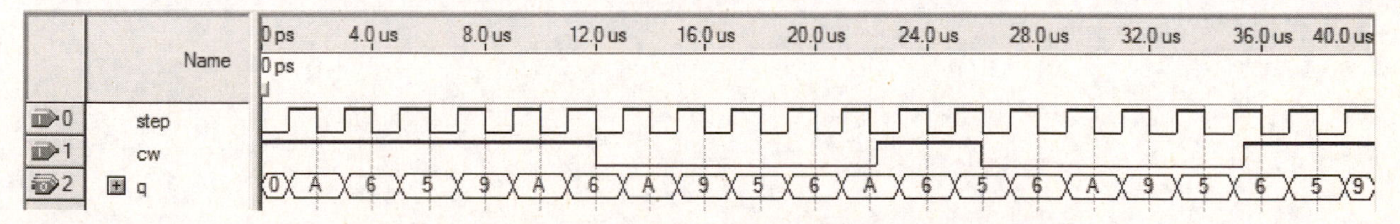

Fig. 14-13 Simulation results for Example 14-4

An alternative VHDL design for Example 14-4 is shown in Fig. 14-14. A VARIABLE named stepper is declared in the PROCESS statement to be an INTEGER with a 0 to 3 RANGE of values. The value of cw, tested with an IF/ELSE statement, will control whether stepper is to be incremented or decremented. The VARIABLE stepper was given the INTEGER data type so that it could be manipulated mathematically by adding 1 to it or subtracting 1 from it. The CASE statement is used to assign the correct bit pattern to the output port q.

```
ENTITY full_step IS
PORT (
      step, cw          :IN BIT;
      q                 :OUT BIT_VECTOR (3 DOWNTO 0)
);
END full_step;

ARCHITECTURE solution2 OF full_step IS
BEGIN
      PROCESS (step)          -- step is clock input
      VARIABLE    stepper     :INTEGER RANGE 0 TO 3;
                  -- integer data type has numerical value
      BEGIN
            IF (step'EVENT AND step = '1') THEN
                  IF (cw = '1')      THEN
                              stepper := stepper + 1;
                  ELSE        stepper := stepper - 1;
                  END IF;
            END IF;

            CASE stepper IS    -- assign bit patterns
                  WHEN 0 =>   q <= "1010" ;
                  WHEN 1 =>   q <= "0110" ;
                  WHEN 2 =>   q <= "0101" ;
                  WHEN 3 =>   q <= "1001" ;
            END CASE;
      END PROCESS;
END solution2;
```

Fig. 14-14 Alternate VHDL file for the stepper motor controller

Example 14-5

Design a sequential circuit that will wait in an idle state for a trigger input (start) to occur and then output a single pulse four clock cycles later. The sequential circuit will then wait for another trigger signal. The state transition diagram for this delay circuit is shown in Fig. 14-15. The state machine should also be self-correcting.

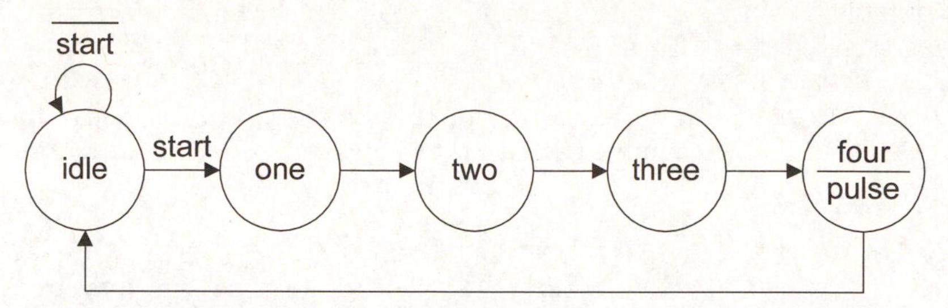

Fig. 14-15 State transition diagram for Example 14-5

A VHDL solution is shown in Fig. 14-16. This example illustrates the use of a new VHDL data type called enumerated. An enumerated data type named count_state is created with the *TYPE count_state IS* statement. The enumerated data type allows us to refer to states by name only (idle, one, two, three, four). An internal VARIABLE named mach is declared to be this user-defined data type called count_state. The bit patterns have not been defined for this VARIABLE because we do not really care what they look like. The compiler will assign bit patterns for each state. The VARIABLE mach will be used to create a "buried" state machine, one that is not connected to the entity's output ports. In the PROCESS, a CASE statement is used to define the sequence of states for our state machine. When the state machine is in the idle state, it will be waiting for the start signal to go HIGH (and the ↑ clock) in order to proceed to the next state (one). After starting the sequence, the state machine will step through each state on each ↑ clock. Upon reaching state four (i.e., *mach = four* is true), the output port pulse will be assigned a 1; otherwise pulse will be a 0. Notice that the only output needed is the signal named pulse.

The results of the simulation for Example 14-5 are shown in Fig. 14-17. The buried state machine (notice its "buried handle" labeled with an "s") has been included in the simulation to verify its correct operation. **See Quartus Notes for the setup procedure.** Only the state names are given by the simulator because the bit values were not user-defined. Note that the first start signal did not initiate the machine sequence because it occurred after the ↑ clock.

```
ENTITY delay IS
PORT (
      clock        :IN BIT;
      start        :IN BIT;
      pulse        :OUT BIT
);
END delay;

ARCHITECTURE vhdl OF delay IS
BEGIN
      PROCESS (clock)
      TYPE count_state IS (idle, one, two, three, four);
            -- enumerated data type called count_state
      VARIABLE mach      :count_state;
      BEGIN
         IF (clock'EVENT AND clock = '1') THEN
             CASE  mach  IS   -- describe mach sequence
                 WHEN idle =>
                     IF (start = '1') THEN
                                  mach := one;
                     ELSE         mach := idle;
                     END IF;
                 WHEN one     =>  mach := two;
                 WHEN two     =>  mach := three;
                 WHEN three   =>  mach := four;
                 WHEN four    =>  mach := idle;
             END CASE;
         END IF;
             -- detect state to assert pulse
         IF (mach = four)     THEN pulse <= '1';
         ELSE                      pulse <= '0';
         END IF;
      END PROCESS;
END vhdl;
```

Fig. 14-16 VHDL file for Example 14-5

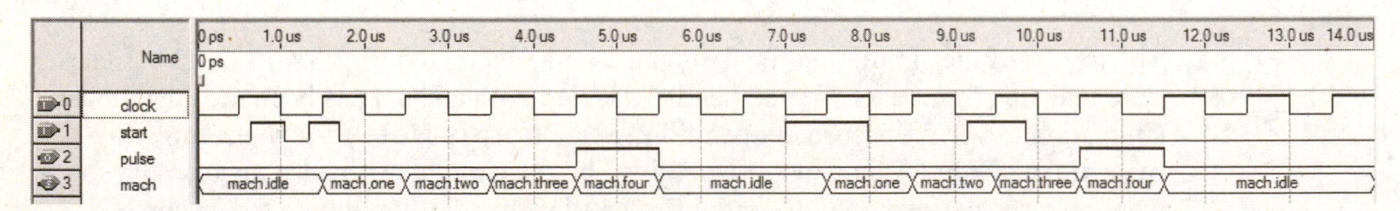

Fig. 14-17 Simulation results for Example 14-5

Example 14-6

Design the state machine (using VHDL) that is defined by the state transition diagram given in Fig. 14-18. The state machine has three inputs—trig, delay(1), and delay(0)—that control the sequence of states. The logic circuit will produce a 3-bit array output named outs that is dependent upon the current state for the machine. The states are named, but the bit patterns for each state are not defined in this design.

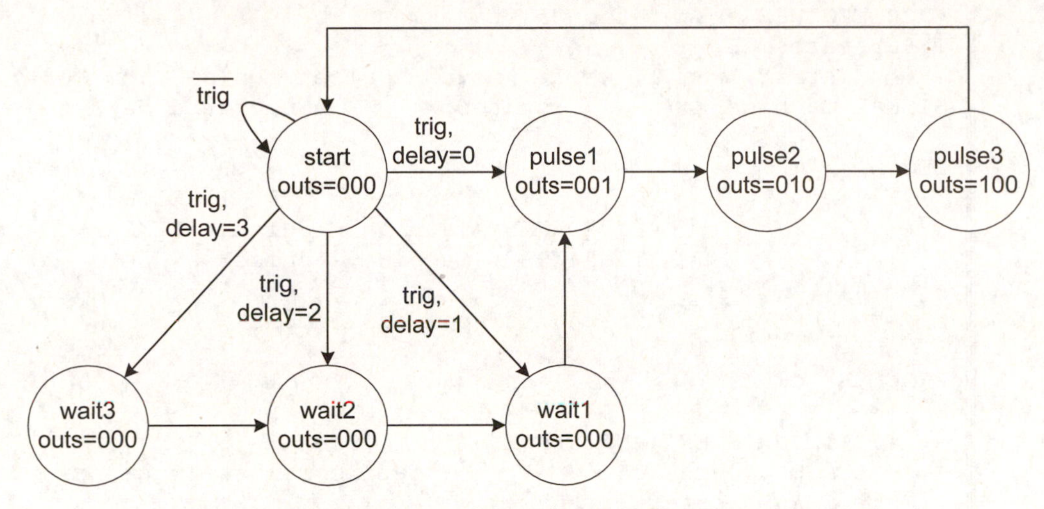

Fig. 14-18 State transition diagram for Example 14-6

An example VHDL solution named state_machine is shown in Fig. 14-19. Seven states are defined for the internal SIGNAL named machine. The states are listed by name only (start, wait1, wait2, wait3, pulse1, pulse2, pulse3) using an enumerated data type called states. The SIGNAL machine is used to create a buried state machine (its outputs are not connected to the output ports for the ENTITY). The bit patterns for machine were left for the compiler to determine because we do not care what they actually look like. Two CASE statements and an IF/ELSE statement are used to describe the operation of machine. The 3-bit output port (outs) values are assigned using a conditional signal assignment (WHEN/ELSE) statement to test for the appropriate state of machine.

Simulation results for Example 14-6 are shown in Fig. 14-20. The buried state machine (machine) has been included in the simulation to make it easier to verify the circuit's operation. Note that the simulation shows the bit values for the enumerated state names for machine.

```
ENTITY state_machine IS
PORT (
      clock, trig         :IN BIT;
      delay               :IN INTEGER RANGE 0 TO 3;
      outs                :OUT BIT_vector (3 DOWNTO 1)
);
END state_machine;

ARCHITECTURE enumerated OF state_machine IS
TYPE states IS
      (start, wait1, wait2, wait3, pulse1, pulse2, pulse3);
SIGNAL machine            :states;    -- enumerated data type

BEGIN
      outs <=     "001" WHEN (machine = pulse1) ELSE
                  "010" WHEN (machine = pulse2) ELSE
                  "100" WHEN (machine = pulse3) ELSE
                  "000";
            -- conditional signal assignment

PROCESS (clock)
BEGIN
   IF (clock'EVENT AND clock = '1') THEN
      CASE machine IS            -- define machine sequence
         WHEN  start  =>
            IF (trig = '1')   THEN  -- wait for trig
               CASE delay IS  -- determine delay
                    WHEN 0    =>    machine <= pulse1;
                    WHEN 1    =>    machine <= wait1;
                    WHEN 2    =>    machine <= wait2;
                    WHEN 3    =>    machine <= wait3;
               END CASE;
            ELSE                    machine <= start;
            END IF;
         WHEN wait1     =>          machine <= pulse1;
         WHEN wait2     =>          machine <= wait1;
         WHEN wait3     =>          machine <= wait2;
         WHEN pulse1    =>          machine <= pulse2;
         WHEN pulse2    =>          machine <= pulse3;
         WHEN pulse3    =>          machine <= start;
      END CASE;
   END IF;
END PROCESS;
END enumerated;
```

Fig. 14-19 VHDL file for Example 14-6

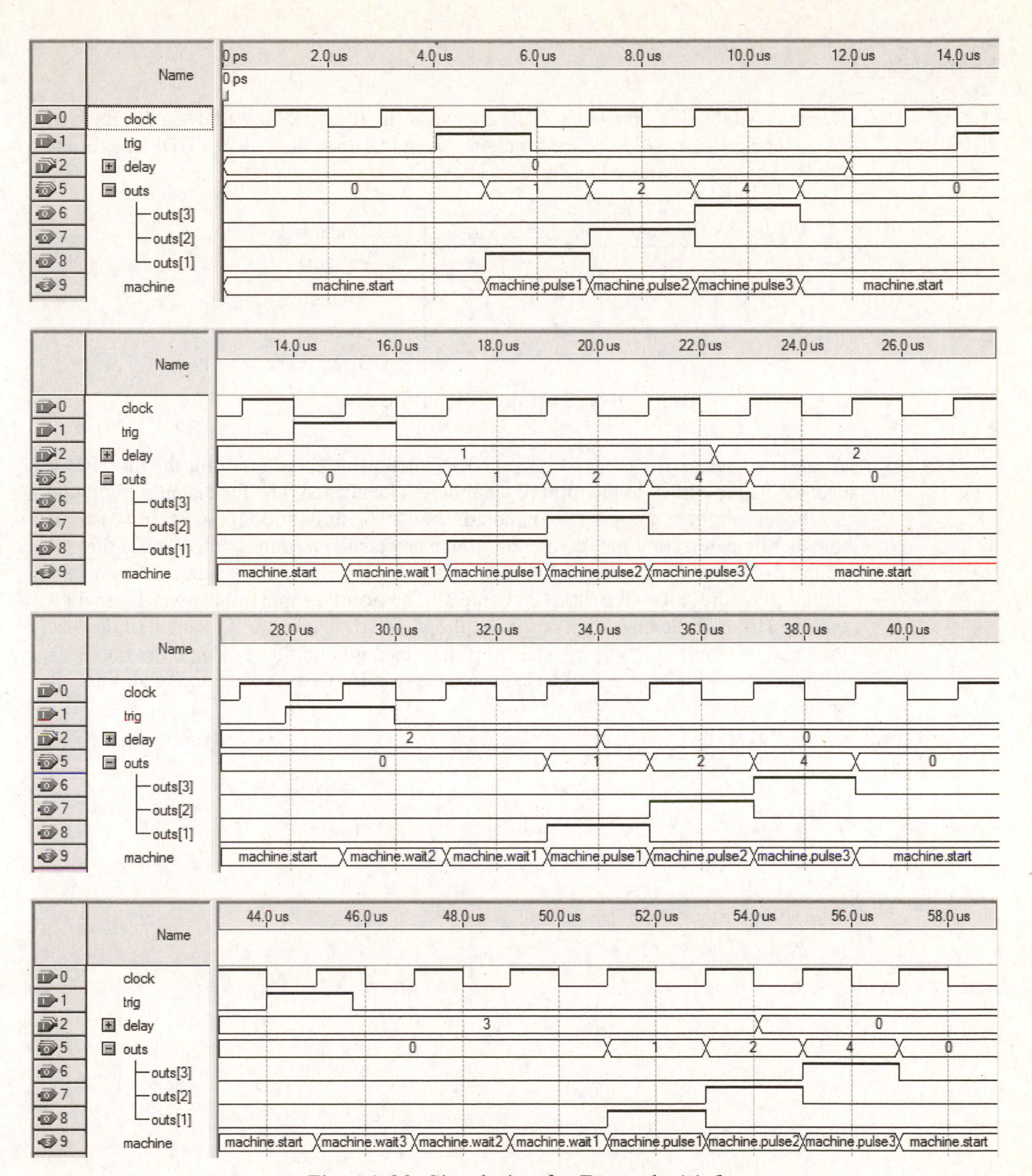

Fig. 14-20 Simulation for Example 14-6

Example 14-7

Design a MOD-1000, recycling BCD counter using the mod10.vhd design in Example 14-2. The counter will have the functions given in Table 14-2. The VHDL file for the mod10 counter is repeated in Fig. 14-21.

clear	load	enable	clock	function
1	X	X	X	CLEAR
0	0	0	↓	HOLD
0	1	X	↓	LOAD
0	0	1	↓	COUNT UP

Table 14-2 Example 14-7 functions

A top-level design file (Fig. 14-22) can be created using the symbol for the mod10.vhd design. Three copies of the mod10 symbol will be needed, one for the ones digit, one for the tens digit, and one for the hundreds digit. The three mod10 stages are cascaded together by connecting the rco output (which detects the terminal state 9 when the counter is enabled) to the next digit's enable control. The enable input on the ones digit will control all of the mod10 counters. The enable input on the tens digit will only be HIGH if the ones digit is a 9 and the MOD-1000 counter is enabled. Likewise, the hundreds digit will only be enabled if the tens digit and the ones digit are both 9 (and the counter is enabled). The load and clear controls are connected to all three counter stages.

```
ENTITY mod10 IS
PORT (
      clock                      :IN BIT;
      load, enable, clear        :IN BIT;
      d                          :IN INTEGER RANGE 0 TO 15;
      q                          :OUT INTEGER RANGE 0 TO 15;
      rco                        :OUT BIT
);
END mod10;

ARCHITECTURE bcd OF mod10 IS
BEGIN
      PROCESS (clock, clear, enable)
                  -- invoke process if clock or clear changes
            VARIABLE  counter        :INTEGER RANGE 0 TO 15;
      BEGIN
            IF (clear = '1') THEN   counter := 0;
                  -- asynchronous clear has priority
            ELSIF (clock'EVENT AND clock = '0') THEN
                  IF (load = '1')  THEN   counter := d;
                        -- synchronous load
                  ELSIF  (enable = '1')  THEN
                        IF (counter = 9)  THEN  -- recycle
                                    counter := 0;
                        ELSE        counter := counter + 1;
                        END IF;
                  END IF;
            END IF;

                  -- rco detects terminal count when enabled
            IF ((counter = 9) AND (enable = '1')) THEN
                        rco <= '1';
            ELSE        rco <= '0';
            END IF;

            q <= counter;       -- output counter to ports
      END PROCESS;
END bcd;
```

Fig. 14-21 VHDL design file for the decade counter from Example 14-2

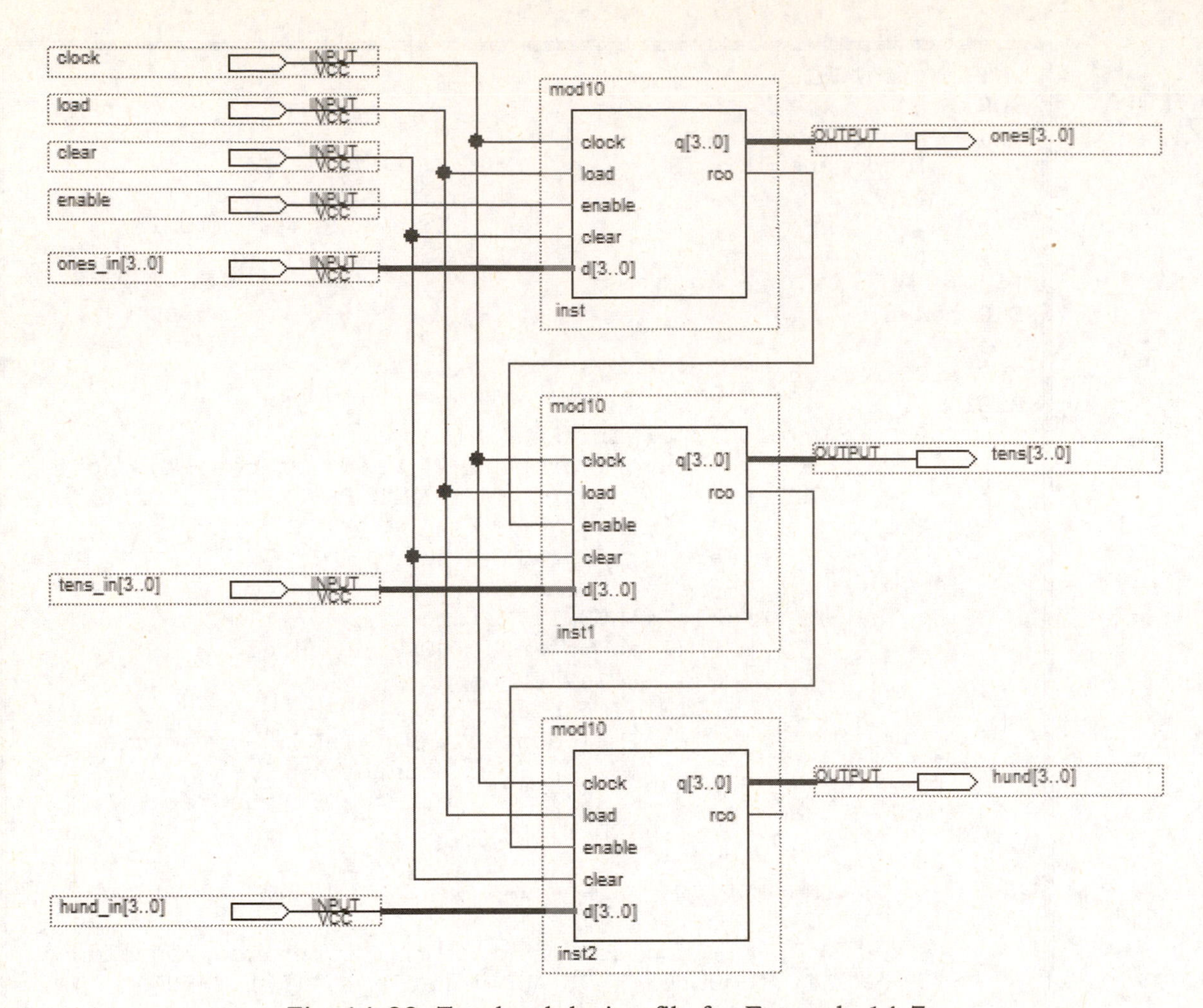

Fig. 14-22 Top-level design file for Example 14-7

An alternate solution is shown in Fig. 14-23. This technique uses a VHDL design file to describe the interconnections between the MOD-10 counters for the top-level description. This design method is said to use a structural style. The mod10.vhd file is declared to be a COMPONENT in the top-level file. The ports for the mod10 entity are given within the COMPONENT declaration. Two additional single-bit signals (riplcary1 and riplcary10) are also declared to make the connections between the ones and tens digit counters and between the tens and hundreds digit counters (see Fig. 14-22). The mod10 counter module is instantiated three times in the top-level file. Each time the counter module is used, it is given a unique label (ones_digit, tens_digit, hund_digit), and the appropriate wiring connection for each port is given in the PORT MAP using named associations. The list of named associations has the COMPONENT port name on the left of the => symbol and the top-level port name on the right. Notice how riplcary1 is used as a "wire" that connects the rco output from the ones_digit to the enable input on the tens_digit. Another "wire" (riplcary10) connects the tens_digit to the hund_digit. The final rco output, produced by the hund_digit module, is not used in our application so it is "connected" (it must be given a named association) in the PORT MAP to OPEN, a keyword that means exactly what it says.

```
ENTITY mod1000v IS
PORT ( clock, load, clear, enable        :IN BIT;
       ones_in, tens_in, hund_in         :IN INTEGER RANGE 0 TO 15;
       ones, tens, hund                  :OUT INTEGER RANGE 0 TO 15
  );
END mod1000v;

ARCHITECTURE structure OF mod1000v IS
SIGNAL riplcary1, riplcary10             :BIT;
COMPONENT mod10
  PORT (       clock                     :IN BIT;
               load, clear, enable       :IN BIT;
               d                         :IN INTEGER RANGE 0 TO 15;
               q                         :OUT INTEGER RANGE 0 TO 15;
               rco                       :OUT BIT     );
END COMPONENT;
BEGIN
  ones_digit          :mod10
       PORT MAP (clock => clock, load => load, clear => clear,
       enable => enable, d => ones_in, q => ones,
       rco => riplcary1);
  tens_digit          :mod10
       PORT MAP (clock => clock, load => load, clear => clear,
       enable => riplcary1, d => tens_in, q => tens,
       rco => riplcary10);
  hund_digit          :mod10
       PORT MAP (clock => clock, load => load, clear => clear,
       enable => riplcary10, d => hund_in, q => hund,
       rco => OPEN);
END structure;
```

Fig. 14-23 Alternate VHDL top-level solution for Example 14-7

Sample simulation results for either design solution of the MOD-1000 counter are shown in Fig. 14-24.

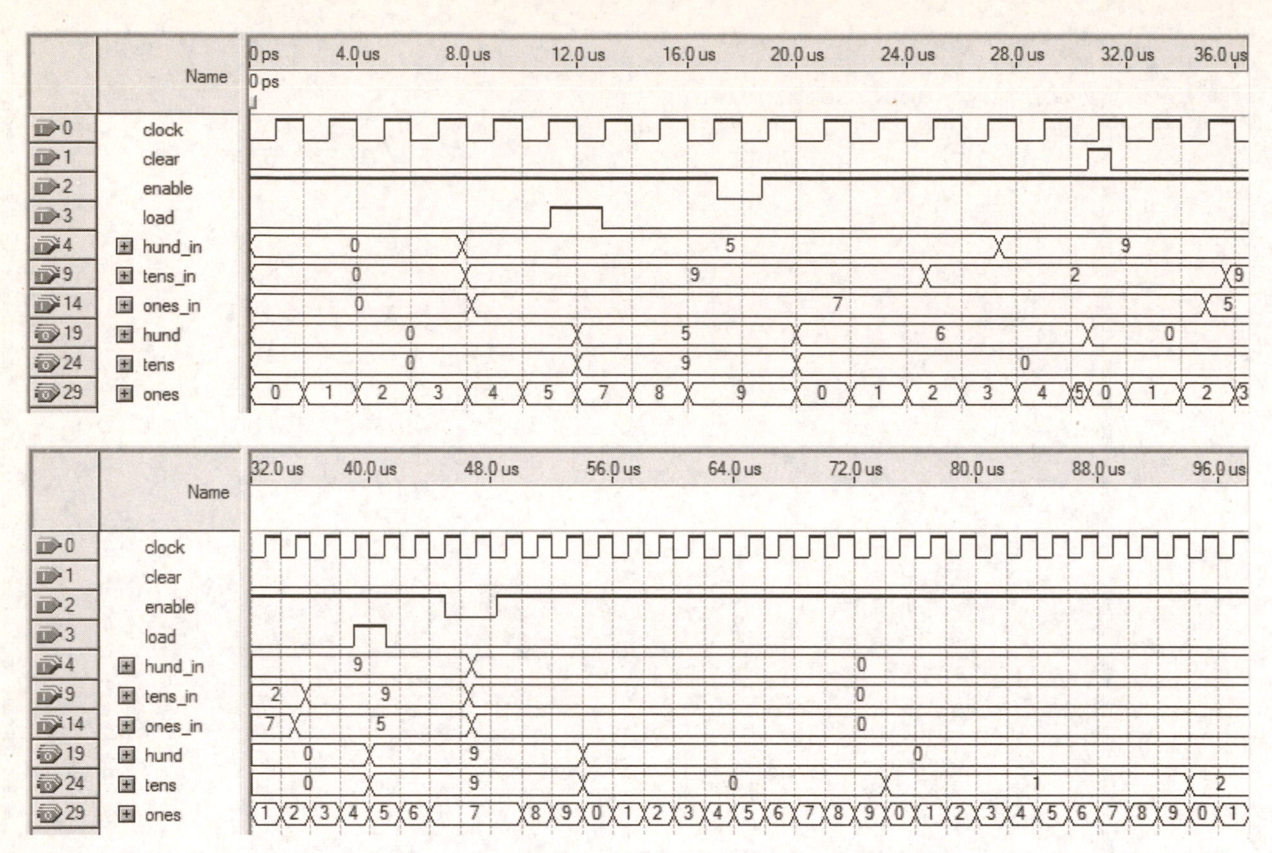

Fig. 14-24 Sample simulation results for Example 14-7

Example 14-8

Design a clock frequency divider that will output a 20-kHz signal with a 50% duty cycle using a 50-MHz input.

The 50-MHz input signal needs to be divided by 2500 to produce an output frequency of 20 kHz. Since the MSB output from a MOD-2500 will not produce a 50% duty cycle, it will be necessary to factor the total modulus (2500) to use a binary counter, which inherently produces 50% duty cycle output signals, that is cascaded with another counter. We have selected a MOD-2 counter and a MOD-1250 counter to produce an overall divide by 2500 ($2 \times 1250 = 2500$) for our design solution (see Fig. 14-25). The individual AHDL design files are shown in Fig. 14-26 and 14-27.

The simulation results are shown in Fig. 14-28. The period measurement of the output signal using the time bars in the simulator is 50.0 μs. The period for a 20-kHz waveform is exactly 50 μs so our design appears to be correct. See Quartus Notes for the setup procedure.

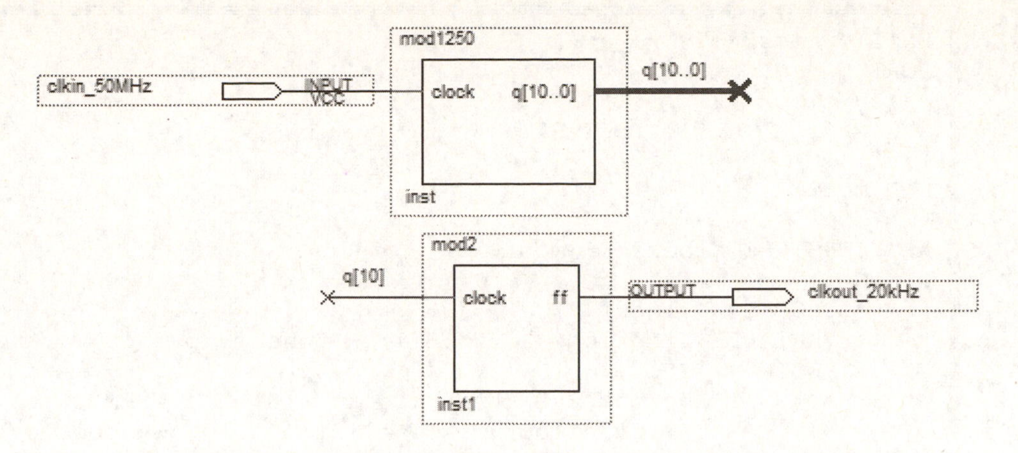

Fig. 14-25 Top-level design file for Example 14-8

VHDL

```
ENTITY  mod1250  IS
PORT (
      clock        :IN BIT;
      q            :OUT INTEGER RANGE 0 TO 1249
);
END mod1250;

ARCHITECTURE block1 OF mod1250 IS
BEGIN
      PROCESS (clock)
      VARIABLE counter   :INTEGER RANGE 0 TO 1249;

      BEGIN
            IF (clock'EVENT AND clock = '1')     THEN
                  IF counter = 1249        THEN
                        -- at terminal state?
                        counter := 0;      -- recycle
                  ELSE
                        counter := counter + 1;
                        -- otherwise, count up
                  END IF;
            END IF;
            q <= counter;
      END PROCESS;
END block1;
```

Fig. 14-26 VHDL design for MOD-1250 counter block

```
ENTITY  mod2  IS
PORT (
      clock        :IN BIT;
      ff           :OUT BIT
);
END mod2;

ARCHITECTURE block2 OF mod2 IS
BEGIN
      PROCESS (clock)
      VARIABLE flipflop       :BIT; -- one flip-flop
      BEGIN
            IF (clock'EVENT AND clock = '1')     THEN
                  flipflop := NOT flipflop;      -- toggle
            END IF;
            ff <= flipflop;
      END PROCESS;
END block2;
```

Fig. 14-27 VHDL design for MOD-2 counter block

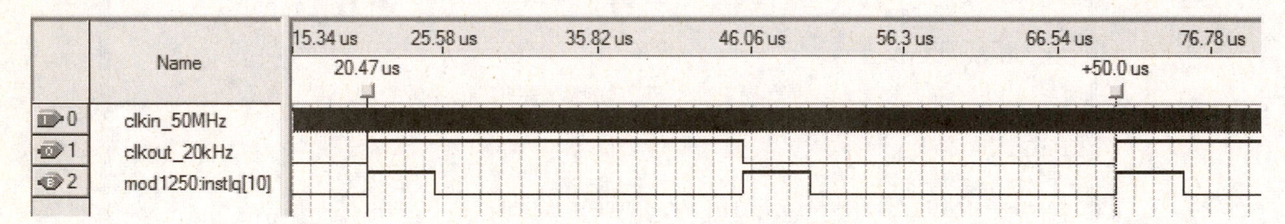

Fig. 14-28 Sample simulation results for Example 14-8

Laboratory Projects

Design the following counter applications using VHDL. Simulate and test the circuit designs.

14V.1 Binary down counter

Design a recycling, 4-bit binary down counter. Counting is controlled by an active-LOW enable named en. The counter can also be synchronously parallel loaded when an input named ld is HIGH. The following function table describes the counter's operation (note the function priorities). Label the parallel inputs data[3..0] and the counter outputs q[3..0]. The circuit also should produce an output signal called last, which goes HIGH whenever the counter reaches its terminal state (0000).

ld	en	operation
1	X	Load
0	0	Count
0	1	Hold

14V.2 Up/down BCD counter

Design a MOD-10, recycling, BCD up/down counter. The counter's function table is given below. Inputs c[1..0] control the counter's function. The counter should also produce an active-LOW ripple carry output signal called carryn. The carry output signal should only be asserted at 9 when counting up or at 0 when counting down.

c1	c0	operation
0	0	Reset
0	1	Count Down
1	0	Count Up
1	1	Hold Count

14V.3 MOD-60 BCD counter

Design a MOD-60, recycling BCD counter. The counter should have an active-LOW, synchronous reset (resn) signal. The count sequence will be 0 through 59_{10} in BCD. Create a hierarchical design with separate VHDL design files for the tens digit and the ones digit.

14V.4 MOD-100 binary counter

Design a MOD-100, recycling binary counter. The counter is triggered with ↓ clocks and should have an active-HIGH count enable called enable. The counter will also output a single active-HIGH pulse (called pulse) during each MOD-100 count sequence. The pulse occurs at the end of the count sequence during either the last 5 or 10 counter states. The pulse width will be controlled by an input signal called w, as shown in the table below. Essentially, w selects the duty cycle of the divide-by-100 output signal. Compare functional and timing simulation results. Why does the timing simulation show glitches?

W	PULSE width	duty cycle
0	5 clock periods	5%
1	10 clock periods	10%

14V.5 Stepper motor sequence controller

Design a half-step sequencer to control a stepper motor. The sequencer should produce the appropriate sequence of states to drive the motor either clockwise (CW) or counterclockwise (CCW) without using a binary counter (i.e., produce the half-step count sequence directly). The stepper direction is controlled by an input signal named cw. The stepper enable is called go. The function table is given below. Make sure the sequencer is self-starting.

GO	CW	function
0	X	HALT
1	0	CCW
1	1	CW

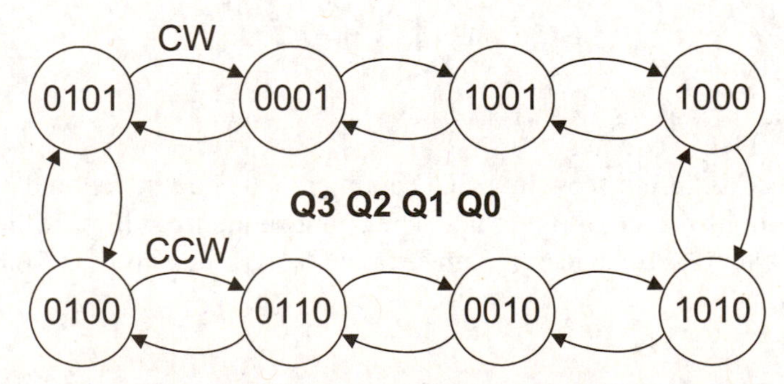

Half-step sequence for stepper motor control

14V.6 Variable frequency divider

Design a variable frequency divider. The frequency divider should divide the input frequency by one of four different factors. The divide-by factor is controlled by two mode controls, as described by the following function table. The mode controls are used to change the modulus of the counter used for the frequency division.

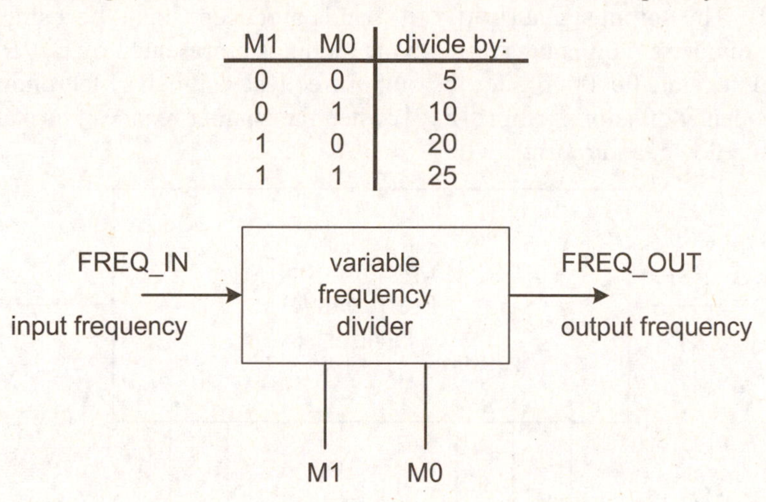

M1	M0	divide by:
0	0	5
0	1	10
1	0	20
1	1	25

14V.7 Digital lock

Design a digital lock circuit. The lock will have a 4-bit data input and an enter signal. Use a state machine with 5 states to keep track of the sequencing through the combination. **The enter signal is a manual clock signal for the state machine.** A sequence of four 4-bit numbers will make up the lock combination (that you will select for your design). Each 4-bit input number will be applied (via switches), and then the manual enter signal (↑clock) will be pulsed. The machine states must be sequenced in the proper order to unlock the lock. The START state is followed by the states STATE1, then STATE2, then STATE3, and then finally DONE, which "unlocks" the lock (indicated by a HIGH logic level on an output named unlock). Use an **enumerated** data type for the sequential machine. If an incorrect input combination is "entered" during any machine state, the state machine will return to START and an output signal named ready will be activated. The state transition diagram for the digital lock is shown below.

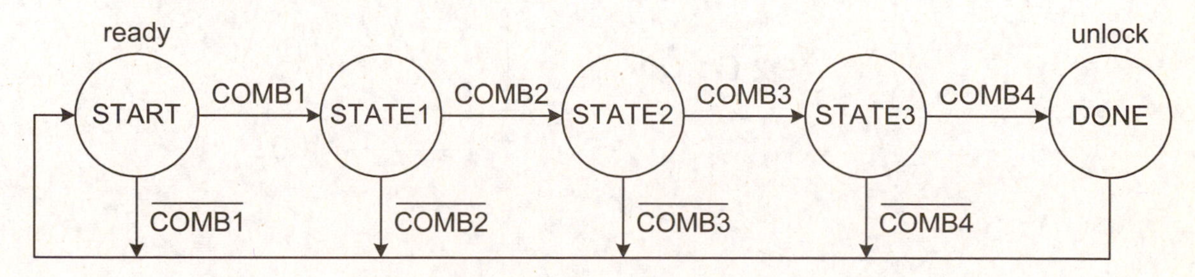

Returns to START if incorrect combination is applied

14V.8 Programmable frequency divider

Design an 8-bit, programmable frequency divider using VHDL. The input signal frequency (freq_in) will be divided by a variable, 8-bit, binary input value (B7 through B0). The output signal (freq_out) that is produced should be a single pulse each time the divider circuit counts through the modulus represented by B7..0. Hint: Detect the counter state 00000001_2 for the output frequency signal, synchronously load in the frequency division factor (b[7..0]) when the counter reaches this state (00000001_2), count down, and repeat.

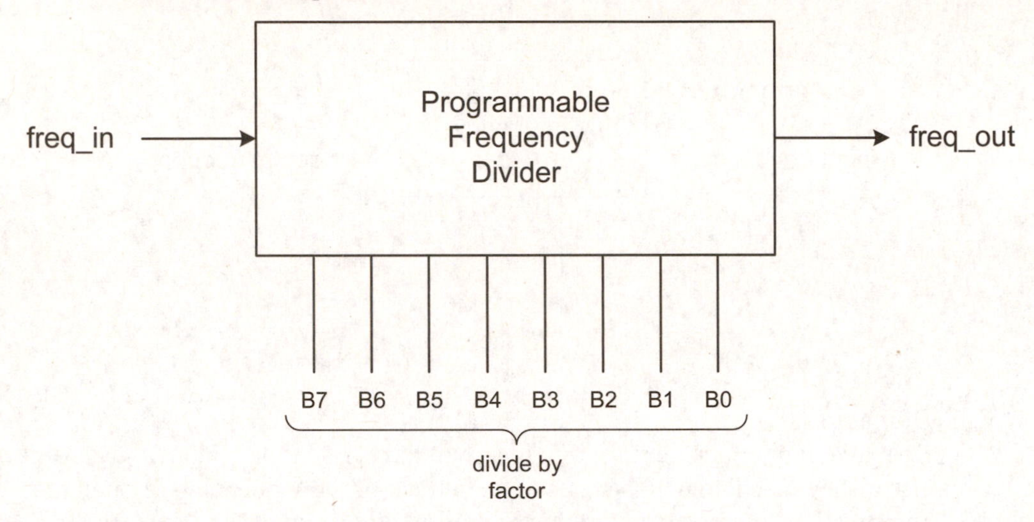

14V.9 State machine

Design the following state machine. Make the design self-correcting by forcing any other state to go to 001_2 on the next ↑ clock.

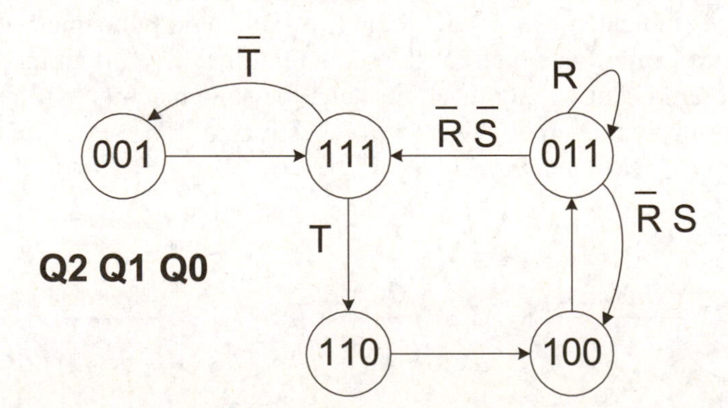

14V.10 Variable pulse delay

Design a state machine that will output a single time-delayed pulse after receiving a triggering signal named start. The state machine can only be triggered to start the timing delay for the output pulse when it is in an idle state. The length of the delay is variable and controlled by an input named delay. The output pulse produced has a duration of one clock cycle and will begin on the fifth rising edge of clock when delay = 0 or on the tenth rising edge of clock when delay = 1. At the end of the variable timing cycle for the state machine, the state machine will return to its idle state and wait for another trigger input on start.

14V.11 Frequency divider

Design a frequency divider circuit that will produce an output signal with a frequency that is equal to the circuit's input signal frequency divided by 100 and has a duty cycle of 50%.

14V.12 Clock frequency divider

Design a clock frequency divider circuit that will output two frequencies, 25 kHz and 10 kHz, when driven by a 50-MHz clock signal. Caution: avoid very-low duty cycle output signals!

14V.13 Gray code counter

Design a 4-bit, up/down, recycling Gray code counter. The count direction is controlled by a signal called dir, as indicated in the following function table. Label the counter outputs q[3..0]. The circuit also should produce an output signal called index, which goes LOW whenever the counter state is 0000.

dir	operation
0	Count Up
1	Count Down

Gray code sequence

(count up sequence: top to bottom; count down sequence: bottom to top)

q3	q2	q1	q0
0	0	0	0
0	0	0	1
0	0	1	1
0	0	1	0
0	1	1	0
0	1	1	1
0	1	0	1
0	1	0	0
1	1	0	0
1	1	0	1
1	1	1	1
1	1	1	0
1	0	1	0
1	0	1	1
1	0	0	1
1	0	0	0

UNIT 15

SHIFT REGISTER APPLICATIONS

Objectives

- To apply shift registers in serial and parallel digital data transfer applications.
- To apply shift registers in counter applications.

Shift Registers

Registers consist of a set of flip-flops used to store and transfer binary data in a digital system. See Quartus Notes available from the publisher's textbook web site at www.pearsonhighered.com/electronics. Registers can be classified according to the types of input and output data movement. With the two basic forms of data transfer, serial and parallel, there are the following categories of registers:

1. Parallel-in/parallel-out (PIPO)
2. Serial-in/serial-out (SISO)
3. Parallel-in/serial-out (PISO)
4. Serial-in/parallel-out (SIPO)

Example 15-1

Create an 8-bit SISO shift register with a CPLD/FPGA. The shift register should shift data left (towards the MSB) and have an asynchronous, active-LOW clear and an active-HIGH enable.

DFFE A schematic design solution using D flip-flops with enables is shown in Fig. 15-1. An 8-bit register using D flip-flops has been created. The SER_IN is fed into the register on the right and data will shift to the left with the SER_OUT coming from the leftmost flip-flop. All flip-flops are clocked synchronously. The CLR input is connected to the asynchronous clear control on each flip-flop and the EN enable input is connected to the ENA control on each DFFE. Simulation results are shown in Fig. 15-2. Buried register nodes for each flip-flop output have been included in the simulation to demonstrate the serial shifting of the register.

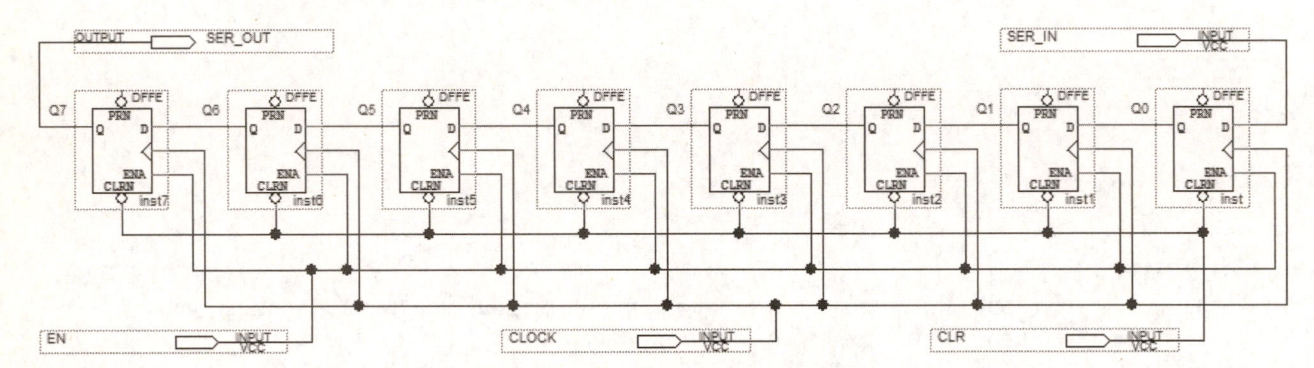

Fig. 15-1 DFFE design solution for Example 15-1

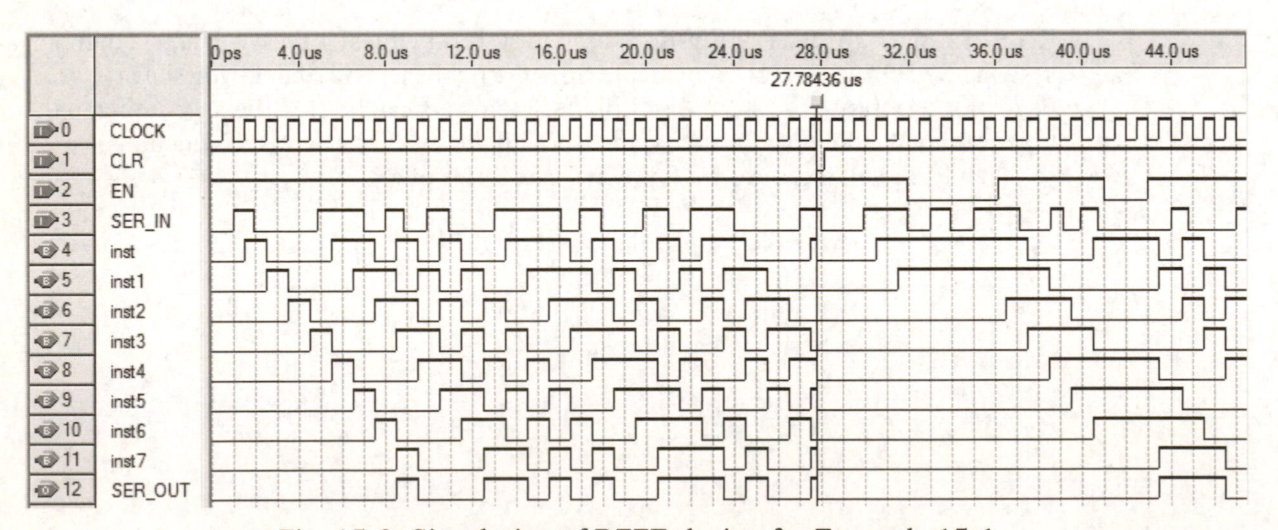

Fig. 15-2 Simulation of DFFE design for Example 15-1

LPM An LPM design solution using LPM_SHIFTREG is shown in Fig. 15-3. An 8-bit shift register with serial shifting input and serial shifting output has been created with the MegaWizard. Enable and asynchronous clear controls have been included as register parameters. Simulation results are shown in Fig. 15-4. Each of the buried register nodes has been included in the simulation to demonstrate the serial shifting of the register.

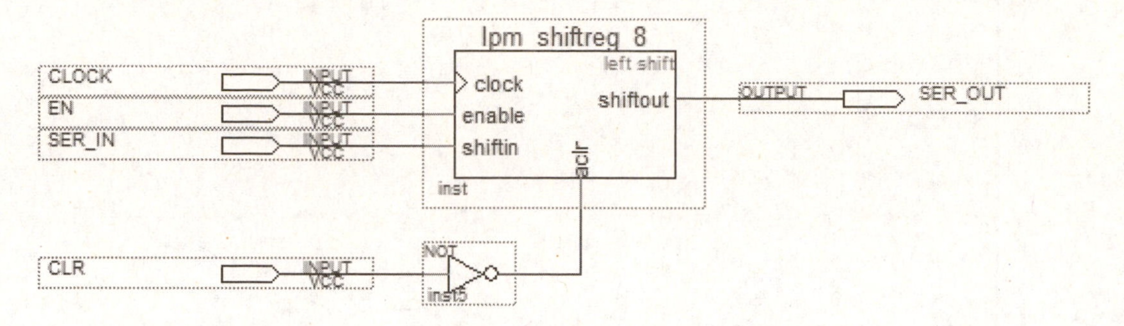

Fig. 15-3 LPM design solution for Example 15-1

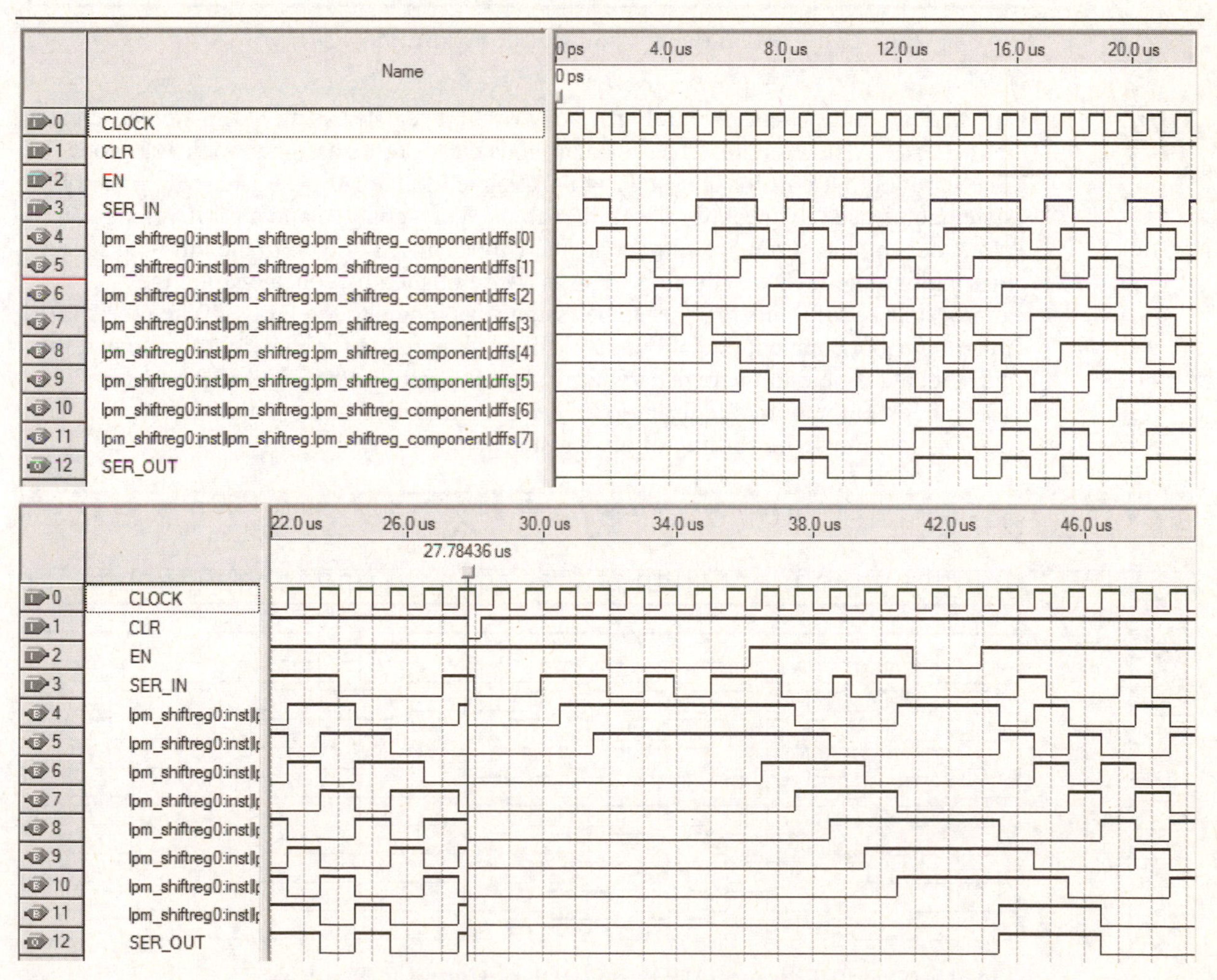

Fig. 15-4 Simulation of LPM design for Example 15-1

```
SUBDESIGN  siso
(
      clock, en, clr, ser_in          :INPUT;
      ser_out                         :OUTPUT;
)
VARIABLE
      q[7..0]                         :DFF;
BEGIN
      q[].clk = clock;
      q[].clrn = clr;
      ser_out = q7.q;            -- output last register bit
      IF (en == VCC)     THEN    -- check for enable active
            q[7..0].d = (q[6..0].q, ser_in);
                                 -- concatenate bits for shift
      ELSE
            q[7..0].d = (q[7..0].q);        -- hold data
      END IF;
END;
```

Fig. 15-5 AHDL design solution for Example 15-1

AHDL

An AHDL solution is shown in Fig. 15-5. An 8-bit register using D flip-flops is created in the VARIABLE section. The asynchronous clear port is assigned to the clr input. The serial output port for the subdesign is assigned to the MSB register output. An IF statement is used to check for the shift enable. When enabled and a positive-edge on clock occurs, the current register contents will be shifted one bit to the left with ser_in moving into the LSB q0 flip-flop. This is accomplished by concatenating the appropriate set of 8 bits together to be assigned as inputs to the DFFs. When the shift function is not enabled, the DFF inputs will be assigned to the same outputs so that the current data is held in the register when clocked. Simulation results are shown in Fig. 15-6. Each of the buried register nodes has been included in the simulation to demonstrate the serial shifting of the register.

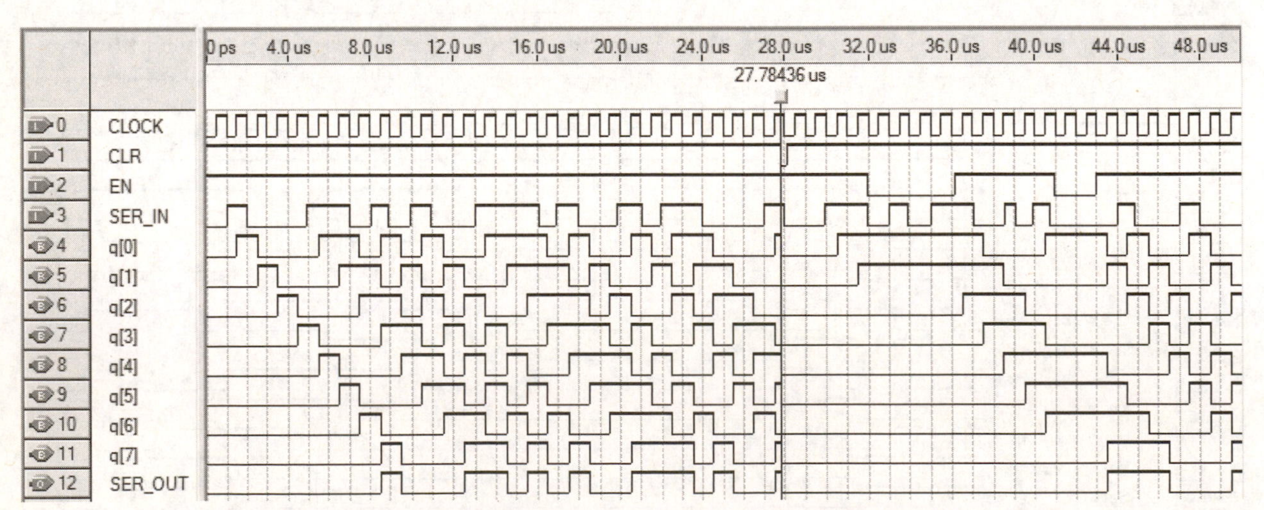

Fig. 15-6 Simulation of AHDL & VHDL designs for Example 15-1

```
ENTITY  siso  IS
PORT ( clock, en, clr, ser_in         :IN BIT;
       ser_out                        :OUT BIT );
END siso;
ARCHITECTURE  vhdl  OF  siso  IS
BEGIN
PROCESS (clock, clr)
      VARIABLE  q                   :BIT_VECTOR (7 DOWNTO 0);
      BEGIN
      ser_out <= q(7);                   -- output MSB register bit

      IF clr = '0'      THEN  q := "00000000";
                                         -- asynchronous clear
      ELSIF (clock'EVENT AND clock = '1') THEN
            IF (en = '1')   THEN         -- check for enable active
                  q := (q(6 DOWNTO 0) & ser_in);
                                         -- concatenate for shift
            END IF;                      -- otherwise hold
      END IF;
END PROCESS;
END vhdl;
```

Fig. 15-7 VHDL solution for Example 15-1

VHDL

A VHDL solution is shown in Fig. 15-7. An 8-bit register is created for the VARIABLE named q. The sensitivity list for the PROCESS includes both clock and the asynchronous clr input. The serial output port for the subdesign is assigned to the MSB register output. An IF statement is used to check for an asynchronous clear first and, if that test fails, for a positive-edge on clock. When clocked, a nested IF checks to see if the shift function is enabled. The current register contents will be shifted one bit to the left with ser_in moving into the LSB q0 flip-flop. This is accomplished by concatenating the appropriate set of 8 bits together to be assigned as inputs to the VARIABLE q. When the shift function is not enabled, it is implied that q does not change. Simulation results are shown in Fig. 15-6. Each of the buried register nodes has been included in the simulation to demonstrate the serial shifting of the register.

Example 15-2

Design an 8-bit universal shift register using an HDL language. The 8-bit shift register can move data serially or in parallel. The output is to be labeled q[0..7] and the parallel data input is d[0..7]. The serial data movement can be either shifting the data to the left (from q7 toward q0) or to the right (from q0 toward q7). The serial data inputs are sr_ser (shift right serial) and sl_ser (shift left serial). The clocked register function is controlled by s1 and s0, as described in the function table in Table 15-1.

s1	s0	Function	$q0_{n+1}$	$q1_{n+1}$	$q2_{n+1}$	$q3_{n+1}$	$q4_{n+1}$	$q5_{n+1}$	$q6_{n+1}$	$q7_{n+1}$
0	0	HOLD	$q0_n$	$q1_n$	$q2_n$	$q3_n$	$q4_n$	$q5_n$	$q6_n$	$q7_n$
0	1	SHIFT RIGHT	sr_ser	$q0_n$	$q1_n$	$q2_n$	$q3_n$	$q4_n$	$q5_n$	$q6_n$
1	0	SHIFT LEFT	$q1_n$	$q2_n$	$q3_n$	$q4_n$	$q5_n$	$q6_n$	$q7_n$	sl_ser
1	1	LOAD	$d0_n$	$d1_n$	$d2_n$	$d3_n$	$d4_n$	$d5_n$	$d6_n$	$d7_n$

Table 15-1 Function table for Example 15-2

```
SUBDESIGN  univ_shift
(
        clk, s[1..0]                        :INPUT;
        d[0..7], sr_ser, sl_ser             :INPUT;
        q[0..7]                             :OUTPUT;
)
VARIABLE
        register[0..7]                      :DFF;     -- 8-bit register
BEGIN
        register[].clk = clk;

        CASE  s[]  IS
              WHEN  0  =>         -- hold
                    register[0..7].d = register[0..7].q;
              WHEN  1  =>         -- shift right
                    register[1..7].d = register[0..6].q;
                    register[0].d = sr_ser;
              WHEN  2  =>         -- shift left
                    register[0..6].d = register[1..7].q;
                    register[7].d = sl_ser;
              WHEN  3  =>         -- parallel load
                    register[0..7].d = d[0..7];
        END CASE;

        q[] = register[].q;
                    -- connect register to output port
END;
```

Fig. 15-8 AHDL solution for Example 15-2

AHDL An AHDL solution is shown in Fig. 15-8. An 8-bit register using D flip-flops is created in the VARIABLE section. The CASE statement determines the necessary input assignments for the register to operate according to the function controls s[1..0]. Each input bit (.d) is assigned in order to the respective output bit (.q). This makes it easy to shift right or left by one bit for each clock cycle when s[1..0] is 01_2 or 10_2, respectively. The appropriate serial input bit is also assigned. The output port for the subdesign is assigned to the register outputs.

```
ENTITY  univ_shift  IS
PORT (
      clk, sr_ser, sl_ser        :IN BIT;
      s                          :IN BIT_VECTOR (1 DOWNTO 0);
      d                          :IN BIT_VECTOR (0 TO 7);
      q                          :OUT BIT_VECTOR (0 TO 7)
);
END univ_shift;

ARCHITECTURE vhdl OF univ_shift ISPORT (
BEGIN
   PROCESS (clk)
      VARIABLE  sr   :BIT_VECTOR (0 TO 7); -- 8-bit register
      BEGIN
         IF (clk'EVENT AND clk = '1')  THEN
            CASE  s  IS
                WHEN  "00"  =>        -- hold
                   sr := sr;
                WHEN  "01"  =>        -- shift right
                   sr(1 TO 7) := sr(0 TO 6);
                   sr(0) := sr_ser;
                WHEN  "10"  =>        -- shift left
                   sr(0 TO 6) := sr(1 TO 7);
                   sr(7) := sl_ser;
                WHEN  "11"  =>        -- parallel load
                   sr:= d;
            END CASE;
         END IF;
      q <= sr;     -- connect register to output port
   END PROCESS;
END vhdl;
```

Fig. 15-9 VHDL solution for Example 15-2

VHDL A VHDL solution is shown in Fig. 15-9. An 8-bit VARIABLE named sr is created to represent the shift register. The CASE statement determines the necessary assignment statements for the register to operate according to the 2-bit function control s. For serial shifting, the appropriate input bit is assigned in order, as specified in the truth table. This makes it easy to shift right or left by one bit for each clock cycle when s is 01 or 10, respectively. The appropriate serial input bit is also assigned. The output port for the entity is assigned to the register outputs.

Simulation results for the HDL designs are shown in Fig. 15-10.

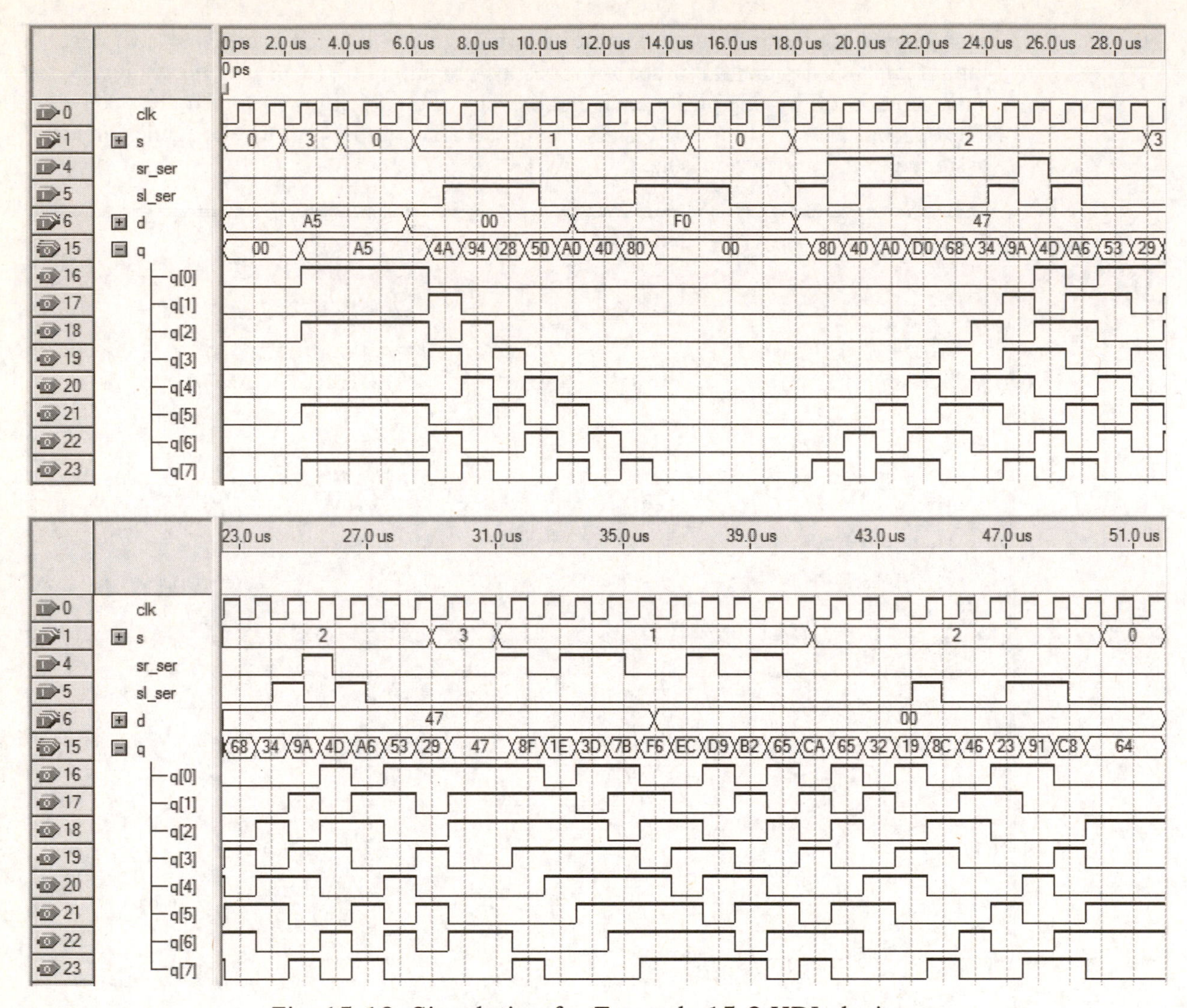

Fig. 15-10 Simulation for Example 15-2 HDL designs

Feedback Shift Register Counters

A count sequence can be generated with a shift register by using the contents of the register to produce a feedback signal for the serial input to the shift register. This type of counter generally requires very little hardware to construct, and the circuitry required to decode the count sequence can be very simple. Ring counters and Johnson counters are examples of feedback shift register counters.

A ring counter has a modulus equal to the number of flip-flops being used in the shift register. There are two variations of the ring counter. One has a single bit that is HIGH and is moved from one flip-flop in the shift register to the next. The other choice is to rotate a single 0 bit throughout the length of the shift register. Decoding of the ring counter is simply done by noting which bit position contains the single 1 (or single 0). Self-starting ring counters can be easily designed by generating a simple feedback signal that is a function of all of the shift register bits except the final bit in the chain. The number of inputs to the feedback function circuit is always one less than the number of flip-flops in the shift register counter.

The Johnson counter (also called a twisted ring counter) uses the flip-flops a little more efficiently to produce a modulus that is two times the number of flip-flops being used in the shift register. The feedback signal needed for the Johnson sequence is accomplished by merely inverting the single output bit of the serial shift register. Multiple sequences are produced with this simple feedback circuit, but the standard Johnson code sequence includes the state where all flip-flops are LOW. The Johnson counter sequence can be decoded by monitoring an appropriate pair of bits for each counter state.

A pseudorandom count sequence can be generated by a linear feedback shift register (LFSR). XNORing two or more flip-flop outputs from the shift register produces the feedback for this simple type of shift register counter. A shift register counter using this feedback arrangement with n flip-flops would have a count modulus of $2^n - 1$. The count modulus for any length LFSR counter can be shortened by causing it to skip an appropriate number of states. Table 15-2 lists the proper inputs to the XNOR feedback gate that is generating the serial input to flip-flop Q0 (the first flip-flop in the shift register chain).

Number of flip-flops	Counter modulus	XNOR gate inputs
2	3	Q1, Q0
3	7	Q2, Q1
4	15	Q3, Q2
5	31	Q4, Q2
6	63	Q5, Q4
7	127	Q6, Q5
8	255	Q7, Q5, Q4, Q3
9	511	Q8, Q4
10	1023	Q9, Q6
11	2047	Q10, Q9
12	4095	Q11, Q10, Q9, Q1

Table 15-2 LFSR counter modulus and feedback signals

Example 15-3

Design a self-starting MOD-4 ring counter using an HDL language. The ring counter should output a single 1 that is rotated through a 4-bit shift register to produce the recycling binary pattern: 1000 → 0100 → 0010 → 0001 → 1000*(repeats)*.

A MOD-4 ring counter will require a 4-bit SIPO register. An active-HIGH output ring counter will have a single 1 stored in the register. The design "trick" is to produce the necessary feedback function for the serial input to the register. The feedback function will be HIGH only if all flip-flop outputs except the last one (the first 3 flip-flops for a 4-bit ring counter) in the register are LOW. All other bit combinations in the first 3 flip-flops should produce a LOW feedback signal. On the other hand, if an active-LOW output ring counter (single 0 output) is desired, the feedback signal will be HIGH unless all flip-flop outputs except the last one are HIGH.

```
SUBDESIGN  ring4
(
      clk                  :INPUT;
      qa, qb, qc, qd       :OUTPUT;
)
VARIABLE
      register[0..3]       :DFF;        -- shift register
      ser                  :NODE;       -- feedback signal
BEGIN
      register[].clk = clk;
-- determine feedback signal
      IF register[0..2].q == B"000" THEN  ser = VCC;
      ELSE                                ser = GND;
      END IF;
-- create ring counter by concatenating bits for input
      register[0..3].d = (ser, register[0..2].q);
-- connect register to output ports
      (qa, qb, qc, qd) = register[];
END;
```

Fig. 15-11 AHDL solution for Example 15-3

AHDL An AHDL design solution is shown in Fig. 15-11. The buried-node feedback function ser is defined with the IF statement. By using concatenation, the serial input is applied to the first flip-flop input while the other three bits of the register are shifted one position. The register outputs are connected to the output ports for the block. This circuit can also be designed using a state machine.

VHDL A VHDL design solution is shown in Fig. 15-12. By using concatenation, the serial input is applied to the first flip-flop input while the other three bits of the register are shifted one position. The buried-node feedback function ser is defined with the NOR logic expression. The register outputs are connected to the output ports for the block.

Simulation results for the HDL designs are shown in Fig. 15-13.

```
ENTITY  ring4  IS
PORT (
      clk                   :IN BIT;
      qa, qb, qc, qd        :OUT BIT
);
END ring4;

ARCHITECTURE vhdl OF ring4 IS
signal      ser            :BIT; -- feedback signal
BEGIN
PROCESS (clk)
      VARIABLE  reg        :BIT_VECTOR (0 TO 3);
                                 -- shift register
      BEGIN
-- create ring counter by concatenating bits for input
            IF (clk'EVENT AND clk = '1')  THEN
                  reg(0 TO 3) := (ser & reg(0 TO 2));
            END IF;
-- determine feedback signal
            IF (reg(0 TO 2) = "000")  THEN  ser <= '1';
            ELSE                            ser <= '0';
            END IF;
-- connect register to output ports
            qa <= reg(0);
            qb <= reg(1);
            qc <= reg(2);
            qd <= reg(3);
END PROCESS;
END vhdl;
```

Fig. 15-12 VHDL solution for Example 15-3

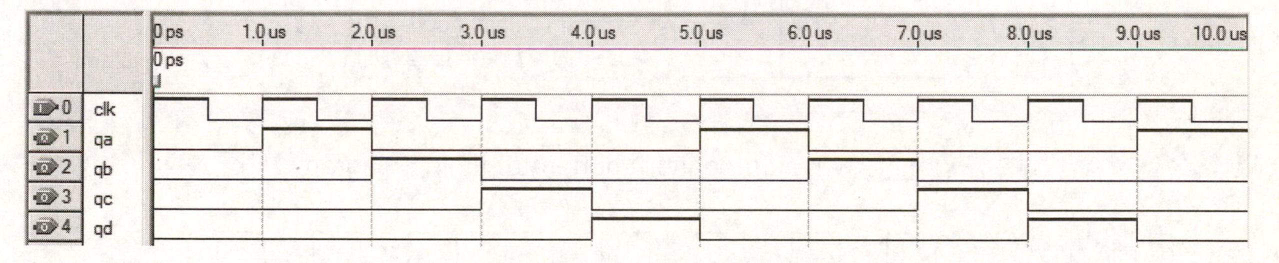

Fig. 15-13 Simulation for Example 15-3 HDL designs

Laboratory Projects

Design the following counter applications using an HDL language. Simulate and test the circuit designs.

15.1 Waveform pattern generator
Design a waveform pattern generator. The parallel-in, serial-out shift register will produce any desired 8-bit waveform pattern such as the one illustrated below. The first bit on each pattern cycle should be the LSB. Test the logic circuit by comparing the CLK and SERIAL_OUT waveforms with a two-channel oscilloscope. **Use the Shift/Load signal to trigger the scope.** Note the change in the SERIAL_OUT waveform as the data input switches are changed. Hints: There will be 8 clock cycles needed for each 8-bit pattern sequence: 1 cycle to parallel-load the 8-bit pattern and 7 more to serially shift the data toward Q0 for serial output. The state machine controls the parallel loading or serial shifting function of the PISO shift register. Design the waveform pattern generator using:

(a) HDL.
(b) LPM megafunctions in a BDF.

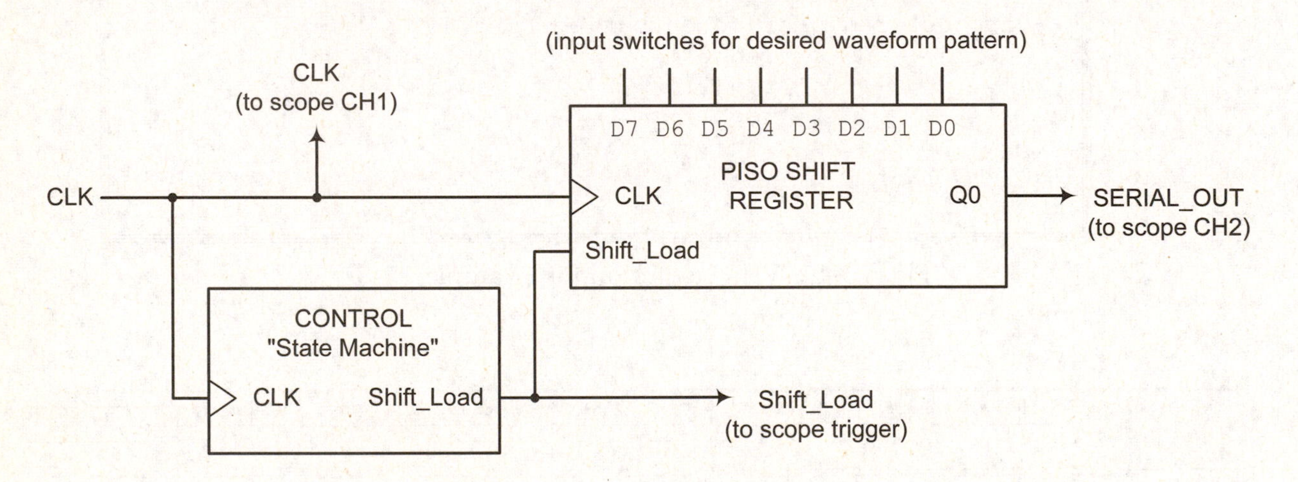

Example waveform pattern generator timing diagram

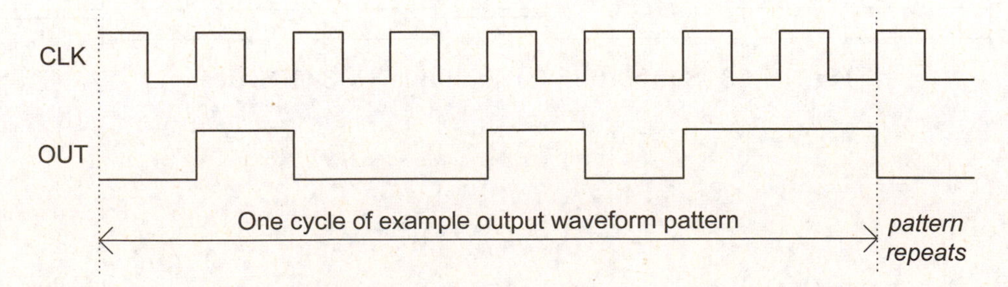

15.2 Serial data buffer

Design a serial data buffer that can be used to receive a serially transmitted 8-bit data word. The CONTROL is a state machine with 9 states, 8 states (shft0 through shft7) in sequence that will enable serial shifting of the SIPO register followed by a state named done. The CONTROL will stay in the done state until the RTS (request to send) input goes HIGH to initiate the data transfer sequence. Then, when clocked, the state machine will proceed to the shft0 state, the first of 8 states in sequence during which the shift enable on the SIPO register is active. The READY signal will be HIGH when the state machine is in the shft0 state, to indicate that the buffer (register) is ready to receive data. Each of the 8 data bits, starting with the LSB, will individually be placed on the serial SER_IN line and clocked into the SIPO register. After 8 clock cycles, all 8 data bits will be received by the SIPO register and the state machine will automatically stop on the done state. While the state machine is in the done state, DAV (data available) will be HIGH. Example operation is shown in the simulation below. Use a **manual** clock to test the circuit.

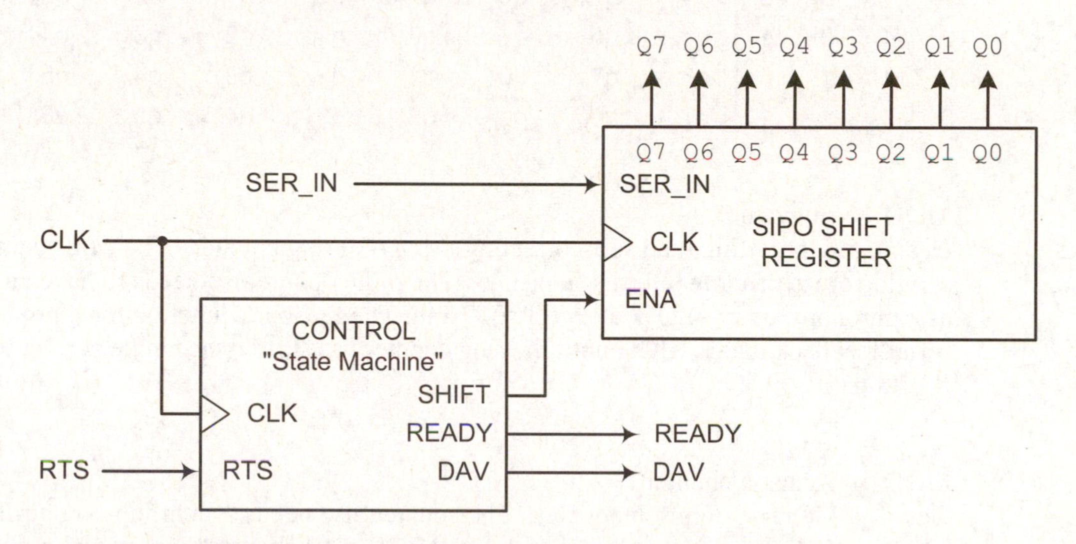

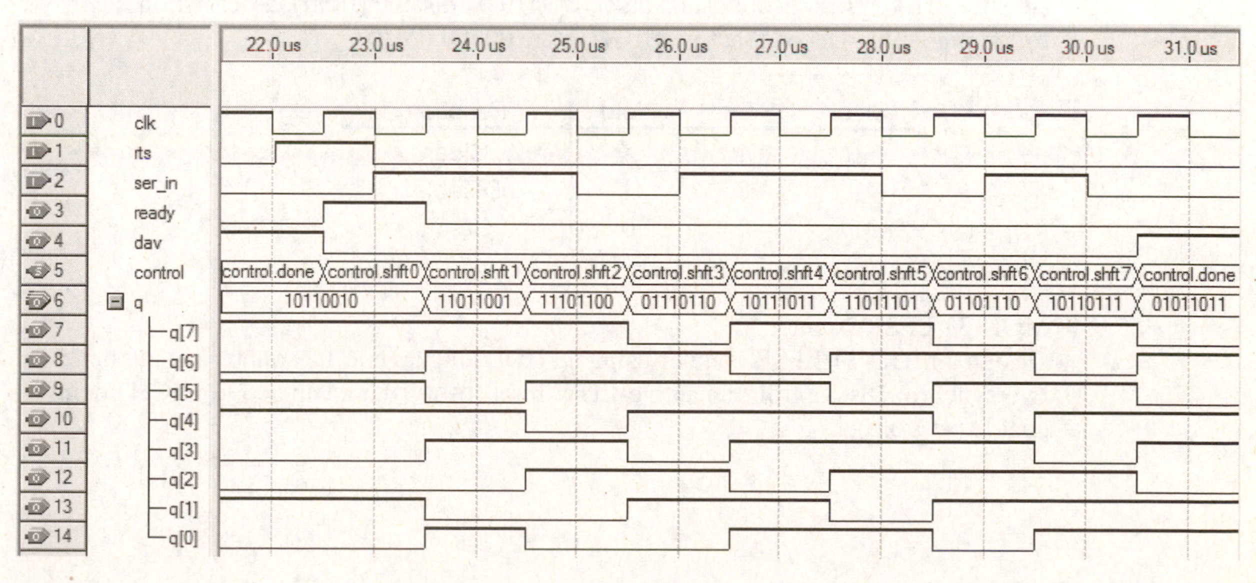

15.3 Barrel shifter

Design an 8-bit, registered barrel shifter using a PLD. The 8-bit barrel shifter can be synchronously parallel-loaded with 8 data inputs (d0 through d7). The register contents can then be cyclically rotated from 0 to 7 places under the control of the select inputs (s[2..0]), as shown in the function table below. The rotated data is stored in the same register when the clock is pulsed.

clk	ld	s2	s1	s0	$q0_{n+1}$	$q1_{n+1}$	$q2_{n+1}$	$q3_{n+1}$	$q4_{n+1}$	$q5_{n+1}$	$q6_{n+1}$	$q7_{n+1}$
↑	1	X	X	X	$d0_n$	$d1_n$	$d2_n$	$d3_n$	$d4_n$	$d5_n$	$d6_n$	$d7_n$
↑	0	0	0	0	$q0_n$	$q1_n$	$q2_n$	$q3_n$	$q4_n$	$q5_n$	$q6_n$	$q7_n$
↑	0	0	0	1	$q1_n$	$q2_n$	$q3_n$	$q4_n$	$q5_n$	$q6_n$	$q7_n$	$q0_n$
↑	0	0	1	0	$q2_n$	$q3_n$	$q4_n$	$q5_n$	$q6_n$	$q7_n$	$q0_n$	$q1_n$
↑	0	0	1	1	$q3_n$	$q4_n$	$q5_n$	$q6_n$	$q7_n$	$q0_n$	$q1_n$	$q2_n$
↑	0	1	0	0	$q4_n$	$q5_n$	$q6_n$	$q7_n$	$q0_n$	$q1_n$	$q2_n$	$q3_n$
↑	0	1	0	1	$q5_n$	$q6_n$	$q7_n$	$q0_n$	$q1_n$	$q2_n$	$q3_n$	$q4_n$
↑	0	1	1	0	$q6_n$	$q7_n$	$q0_n$	$q1_n$	$q2_n$	$q3_n$	$q4_n$	$q5_n$
↑	0	1	1	1	$q7_n$	$q0_n$	$q1_n$	$q2_n$	$q3_n$	$q4_n$	$q5_n$	$q6_n$

15.4 MOD-10 ring counter

Design a self-starting (and self-correcting) MOD-10 ring counter. The ring counter should **rotate a single 0** for its sequence. The ring counter also needs to have an asynchronous, active-LOW clear (clrn). Hint: The feedback function must produce a serial feedback that is HIGH until the ring counter has shifted the single zero to the last flip-flop.

15.5 MOD-10 Johnson counter

Design a MOD-10 Johnson counter. The counter also needs synchronous controls for an active-HIGH reset (reset) and an active-HIGH disable (hold), as shown in the following table. Decode (active-HIGH) state 00000 (zero).

Clock	Reset	Hold	Function
↑	1	X	Clear
↑	0	1	Disable
↑	0	0	Count

15.6 MOD-31 LFSR counter

Design a MOD-31 LFSR counter using an HDL language. The counter should have two synchronous controls, an active-LOW reset (resetn), and an active-HIGH count enable (enable).

15.7 Parallel data pipeline

Design a 4-bit data pipeline circuit. A 4-bit data word (D3 D2 D1 D0) will be entered in parallel into the data pipeline when the LOAD control is HIGH and the registers are clocked. Use a **manual** clock to load in one data word at a time into the pipeline. The data pipeline is five 4-bit words deep. Create a **hierarchical** design by defining a 4-bit register with HDL code and then connecting 5 copies in a BDF.

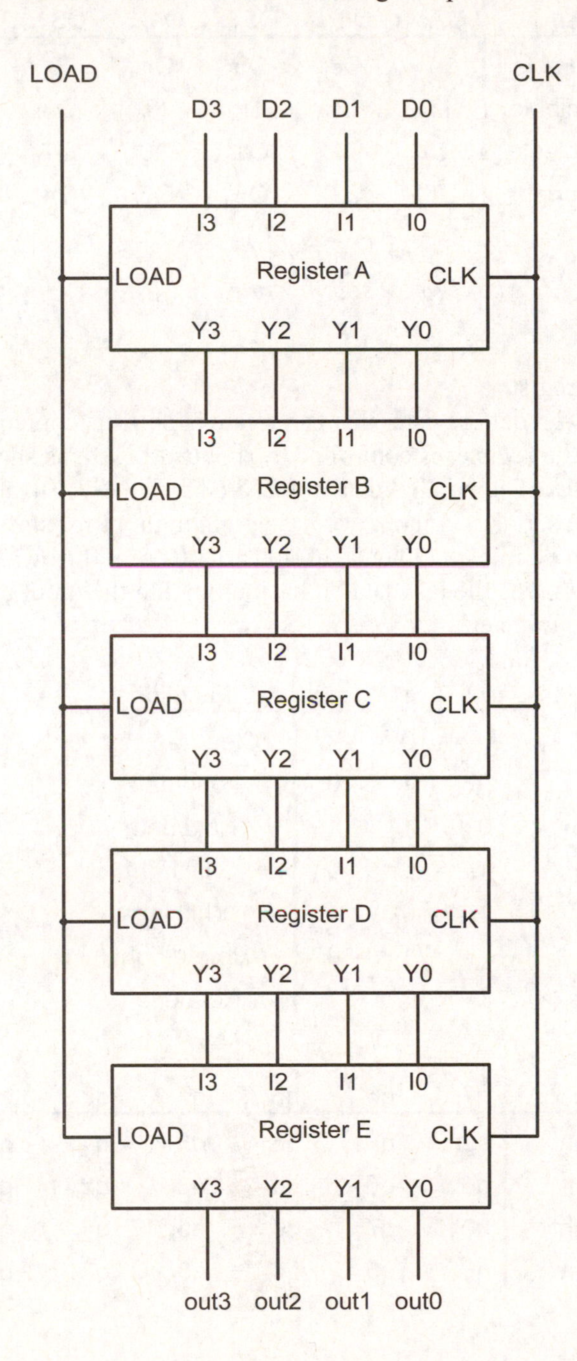

15.8 Special-purpose data register

Design an 8-bit data register that operates according to the following function table. The register function is selected with the controls s[1..0]. The 4-bit data input is labeled i[3..0], and the register output is labeled q[7..0].

s1	s0	Operation	$Q7_{n+1}$	$Q6_{n+1}$	$Q5_{n+1}$	$Q4_{n+1}$	$Q3_{n+1}$	$Q2_{n+1}$	$Q1_{n+1}$	$Q0_{n+1}$
0	0	Load L.S. nibble	$Q7_n$	$Q6_n$	$Q5_n$	$Q4_n$	$I3_n$	$I2_n$	$I1_n$	$I0_n$
0	1	Load M.S. nibble	$I3_n$	$I2_n$	$I1_n$	$I0_n$	$Q3_n$	$Q2_n$	$Q1_n$	$Q0_n$
1	0	Swap nibbles	$Q3_n$	$Q2_n$	$Q1_n$	$Q0_n$	$Q7_n$	$Q6_n$	$Q5_n$	$Q4_n$
1	1	Rotate data right	$Q0_n$	$Q7_n$	$Q6_n$	$Q5_n$	$Q4_n$	$Q3_n$	$Q2_n$	$Q1_n$

Note: L.S. = least significant
M.S. = most significant

15.9 Shift/rotate register

Design an 8-bit data register that can rotate or shift its contents in either direction. The function of the register is controlled by inputs m[2..0], as shown in the following function table. The parallel data inputs are labeled d[7..0], and the register outputs are q[7..0]. The serial data input ser is a common input for shifting in either direction. The data can be moved either right (toward q0) or left (toward q7) in the register. A data rotation feeds the last data bit back around to the input of the shift register instead of losing it out the end.

m2	m1	m0	Function
0	0	X	Clear register
0	1	0	Load data
0	1	1	Hold data
1	0	0	Shift right
1	0	1	Shift left
1	1	0	Rotate right
1	1	1	Rotate left

Shift/rotate	Direction	$q7_{n+1}$	$q6_{n+1}$	$q5_{n+1}$	$q4_{n+1}$	$q3_{n+1}$	$q2_{n+1}$	$q1_{n+1}$	$q0_{n+1}$
Shift	Right	ser	$q7_n$	$q6_n$	$q5_n$	$q4_n$	$q3_n$	$q2_n$	$q1_n$
Shift	Left	$q6_n$	$q5_n$	$q4_n$	$q3_n$	$q2_n$	$q1_n$	$q0_n$	ser
Rotate	Right	$q0_n$	$q7_n$	$q6_n$	$q5_n$	$q4_n$	$q3_n$	$q2_n$	$q1_n$
Rotate	Left	$q6_n$	$q5_n$	$q4_n$	$q3_n$	$q2_n$	$q1_n$	$q0_n$	$q7_n$

UNIT 16

SYNCHRONOUS COUNTER DESIGN WITH FLIP-FLOPS

Objectives

- To design synchronous counters with specified sequences implemented with JK or D flip-flops using Karnaugh mapping.
- To verify synchronous counter designs using logic simulation and circuit testing.

Synchronous Counter Design

Synchronous sequential circuits may be designed by developing a transition table from the desired state sequence. A transition table is used to identify the synchronous inputs that must be applied to each flip-flop to produce the specified count sequence. The Boolean expression for each flip-flop input can be derived by Karnaugh mapping the transition table information. The set of logic equations describes the necessary input circuitry for each flip-flop. This procedure can be applied to any type of flip-flop and any desired count sequence (as long as the logic simplification is manageable with Karnaugh mapping).

Example 16-1

Use flip-flops to design a synchronous, MOD-7 counter with the sequence of states given in the state transition diagram of Fig. 16-1. Create two design solutions, one using JK flip-flops and the other using D flip-flops.

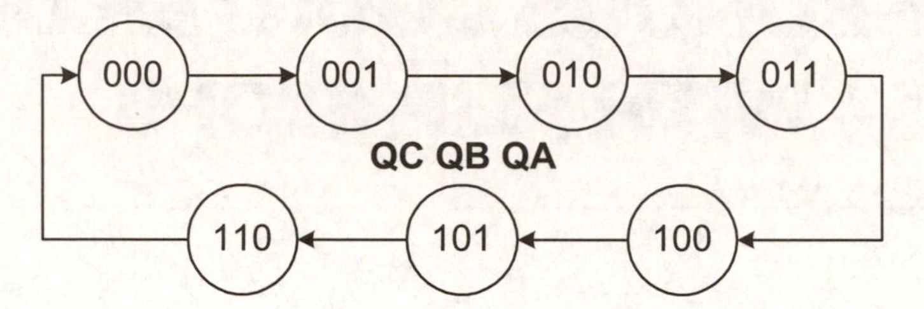

Fig. 16-1 MOD-7 counter state transition diagram for Example 16-1

First, complete a present state/next state table, as illustrated in Table 16-1, to describe the desired sequence from the given state diagram.

Present State			Next State		
QC_n	QB_n	QA_n	QC_{n+1}	QB_{n+1}	QA_{n+1}
0	0	0	0	0	1
0	0	1	0	1	0
0	1	0	0	1	1
0	1	1	1	0	0
1	0	0	1	0	1
1	0	1	1	1	0
1	1	0	0	0	0
1	1	1	X	X	X

Table 16-1 Present state/next state table for Example 16-1

Next, for the JK design, list the state transitions for each flip-flop so that the desired sequence is created, and determine the required flip-flop inputs that will produce this sequence. This information is recorded in an excitation table (see Table 16-2).

Present States			State Transitions			Flip-flop Inputs					
QC_n	QB_n	QA_n	$QC_n \rightarrow QC_{n+1}$	$QB_n \rightarrow QB_{n+1}$	$QA_n \rightarrow QA_{n+1}$	JC	KC	JB	KB	JA	KA
0	0	0	$0 \rightarrow 0$	$0 \rightarrow 0$	$0 \rightarrow 1$	0	X	0	X	1	X
0	0	1	$0 \rightarrow 0$	$0 \rightarrow 1$	$1 \rightarrow 0$	0	X	1	X	X	1
0	1	0	$0 \rightarrow 0$	$1 \rightarrow 1$	$0 \rightarrow 1$	0	X	X	0	1	X
0	1	1	$0 \rightarrow 1$	$1 \rightarrow 0$	$1 \rightarrow 0$	1	X	X	1	X	1
1	0	0	$1 \rightarrow 1$	$0 \rightarrow 0$	$0 \rightarrow 1$	X	0	0	X	1	X
1	0	1	$1 \rightarrow 1$	$0 \rightarrow 1$	$1 \rightarrow 0$	X	0	1	X	X	1
1	1	0	$1 \rightarrow 0$	$1 \rightarrow 0$	$0 \rightarrow 0$	X	1	X	1	0	X
1	1	1	$1 \rightarrow X$	$1 \rightarrow X$	$1 \rightarrow X$	X	X	X	X	X	X

Table 16-2 Excitation table for Example 16-1

Record the information for each flip-flop J and K input in separate K maps (see Fig. 16-2) and determine the appropriate, simplified Boolean expressions. Note that "don't care" output conditions (Xs in the K maps) may be defined as either 0s or 1s and may be used to simplify the expressions. The schematic for the synchronous circuit design (see Fig. 16-3) can now be drawn from the J and K flip-flop input equations determined by K mapping.

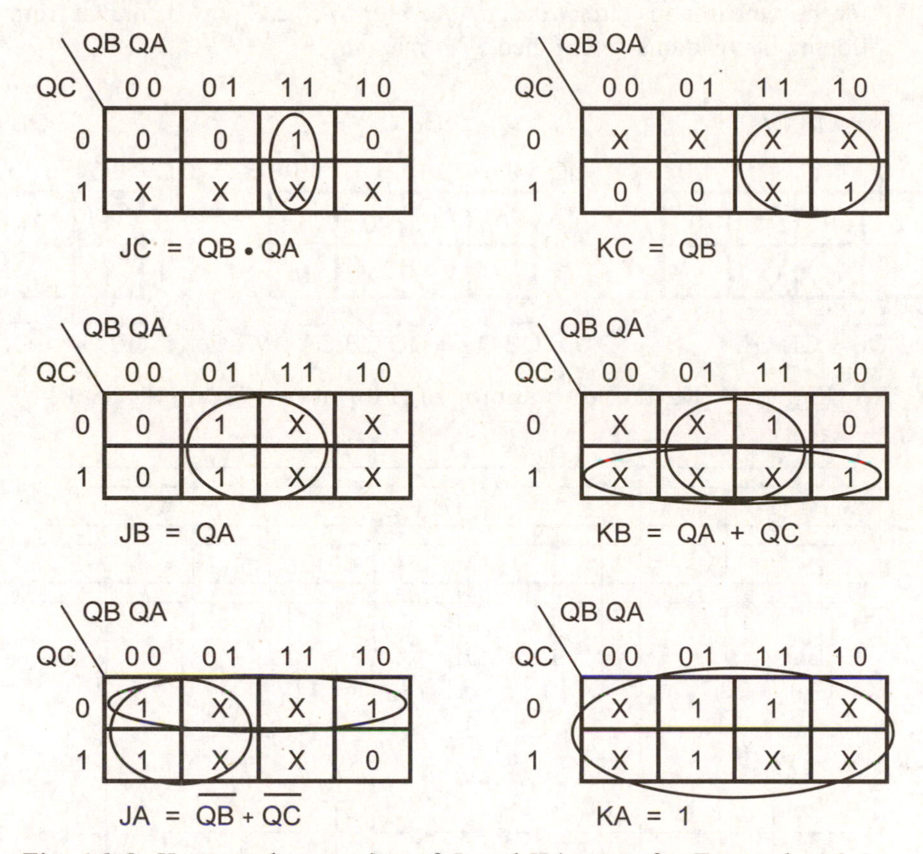

Fig. 16-2 Karnaugh mapping of J and K inputs for Example 16-1

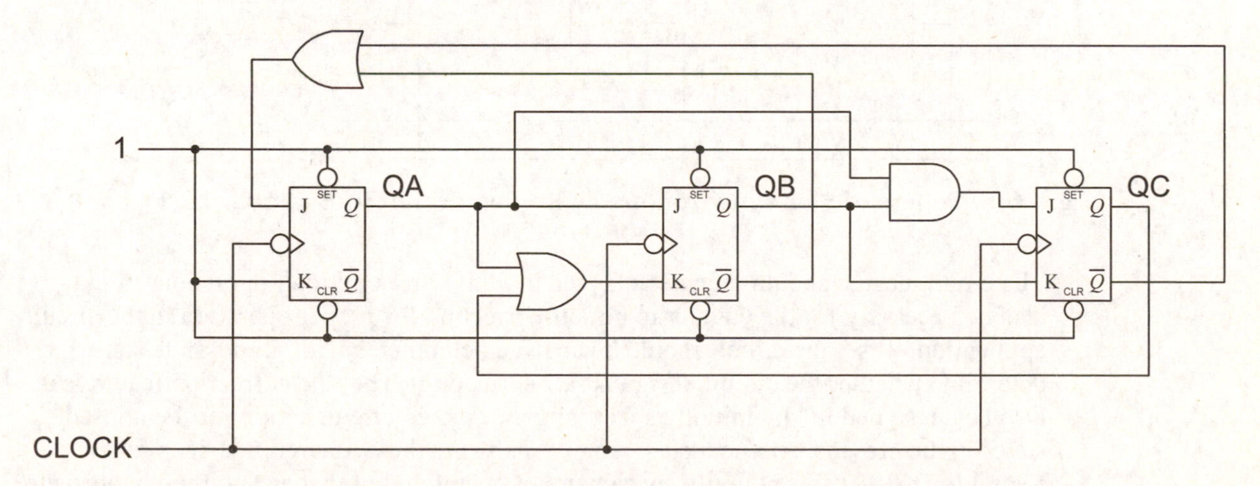

Fig. 16-3 Schematic for synchronous JK flip-flop design in Example 16-1

For the D flip-flop design, we only need to consult the present state/next state table shown in Table 16-1. The D inputs needed to produce the specified next state for each present state are the same as the next state bits. Thus a combinational circuit that produces the necessary corresponding next bit for each state must be designed for each flip-flop's D input. Record this information for each flip-flop in a separate K map (see Fig. 16-4) and determine the appropriate simplified Boolean expression. The schematic for the synchronous circuit design (see Fig. 16-5) can now be drawn from the D flip-flop input equations determined by K mapping.

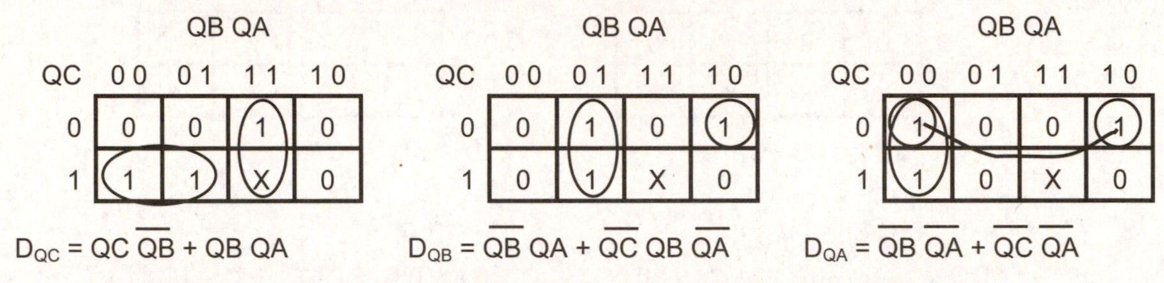

Fig. 16-4 Karnaugh mapping of D inputs for Example 16-1

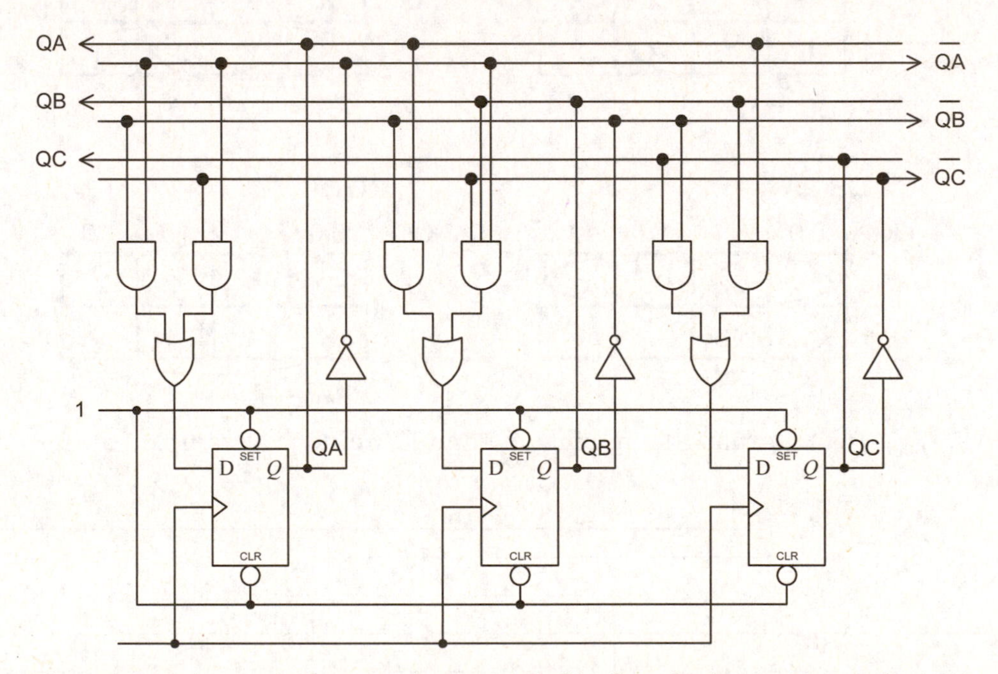

Fig. 16-5 Schematic for synchronous D flip-flop design in Example 16-1

The circuit design should then be analyzed to verify proper circuit operation. It may also be necessary for the counter to be self-correcting for proper operation in the circuit application. If so, the circuit should be analyzed completely (for all possible states) to determine whether the circuit design is self-correcting. The self-correction feature can also be "designed in" by initially specifying the desired circuit action for the unused states in the present state/next state table. Analyzing the two circuit designs for Example 16-1 verifies not only correct circuit operation but that each circuit happens to be self-correcting because the unused state 111 returns into the count sequence loop.

Laboratory Projects

Design each of the following synchronous counters using JK or D flip-flops. Do not use the asynchronous flip-flop inputs (PRE or CLR) in the circuit designs. Construct and test each circuit design. Use an oscilloscope to display the timing diagram for each counter. Functionally simulate the circuits with Quartus.

16.1 MOD-6 synchronous counter design

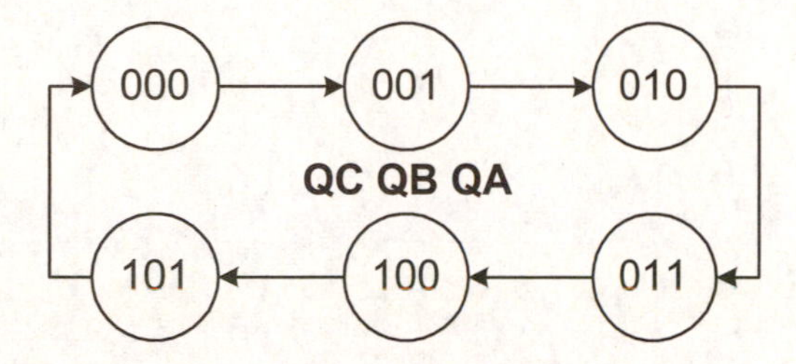

16.2 MOD-5 synchronous down-counter design

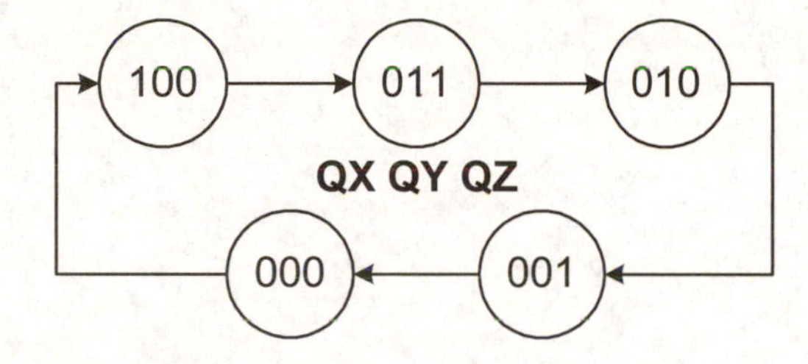

16.3 Synchronous 8421 BCD counter design

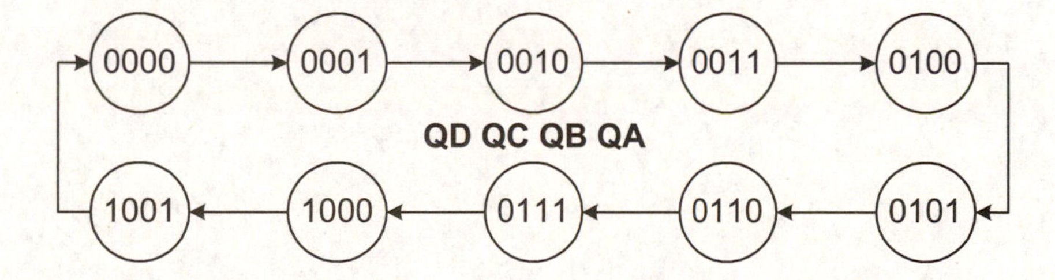

16.4 Synchronous 8421 BCD down-counter design

Design a synchronous 8421 BCD down-counter. The outputs are QD QC QB QA. QA is the LSB.

16.5 Synchronous MOD-11 counter

Design a recycling, synchronous MOD-11 counter. The count sequence should be 0000 through 1010. The outputs are QD QC QB QA (QA = LSB).

UNIT 17

DECODER AND DISPLAY APPLICATIONS

Objectives

- To apply decoding circuits in digital system applications.
- To implement digital displays using 7-segment devices.

Decoder Circuits

A decoder is a common logic function that is used to detect a particular combination of bits applied to the input of the circuit and to display that information in a specified fashion. Logic gates can be used to design any type of decoder circuit. Decoders are available as standard IC chips and also as maxplus2 functions. For example, the 74138 (shown in Fig. 17-1) is a 3-line-to-8-line decoder. It is being used to decode a MOD-8 binary counter (created as an LPM). This decoder will identify which of the 8 possible counter states is currently applied to its 3 data input lines (C B A). The 74138 also has 3 enable inputs (2 active-LOW G2A G2B and 1 active-HIGH G1), which adds greatly to its flexibility and usefulness. Note the Altera signal naming convention of ending the name with an "n" to indicate active-LOW logic levels. When enabled, only one of the decoder's 8 outputs will be active (LOW), which will indicate the counter's output state. The truth table for the 74138 is shown in Table 17-1. Simulation results are shown in Fig. 17-2.

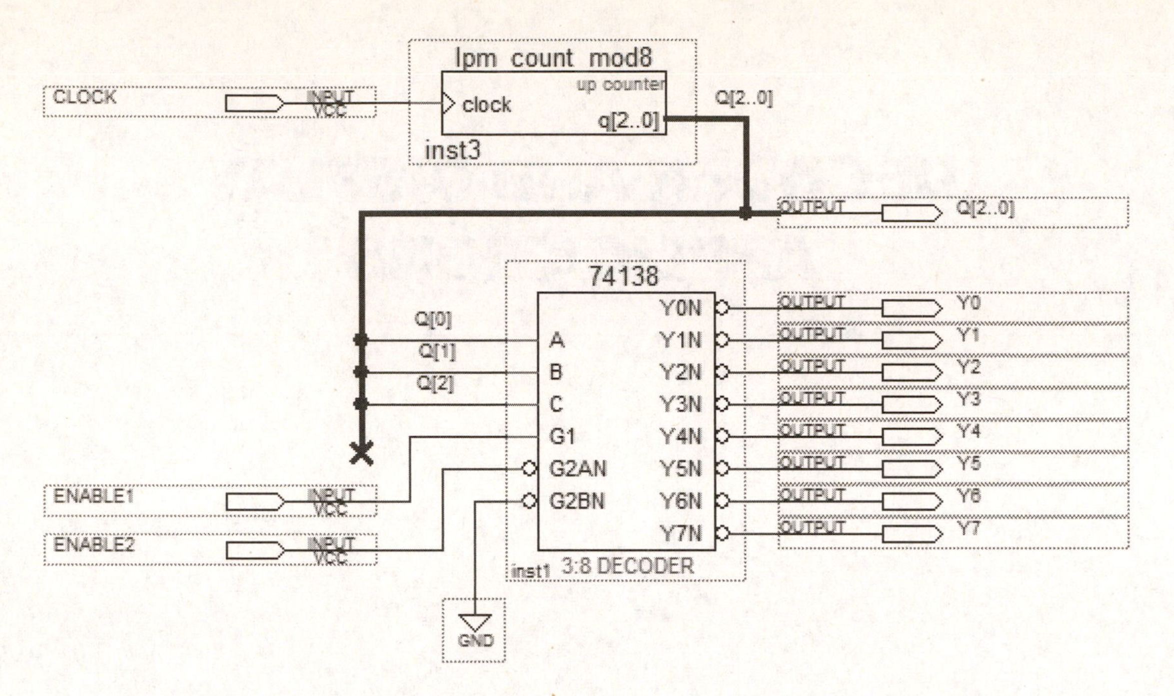

Fig. 17-1 A 74138 decoding a MOD-8 binary counter

G1	G2A	G2B	C	B	A	Y0	Y1	Y2	Y3	Y4	Y5	Y6	Y7
1	0	0	0	0	0	0	1	1	1	1	1	1	1
1	0	0	0	0	1	1	0	1	1	1	1	1	1
1	0	0	0	1	0	1	1	0	1	1	1	1	1
1	0	0	0	1	1	1	1	1	0	1	1	1	1
1	0	0	1	0	0	1	1	1	1	0	1	1	1
1	0	0	1	0	1	1	1	1	1	1	0	1	1
1	0	0	1	1	0	1	1	1	1	1	1	0	1
1	0	0	1	1	1	1	1	1	1	1	1	1	0
0	X	X	X	X	X	1	1	1	1	1	1	1	1
X	1	X	X	X	X	1	1	1	1	1	1	1	1
X	X	1	X	X	X	1	1	1	1	1	1	1	1

Table 17-1 Truth table for a 74138 decoder

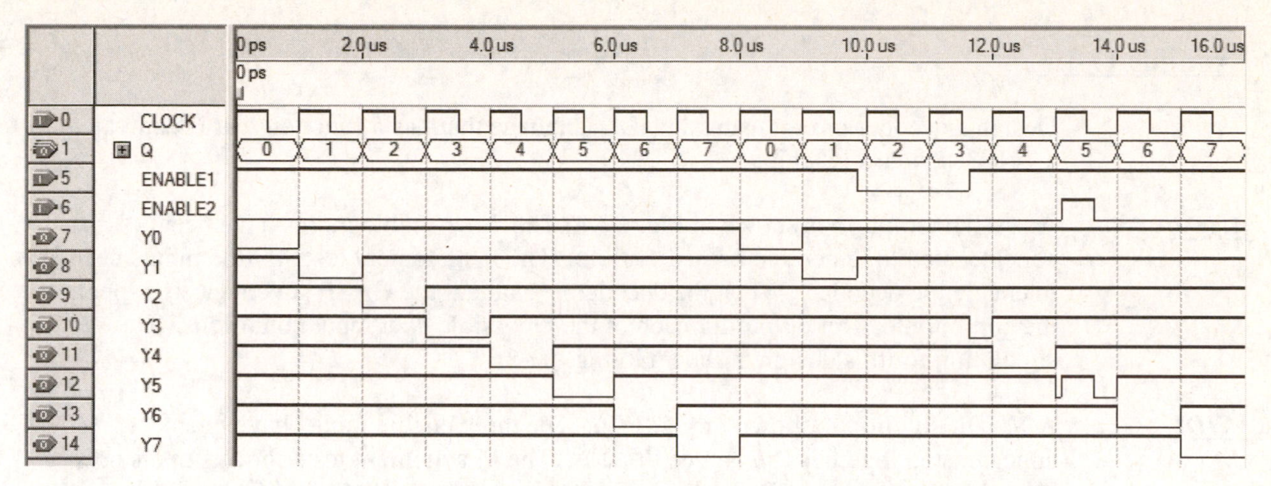

Fig. 17-2 Simulation of MOD-8 binary counter and 74138 decoder

Decoder Megafunction

Quartus also has a decoder megafunction called LPM_DECODE that can be used for applications. An example 3-bit decoder with enable is shown in Fig. 17-3.

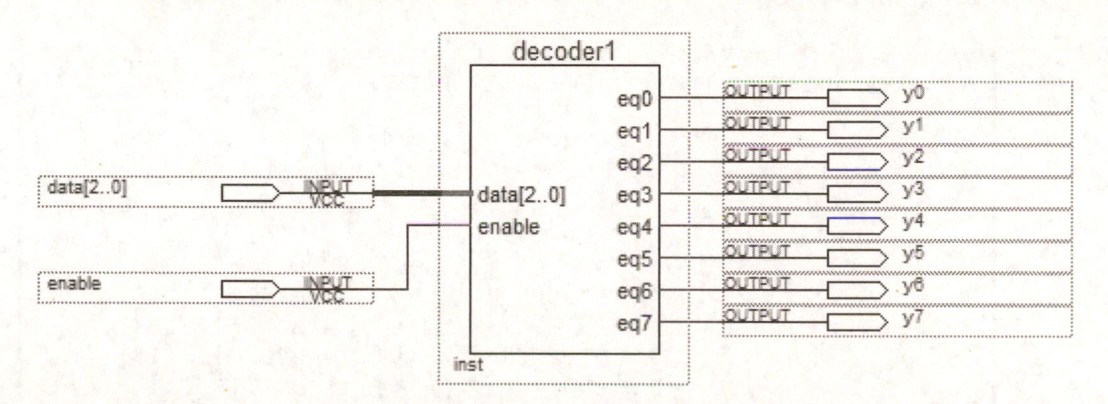

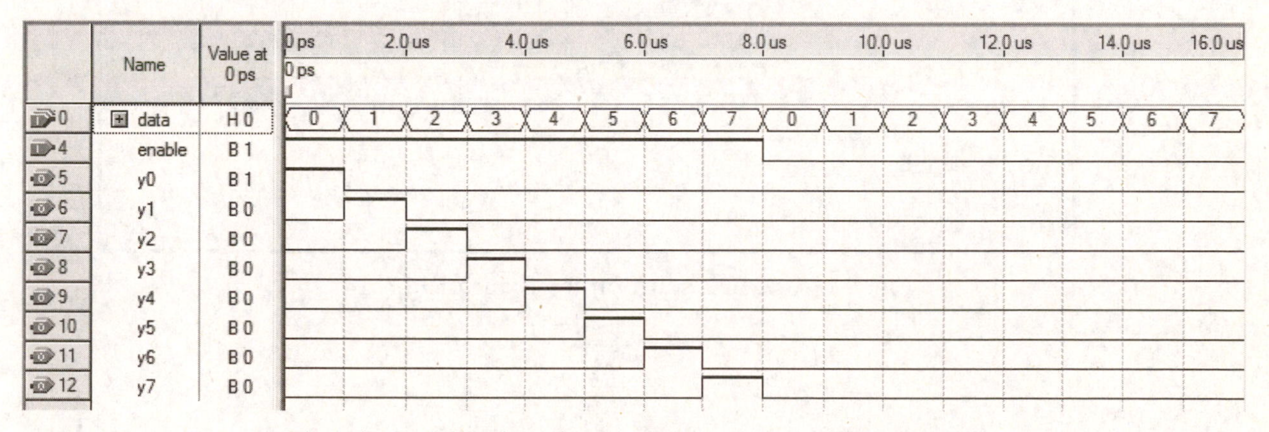

Fig. 17-3 3-bit decoder megafunction and simulation

Example 17-1

Design a decoder circuit using an HDL language that has a function that is equivalent to a 74138 standard IC chip.

AHDL An AHDL solution is shown in Fig. 17-4. The three enable inputs have been concatenated together and named enable. The IF statement tests for the inputs needed to enable the decoder. When the decoder is enabled, the CASE statement will specify the appropriate 8-bit output for each of the 8 possible data input conditions. The simulation for this design is shown in Fig. 17-5.

VHDL A VHDL solution is shown in Fig. 17-6. The three enable inputs have been concatenated together and named enable. The IF statement tests for the inputs needed to enable the decoder. When the decoder is enabled, the CASE statement will specify the appropriate 8-bit output for each of the 8 possible data input conditions. The simulation for this design is shown in Fig. 17-5.

```
SUBDESIGN  octal_decoder
(
      c, b, a, g1, g2a, g2b          :INPUT;
      y[0..7]                        :OUTPUT;
)
VARIABLE
      enable[1..3]           :NODE;
      bin[2..0]              :NODE;
BEGIN
      bin[] = (c, b, a);             -- concatenate data in
      enable[] = (g1, g2a, g2b);     -- concatenate enables
      IF enable[] == B"100"   THEN   -- enabled
            CASE bin[] IS
                  WHEN B"000" =>     y[] = B"01111111";
                  WHEN B"001" =>     y[] = B"10111111";
                  WHEN B"010" =>     y[] = B"11011111";
                  WHEN B"011" =>     y[] = B"11101111";
                  WHEN B"100" =>     y[] = B"11110111";
                  WHEN B"101" =>     y[] = B"11111011";
                  WHEN B"110" =>     y[] = B"11111101";
                  WHEN B"111" =>     y[] = B"11111110";
            END CASE;
      ELSE                           y[] = B"11111111";
                                     -- disabled
      END IF;END;
```

Fig. 17-4 AHDL solution for Example 17-1

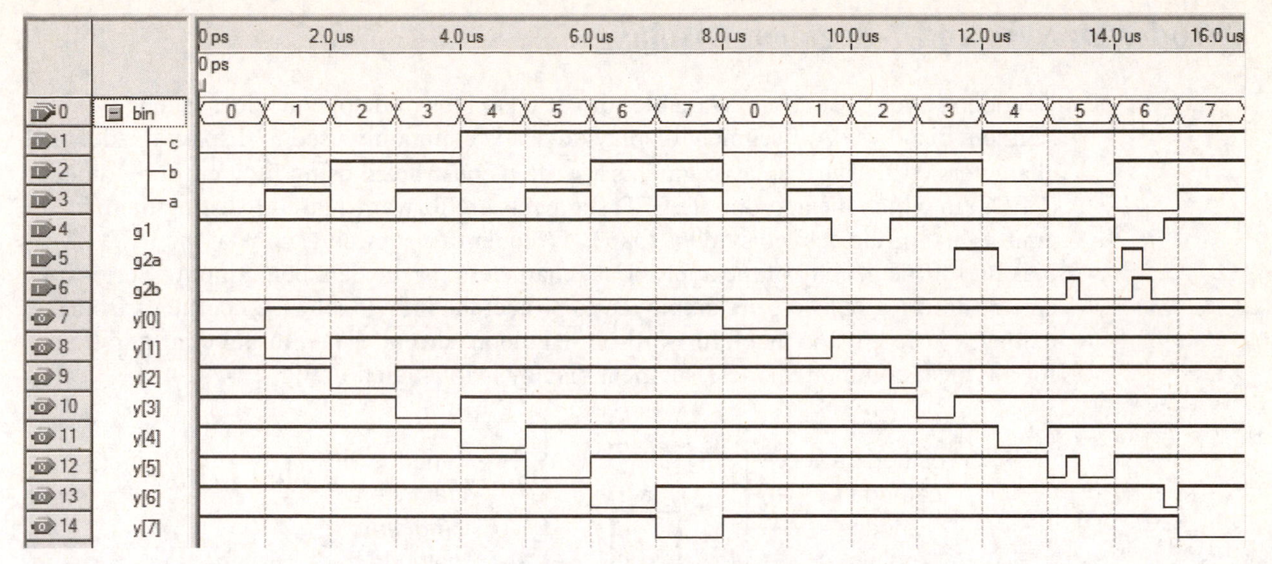

Fig. 17-5 Simulation results for Example 17-1

```
ENTITY  octal_decoder  IS
PORT (
   c, b, a, g1, g2a, g2b          :IN BIT;
   y                              :OUT BIT_VECTOR (0 TO 7)
);
END octal_decoder;

ARCHITECTURE vhdl OF octal_decoder IS
BEGIN
PROCESS (c, b, a, g1, g2a, g2b)
VARIABLE enable             :BIT_VECTOR (1 TO 3);
VARIABLE bin                :BIT_VECTOR (2 DOWNTO 0);
BEGIN
   bin := (c & b & a);                 -- concatenate data in
   enable := (g1 & g2a & g2b);         -- concatenate enables
   IF (enable = "100")  THEN           -- enabled
      CASE bin IS
         WHEN "000"     =>     y <= B"01111111";
         WHEN "001"     =>     y <= B"10111111";
         WHEN "010"     =>     y <= B"11011111";
         WHEN "011"     =>     y <= B"11101111";
         WHEN "100"     =>     y <= B"11110111";
         WHEN "101"     =>     y <= B"11111011";
         WHEN "110"     =>     y <= B"11111101";
         WHEN "111"     =>     y <= B"11111110";
      END CASE;
   ELSE                                y <= B"11111111";
   END IF;                                  -- disabled
END PROCESS;
END vhdl;
```

Fig. 17-6 VHDL solution for Example 17-1

Decoder/Drivers and 7-Segment Displays

Decoder/driver circuits are available for various kinds of display devices, such as 7-segment displays. A 7-segment display device is commonly used to display the decimal characters 0–9. The display segments are often constructed using light emitting diodes (LEDs) in which the appropriate LED segments are forward-biased (causing them to emit light) for the desired symbol shape. Decoder/driver circuits control the LED biasing for the display of the appropriate characters for the data being input. Series current-limiting resistors are employed to protect the individual LED segments from damage caused by too much forward-biased diode current. The pin-out configuration for a typical common-anode 7-segment display is illustrated in Fig. 17-7.

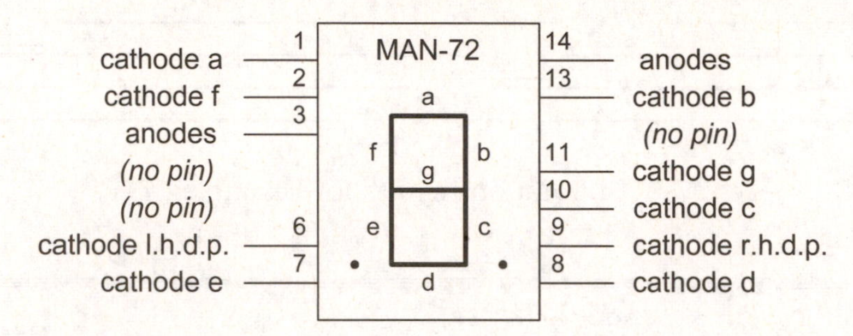

Fig. 17-7 MAN72 (or equivalent) 7-segment common-anode display (top view)

Standard BCD-to-7-segment decoder/driver chips are available to provide the necessary biasing signals for a 7-segment LED display device to produce the decimal characters 0 through 9. Since common-anode and common-cathode 7-segment devices are available, an appropriate decoder/driver must be selected to match the display type. The 7447 shown in Fig. 17-8 is designed to be used with a common-anode type device. The BCD input is applied to DCBA (A = lsb). The segment driver outputs (labeled a through g) are active-LOW because they will be connected to the LED's cathode pin on the corresponding segment. Note that a series resistor is needed for each segment of the display to limit the amount of LED current to a safe level. The active-LOW lamp test (LT) input is used to turn on all 7 segments to check if any are burned out. The active-LOW ripple-blanking input (RBI) is used to blank (turn off) the display if the BCD input is 0000. Both of these functions have been disabled in this schematic by tying them HIGH. The blanking input/ripple-blanking output pin (BI/RBO) can be used as either an input or output control. If a LOW is applied to the pin, the display will be blanked regardless of the BCD input value. Used as an output, this pin will allow all leading zeros in a multi-digit display to be blanked.

Quartus also has an equivalent 7447 function available, although the maxplus2 version has separate BI input and RBO output ports.

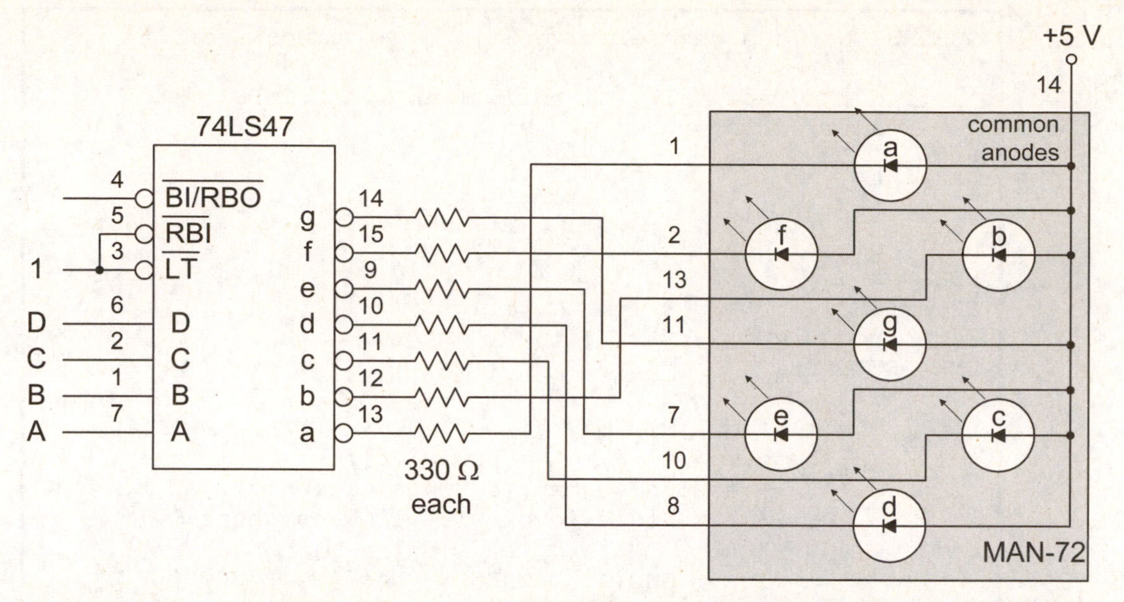

Fig. 17-8 Standard 7447 decoder/driver and 7-segment LED display circuit

Example 17-2

Design a 7-segment, common-anode, decoder/driver function using an HDL language. The circuit should have the features that are found in a standard 7447 chip (lamp test, blanking, ripple-blanking input, and ripple-blanking output). The 7447 produces a LOW output that drives the appropriate LED segment cathodes to create the desired decimal characters. Consult a data sheet for the function table details on a 7447 chip.

We will split the BI/RBO input/output pin found on the 7447 into two separate signals (blank and rbo) to design the equivalent block in a PLD. A standard 7447 will produce non-numeric characters for any data input that is not a valid BCD number. Instead, we will blank the 7-segment display for any invalid BCD input. The BCD data inputs will be named di ci bi ai and concatenated together using a variable named bin[3..0]. The 7-segment outputs will be grouped together using a variable name seg[0..6].

AHDL An AHDL solution is shown in Fig. 17-9. The IF statement tests for the higher priority, active-LOW lt, blank, and rbi inputs first. The lt control will turn on all segments by outputting a LOW for each LED segment's cathode. The blank control (equivalent to BI on the 7447) will turn off all segments by sending a HIGH to each cathode. The rbi control will blank the display only when the BCD data input is equal to 0000. When all of the control inputs are disabled, the CASE statement will specify the appropriate segment pattern output for each of the valid BCD data input conditions. Invalid BCD inputs will result in a blank display. The simulation for this design is shown in Fig. 17-10.

```
SUBDESIGN  7seg_decoderdriver        -- common anode
(
     di, ci, bi, ai, lt, rbi, blank       :INPUT;
     a, b, c, d, e, f, g, rbo             :OUTPUT;
)
VARIABLE
     seg[0..6]                  :NODE;
     bin[3..0]                  :NODE;
BEGIN
DEFAULTS
     rbo = VCC;
END DEFAULTS;

     bin[] = (di, ci, bi, ai);    -- concatenate data in
     (a, b, c, d, e, f, g) = seg[0..6];
          -- assign segment outputs to individual ports
          -- low logic level will light segment

     IF    (lt == GND)      THEN    seg[] = B"0000000";
     ELSIF (blank == GND)   THEN    seg[] = B"1111111";
     ELSIF (rbi == GND & bin[] == 0)  THEN  rbo = GND;
                                    seg[] = B"1111111";
     ELSE
          CASE bin[] IS      -- segments     abcdefg
             WHEN 0          =>    seg[] = B"0000001";
             WHEN 1          =>    seg[] = B"1001111";
             WHEN 2          =>    seg[] = B"0010010";
             WHEN 3          =>    seg[] = B"0000110";
             WHEN 4          =>    seg[] = B"1001100";
             WHEN 5          =>    seg[] = B"0100100";
             WHEN 6          =>    seg[] = B"1100000";
             WHEN 7          =>    seg[] = B"0001111";
             WHEN 8          =>    seg[] = B"0000000";
             WHEN 9          =>    seg[] = B"0001100";
             WHEN OTHERS     =>    seg[] = B"1111111";
          END CASE;
     END IF;
END;
```

AHDL

Fig. 17-9 AHDL solution for Example 17-2

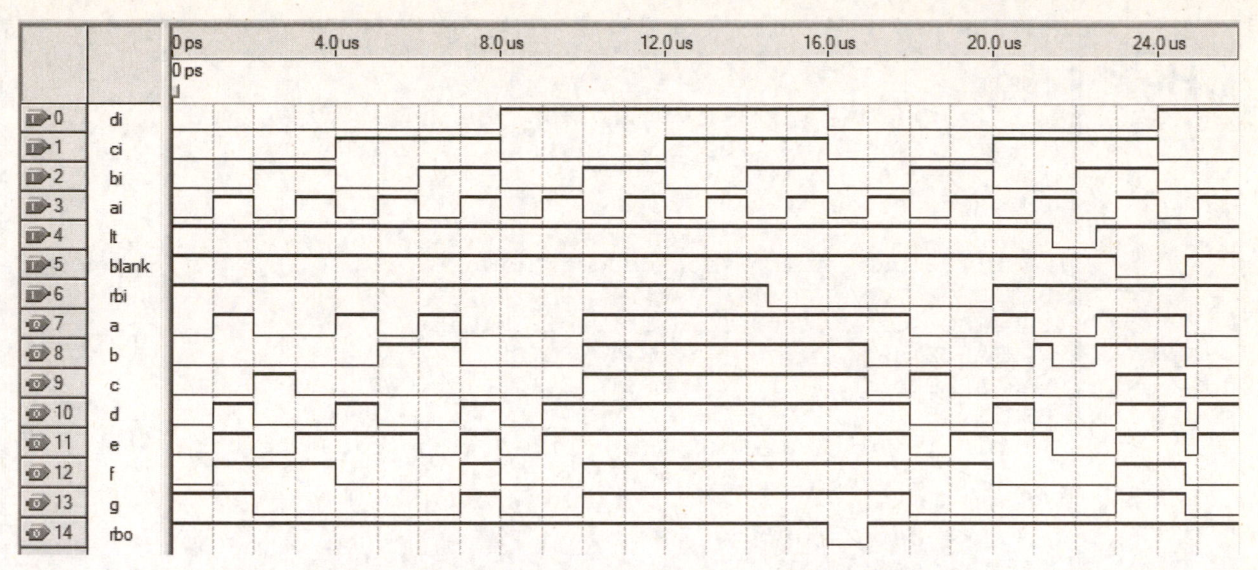

Fig. 17-10 Simulation results for Example 17-2

VHDL

A VHDL solution is shown in Fig. 17-11. The IF statement tests for the higher priority, active-LOW lt, blank, and rbi inputs first. The lt control will turn on all segments by outputting a LOW for each LED segment's cathode. The blank control (equivalent to BI on the 7447) will turn off all segments by sending a HIGH to each cathode. The rbi control will blank the display only when the BCD data input is equal to 0000. When all of the control inputs are disabled, the CASE statement will specify the appropriate segment pattern output for each of the valid BCD data input conditions. Invalid BCD inputs will result in a blank display. The simulation for this design is shown in Fig. 17-10.

```
ENTITY  seg_decoderdriver  IS                  -- common anode
PORT (
   di, ci, bi, ai, lt, rbi, blank      :IN BIT;
   a, b, c, d, e, f, g, rbo            :OUT BIT
);
END seg_decoderdriver;

ARCHITECTURE vhdl OF seg_decoderdriver IS
BEGIN
PROCESS (di, ci, bi, ai, lt, rbi, blank)
VARIABLE seg            :BIT_VECTOR (0 TO 6);
VARIABLE bin            :BIT_VECTOR (3 DOWNTO 0);
BEGIN
   bin := (di & ci & bi & ai);      -- concatenate data in
   a <= seg(0);
   b <= seg(1);
   c <= seg(2);
   d <= seg(3);
   e <= seg(4);
   f <= seg(5);
   g <= seg(6);

   IF    (lt = '0')      THEN  seg := "0000000"; rbo <= '1';
   ELSIF (blank = '0')   THEN  seg := "1111111"; rbo <= '1';
   ELSIF (rbi = '0' AND bin = "0000")
                         THEN  seg := "1111111"; rbo <= '0';
   ELSE
        CASE bin IS      -- segments   abcdefg
             WHEN "0000"   =>   seg := "0000001"; rbo <= '1';
             WHEN "0001"   =>   seg := "1001111"; rbo <= '1';
             WHEN "0010"   =>   seg := "0010010"; rbo <= '1';
             WHEN "0011"   =>   seg := "0000110"; rbo <= '1';
             WHEN "0100"   =>   seg := "1001100"; rbo <= '1';
             WHEN "0101"   =>   seg := "0100100"; rbo <= '1';
             WHEN "0110"   =>   seg := "1100000"; rbo <= '1';
             WHEN "0111"   =>   seg := "0001111"; rbo <= '1';
             WHEN "1000"   =>   seg := "0000000"; rbo <= '1';
             WHEN "1001"   =>   seg := "0001100"; rbo <= '1';
             WHEN OTHERS =>     seg := "1111111"; rbo <= '1';
        END CASE;
   END IF;
END PROCESS;
END vhdl;
```

Fig. 17-11 VHDL solution for Example 17-2

Example 17-3

Design a memory chip decoder using a PLD. The decoder should produce individual chip select signals (cs6 through cs0) for seven different memory chips. The decoder's function table and specific address range for each chip is given in Table 17-2. The 16-bit addresses (a15 through a0) are given in hexadecimal in the table.

Address Range (hexadecimal)	Chip Selects cs6	cs5	cs4	cs3	cs2	cs1	cs0
0000–07FF	1	1	1	1	1	1	0
0800–0FFF	1	1	1	1	1	0	1
1000–17FF	1	1	1	1	0	1	1
1800–1FFF	1	1	1	0	1	1	1
8000–8FFF	1	1	0	1	1	1	1
A000–AFFF	1	0	1	1	1	1	1
C000–FFFF	0	1	1	1	1	1	1

Table 17-2 Decoder function table for Example 17-3

AHDL An AHDL solution is given in Fig. 17-12. An IF/THEN statement is used to specify the appropriate hexadecimal address range that will activate (with a LOW output) the desired chip select output signal. The compiler will automatically determine which of the address input bits need to be decoded to distinguish the address ranges from each other.

```
SUBDESIGN  mem_decoder
(
        a[15..0]     :INPUT;
        cs[6..0]     :OUTPUT;
)
BEGIN
IF    a[] >= H"0000" & a[] <= H"07FF" THEN cs[] = B"1111110";
ELSIF a[] >= H"0800" & a[] <= H"0FFF" THEN cs[] = B"1111101";
ELSIF a[] >= H"1000" & a[] <= H"17FF" THEN cs[] = B"1111011";
ELSIF a[] >= H"1800" & a[] <= H"1FFF" THEN cs[] = B"1110111";
ELSIF a[] >= H"8000" & a[] <= H"8FFF" THEN cs[] = B"1101111";
ELSIF a[] >= H"A000" & a[] <= H"AFFF" THEN cs[] = B"1011111";
ELSIF a[] >= H"C000" & a[] <= H"FFFF" THEN cs[] = B"0111111";
ELSE                                       cs[] = B"1111111";
END IF;
END;;
```

Fig. 17-12 AHDL solution for Example 17-3

VHDL A VHDL solution is given in Fig. 17-13. The 16-bit address port is declared to be an INTEGER data type with a value range of 0 through 65535 ($2^{16} - 1$). The 7-bit chip select output port is a BIT_VECTOR data type. Each chip select output signal will be LOW when the appropriate address range is applied to the decoder. An IF/THEN statement is used in the PROCESS to determine the value of a and specify the appropriate output for cs. The integer values for each of the given address ranges is given in hexadecimal. A hexadecimal integer value is specified by preceding the value with the base (16) and enclosing it with a pair of # characters. The compiler will automatically determine which of the address input bits need to be decoded to distinguish the address ranges from each other.

```
ENTITY  mem_decoder  IS
PORT (    a           :IN INTEGER RANGE 0 TO 65535;
          cs          :OUT BIT_VECTOR (6 DOWNTO 0) );
END mem_decoder;

ARCHITECTURE decoder OF mem_decoder IS
BEGIN
   PROCESS (a)
   BEGIN
            -- determine input address range in hex
      IF    a >= 16#0000# AND a <= 16#07FF#  THEN  cs <= "1111110";
      ELSIF a >= 16#0800# AND a <= 16#0FFF#  THEN  cs <= "1111101";
      ELSIF a >= 16#1000# AND a <= 16#17FF#  THEN  cs <= "1111011";
      ELSIF a >= 16#1800# AND a <= 16#1FFF#  THEN  cs <= "1110111";
      ELSIF a >= 16#8000# AND a <= 16#8FFF#  THEN  cs <= "1101111";
      ELSIF a >= 16#A000# AND a <= 16#AFFF#  THEN  cs <= "1011111";
      ELSIF a >= 16#C000# AND a <= 16#FFFF#  THEN  cs <= "0111111";
      ELSE                                         cs <= "1111111";
      END IF;
   END PROCESS;
END decoder;
```

Fig. 17-13 VHDL solution for Example 17-3

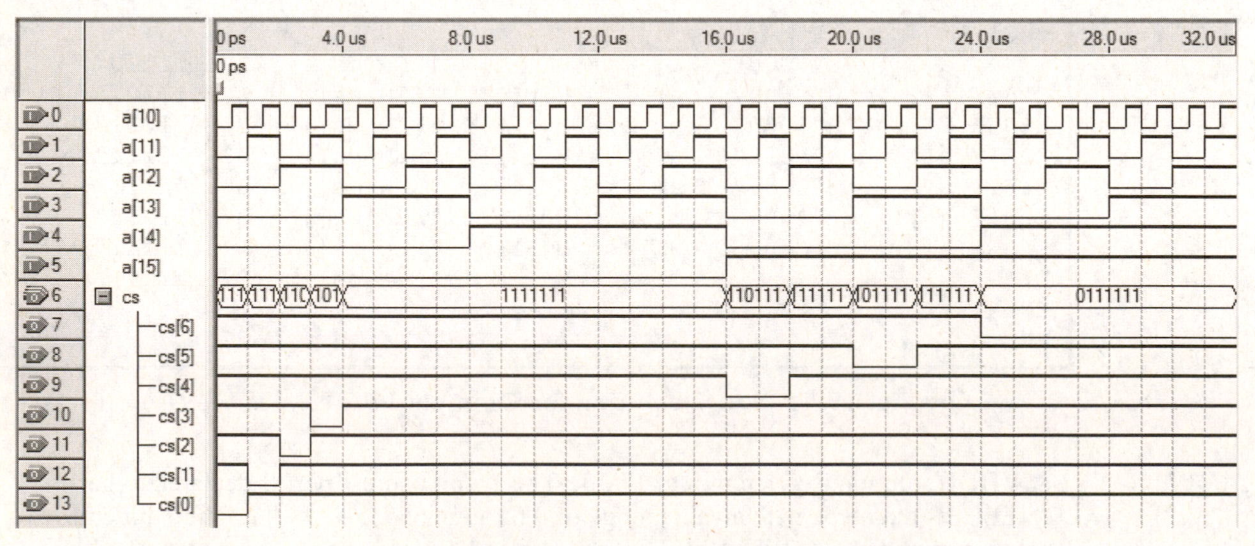

Fig. 17-14 Simulation results for Example 17-3

The simulation results for either HDL solution are shown in Fig. 17-14. The compiler will produce a warning that states "Design contains 11 input pin(s) that do not drive logic" indicating that inputs a10 through a0 do not matter in determining the circuit's output. Inputs a15 through a11, which control the decoder's function, plus one input (a10) that does not affect the output have been included in the simulation. Note in the simulation that a10 does not matter.

Example 17-4

Design a binary-to-BCD-code converter using a PLD. The code converter will have a 5-bit input (00000_2 through 11111_2) and a 2-digit output (00 through 31_{10} in BCD).

AHDL

An AHDL solution is given in Fig. 17-15. An IF/THEN statement determines if the binary input is greater than or equal to 30 (30 or 31), greater than or equal to 20 (20-29), greater than or equal to 10 (10-19), or otherwise (0-9) and then assigns the appropriate BCD value for most significant digit (msd). Because an IF statement automatically establishes priority, there will be no overlap of tested values. The BCD value for the least significant digit (lsd) can be calculated by subtracting the appropriate decade range from the binary input value. The only "trick" to this calculation is that AHDL must have the same number of bits on each side of the equals sign. Because the binary input (bin[4..0]) is 5 bits, we have created a 5-bit variable named ones[4..0] that will be assigned the calculation's result. The largest difference will be nine, so we can assign the least significant 4 bits of ones[] to the output port lsd[3..0]. The simulation results are shown in Fig. 17-16.

```
SUBDESIGN  bin2bcd
     (bin[4..0]                :INPUT;
      msd[1..0], lsd[3..0]     :OUTPUT;)
VARIABLE
      ones[4..0]               :NODE;
                  -- node to capture the ones digit
BEGIN
      lsd[3..0] = ones[3..0];        -- output LSD digit
      IF bin[] >= 30    THEN
            msd[] = 3;               -- set MSD value
            ones[] = bin[] - 30;     -- calculate LSD value
      ELSIF bin[] >= 20 THEN
            msd[] = 2;
            ones[] = bin[] - 20;
      ELSIF bin[] >= 10 THEN
            msd[] = 1;
            ones[] = bin[] - 10;
      ELSE
            msd[] = 0;
            ones[] = bin[];
      END IF;
END;
```

Fig. 17-15 AHDL solution for Example 17-4

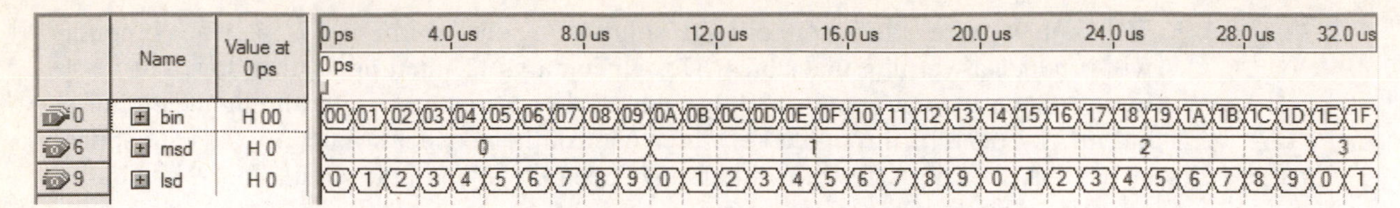

Fig. 17-16 Simulation results for Example 17-4

```
ENTITY  bin2bcd  IS
PORT ( bin          :IN  INTEGER RANGE 0 TO 31;
       msd          :OUT INTEGER RANGE 0 TO 3;
       lsd          :OUT INTEGER RANGE 0 TO 9 );
END bin2bcd;

ARCHITECTURE  vhdl  OF  bin2bcd  is
BEGIN
   PROCESS  (bin)
   VARIABLE tens   :INTEGER RANGE 0 TO 3;
   BEGIN
            -- determine most significant digit
      IF     bin >= 30        THEN  tens := 3;
      ELSIF  bin >= 20        THEN  tens := 2;
      ELSIF  bin >= 10        THEN  tens := 1;
      ELSE                          tens := 0;
      END IF;
      msd <= tens;
            -- determine least significant digit
      lsd <= bin - 10 * tens;
   END PROCESS;
END vhdl;
```

Fig. 17-17 VHDL solution for Example 17-4

VHDL A VHDL solution is given in Fig. 17-17. The input and output ports are declared to be an INTEGER data type. An IF/THEN statement determines if the binary input is greater than or equal to 30 (30 or 31), greater than or equal to 20 (20-29), greater than or equal to 10 (10-19), or otherwise (0-9) and then assigns the appropriate BCD value for most significant digit to a local variable named tens. Because an IF statement automatically establishes priority, there will be no overlap of tested values. The output port msd will be assigned the value of tens. The BCD value for the least significant digit (lsd) can be calculated by subtracting the appropriate decade range from the binary input value. The simulation results are shown in Fig. 17-16.

Schematic An interesting alternative schematic solution is shown in Fig. 17-18. A 74185 binary-to-bcd converter maxplus2 function is used in the design file. This function can handle up to a 6-bit binary input. The input labeled GN is an active-LOW enable. The additional input bit is applied to the E input, and another output bit would be produced at Y6. Using multiple 74185 functions can create larger binary-to-BCD converters, but the circuit arrangement is not obvious. For example, Fig. 17-19 shows an 8-bit binary-to-BCD converter.

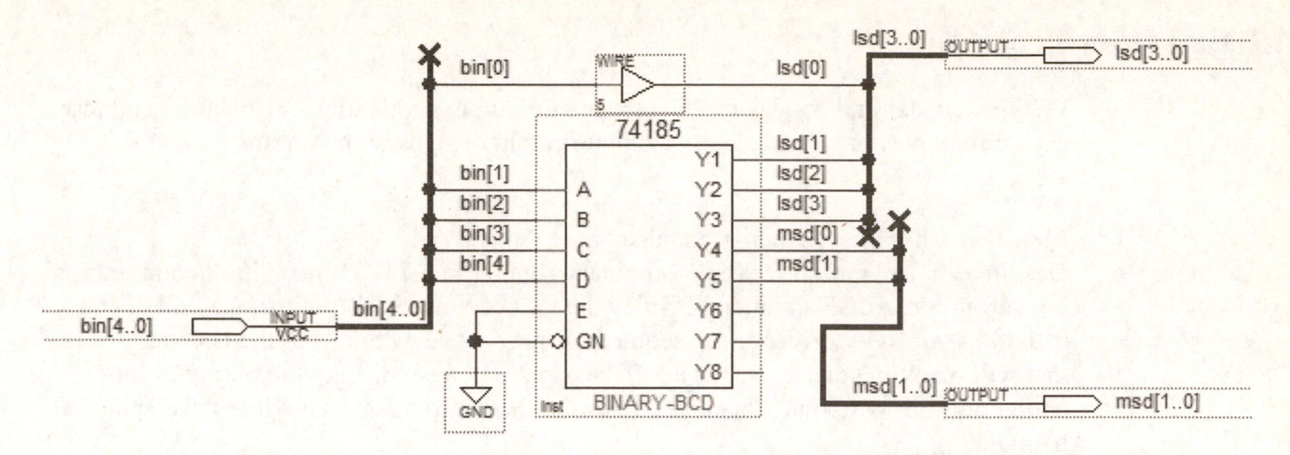

Fig. 17-18 Schematic solution using the 74185 maxplus2 function for Example 17-4

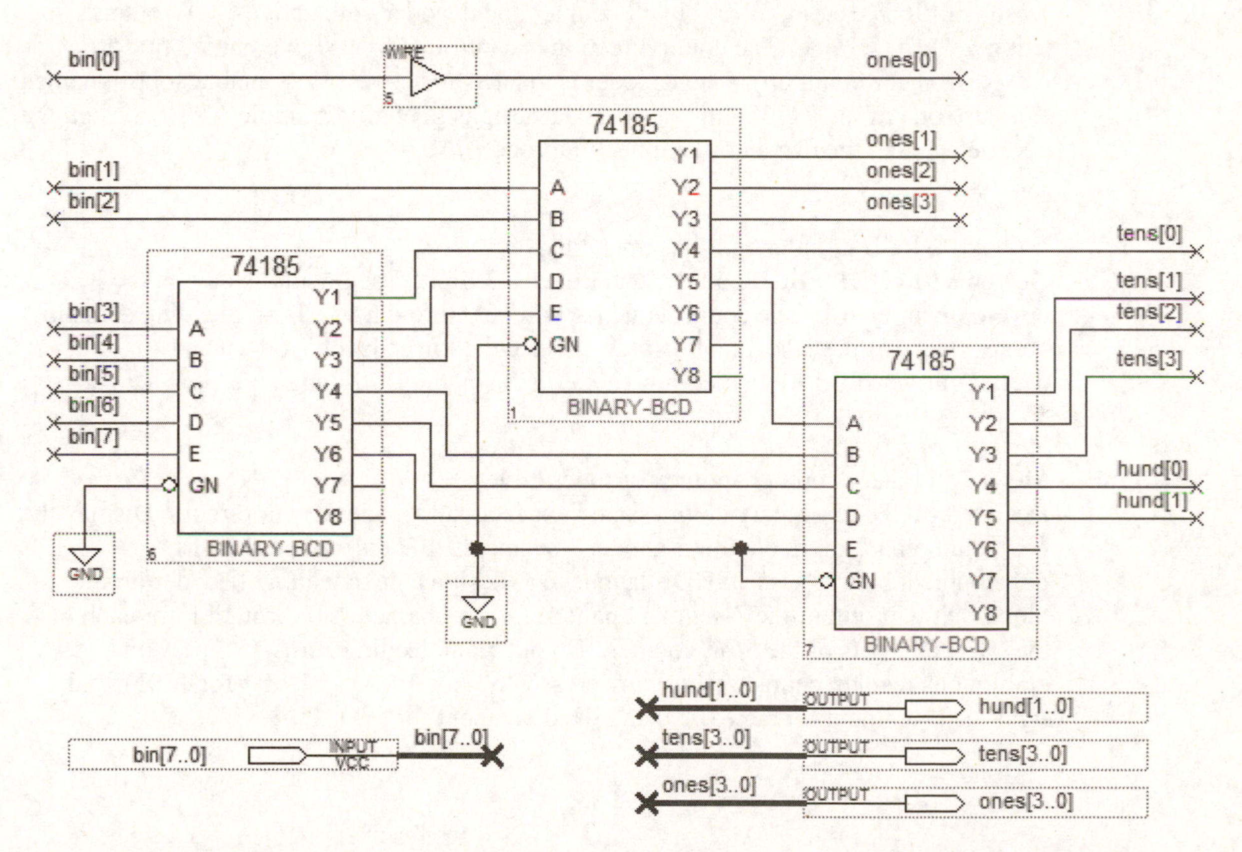

Fig. 17-19 An 8-bit binary-to-BCD converter using the 74185 maxplus2 function

Laboratory Projects

Design decoder and display circuits for the following applications. Construct and test each circuit design. Functionally simulate circuit designs with Quartus.

17.1 Modified 4-bit binary counter and display
Design a control circuit for the 4-bit binary counter in a 74163 maxplus2 counter to convert it into a BCD counter. Display the BCD counter's output using a 7447 maxplus2 decoder/driver and 7-segment display. Add a control to the decoder/driver that will blank the display when a LOW is applied. Also display the output of the counter on LEDs so that you can monitor the count progress even while the display is blanked.

17.2 MOD-16 counter and decoder
Design a MOD-16, up/down, binary counter and decoder with active-LOW outputs using a CPLD/FPGA. The count direction is controlled by a signal named up_dn (up_dn = 1 for count up). The decoder is to have an active-LOW enable control named en. Use one of the 74138-equivalent HDL designs given in Example 17-1 to design the decoder. Label the decoder's outputs y0 through y15.

17.3 MOD-100 BCD counter and 7-segment display
Design a MOD-100 BCD counter that drives a 2-digit, 7-segment display on the PLD development board. The count sequence will be 0 through 99. Blank leading zeros in the tens digit of the 7-segment display. Create the lower-level Mod-10 and decoder/driver blocks in a hierarchical design using an HDL language.

17.4 MOD-256 binary counter and hex decoder/driver
Design a MOD-256 binary counter and hexadecimal decoder/driver circuit. Display the hex count on a 2-digit, common-anode, 7-segment LED display on the PLD development board. Use an HDL language to design a hexadecimal decoder/driver block that will output the 7-segment pattern for the characters 0 through F for each of the two hex digits of the 8-bit counter. Do not blank leading zeros. To prevent ambiguous display characters, use lowercase "**b**" (for 1011) and "**d**" (for 1101), and light the a-segment for a **6** (0110) and the d-segment for a **9** (1001).

17.5 BCD decoder

Use an HDL language to design a BCD decoder circuit with active-HIGH outputs (to drive individual LEDs). The decoder should also detect any invalid BCD input that is applied to the decoder and produce an active-HIGH output signal named err.

17.6 MOD-64 counter with binary-to-BCD converter and display

Design a MOD-64 binary counter and binary-to-BCD converter using an FPGA/CPLD. The count sequence will be 00 through 63 and will be displayed in binary (on LEDs) and decimal (on a 2-digit, 7-segment display). The binary-to-BCD converter will produce the equivalent BCD value for the 6-bit binary count. Do not blank leading zeros on the 7-segment display.

17.7 Memory decoder

Design a memory chip decoder using an FPGA/CPLD. The decoder should produce individual chip select signals for six memory chips. The decoder should also be enabled by either of two active-LOW, memory-control strobe signals (memr and memw). The memory should be disabled if the two strobe signals are active simultaneously. The specific address range to be decoded for each memory chip is given below. The 16-bit addresses are given in hexadecimal in the table. Hint: Create a "buried" node whose logic equation describes the function of the memory-control signals.

Address Range	Chip Selects					
(hexadecimal)	CS5	CS4	CS3	CS2	CS1	CS0
0000–03FF	1	1	1	1	1	0
0400–07FF	1	1	1	1	0	1
0800–0BFF	1	1	1	0	1	1
0C00–0FFF	1	1	0	1	1	1
E800–EBFF	1	0	1	1	1	1
EC00–EFFF	0	1	1	1	1	1

17.8 State decoder

Design a logic circuit using an FPGA/CPLD that will decode the 9 specified states given in the following truth table. The inputs to the decoder are in binary and represent 0 to 99_{10}. The outputs are labeled y[9..1].

State (decimal)	y9	y8	y7	y6	y5	y4	y3	y2	y1
10	0	0	0	0	0	0	0	0	1
20	0	0	0	0	0	0	0	1	0
30	0	0	0	0	0	0	1	0	0
40	0	0	0	0	0	1	0	0	0
50	0	0	0	0	1	0	0	0	0
60	0	0	0	1	0	0	0	0	0
70	0	0	1	0	0	0	0	0	0
80	0	1	0	0	0	0	0	0	0
90	1	0	0	0	0	0	0	0	0

UNIT 18

ENCODER APPLICATIONS

Objective

- To apply encoder circuits in digital system applications.

Encoder Circuits

Input data to a logic system often comes from switches. This switch input information must be encoded into a representative binary form that can be processed by the system. Encoders have several data lines for the inputs that are being applied. The encoded output will be a binary number that will be used by the digital system to represent the specific input condition. Simple types of encoders can handle only one active input at a time. A priority encoder has an established order of precedence for the input lines so that the input with the highest priority will be encoded when multiple inputs are active simultaneously.

In Fig. 18-1, the eight pushbutton switches are encoded using a 74148, a standard IC encoder chip and maxplus2 function. Each of the switch inputs to the 74148 (labeled 0 through 7) are active-LOW. Every SPST switch must be connected to a pull-up resistor to provide a logic one input to the 74148 when the normally open switch is not pressed. Remember that an open input to any logic device is not desirable. The 74148 is a priority encoder, so the greatest input value will be encoded if multiple inputs are active simultaneously. The truth table for the 74148 is shown in Table 18-1. The inverters have been added to the outputs from the 74148 because the encoded output that is produced by this chip happens to be the inverted binary value. The enable input (EI) and enable output (EO) pins can be used to encode more than 8 switches by cascading multiple chips together. The encoder chip has been permanently enabled with EI tied LOW in the schematic of Fig. 18-1. The active-LOW strobe output pin (GS) will indicate that an input has been applied and the encoded result is available at the output.

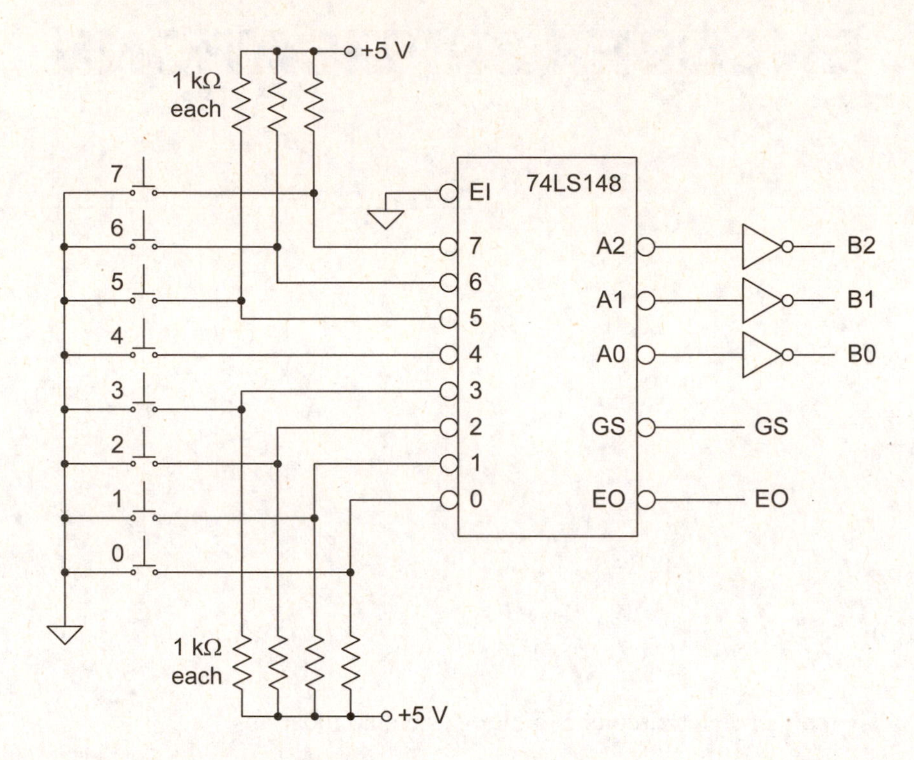

Fig. 18-1 Encoding pushbuttons with a 74148, 8-line-to-3-line priority encoder

EI	0	1	2	3	4	5	6	7	EO	GS	A2	A1	A0
1	X	X	X	X	X	X	X	X	1	1	1	1	1
0	1	1	1	1	1	1	1	1	0	1	1	1	1
0	0	1	1	1	1	1	1	1	1	0	1	1	1
0	X	0	1	1	1	1	1	1	1	0	1	1	0
0	X	X	0	1	1	1	1	1	1	0	1	0	1
0	X	X	X	0	1	1	1	1	1	0	1	0	0
0	X	X	X	X	0	1	1	1	1	0	0	1	1
0	X	X	X	X	X	0	1	1	1	0	0	1	0
0	X	X	X	X	X	X	0	1	1	0	0	0	1
0	X	X	X	X	X	X	X	0	1	0	0	0	0

Table 18-1 Truth table for a 74148 priority encoder

Example 18-1

Design a 16-input priority encoder circuit using the 74148 maxplus2 function in a PLD. The encoder is to be enabled by an active-LOW signal named en. The encoder should also produce an active-HIGH output signal (called flag) to indicate that one or more of the inputs is active.

A solution is shown in Fig. 18-2. There are 16 active-LOW inputs that will be encoded into the corresponding binary output value (out[3..0]) when en is LOW. If any of the higher eight inputs (sw15 through sw8) is active, the resulting HIGH output for out3 (produced by the EON output from the 74184 that handles inputs sw15 through sw8) will disable the other 74148, which handles the encoding for the lower eight inputs. The active-LOW inputs on the ORs (equivalent to a NAND function according to DeMorgan's theorem and called the BOR primitive in Quartus) will detect if an active-LOW output occurs for either 74148. Simulation results for this design are shown in Fig. 18-3.

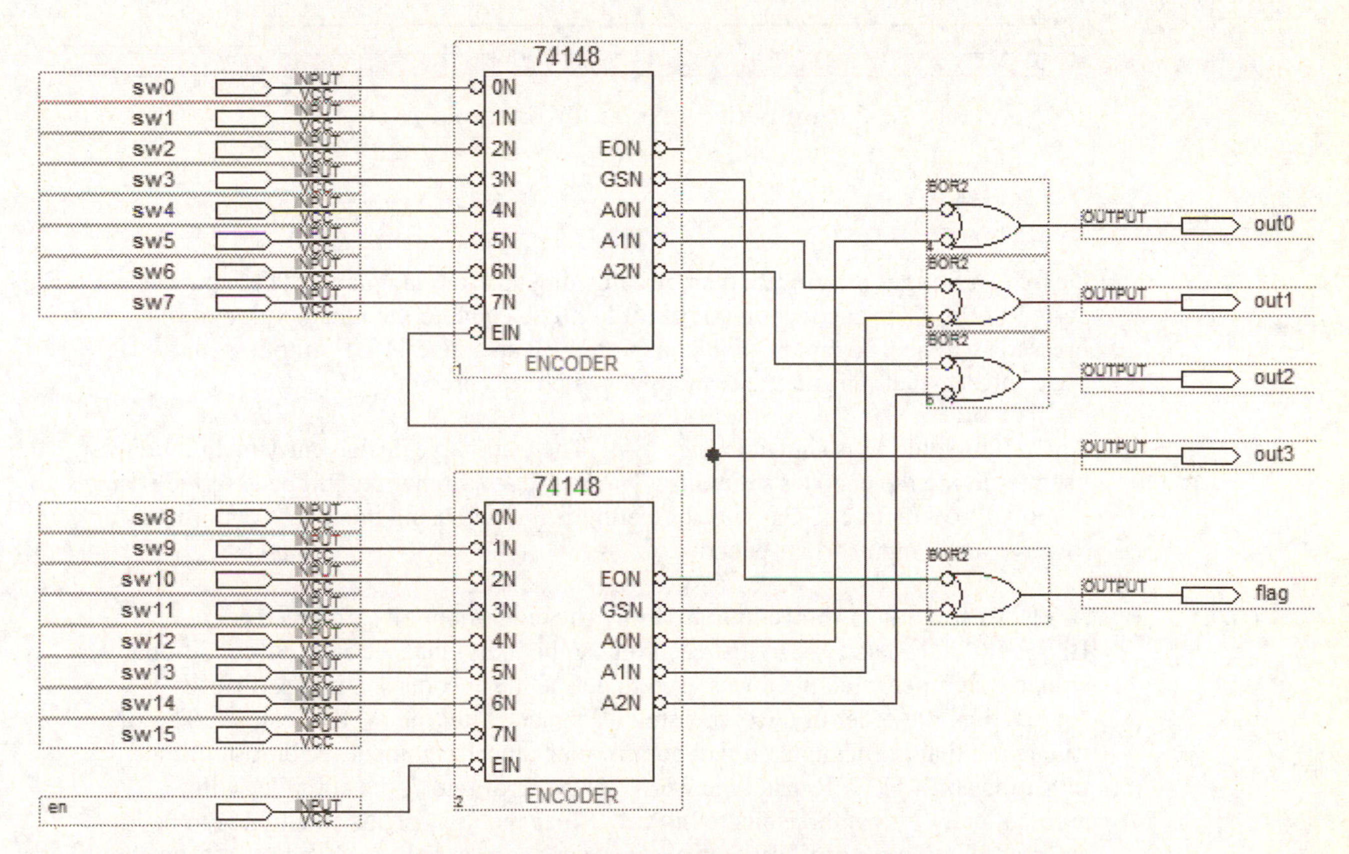

Fig. 18-2 Schematic for Example 18-1

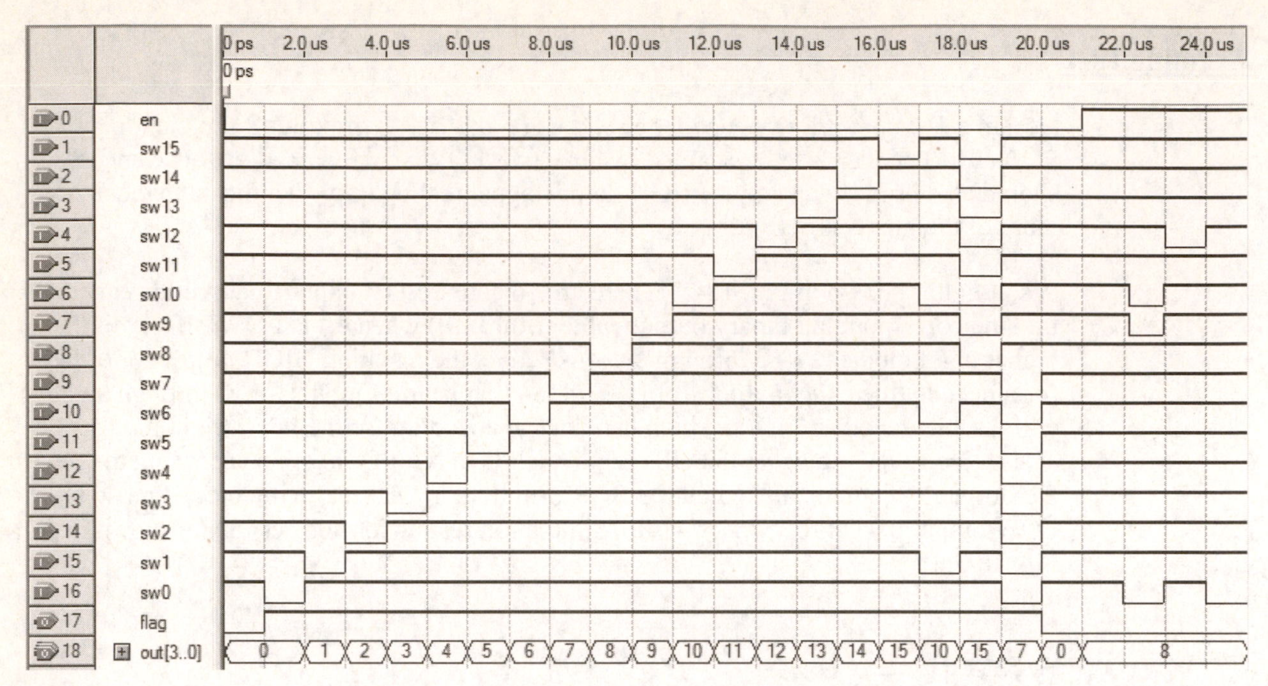

Fig. 18-3 Simulation results for Example 18-1

Example 18-2

Design an 8-input priority encoder circuit using an HDL language. The inputs are active-LOW. The encoder output binout will be equal to the binary value of the pressed switch. The circuit should also produce an active-HIGH output signal (key_press) that indicates that an input switch was pressed.

AHDL An AHDL solution is shown in Fig. 18-4. The truth table design entry technique was used to define the encoder's outputs. The key_press function will be asserted if any input is LOW (active). The priority feature is defined using don't-care conditions for lower priority inputs to the encoder.

VHDL A VHDL solution is shown in Fig. 18-5. The eight input bits are given a BIT_VECTOR data type so that each bit can be individually tested. A conditional signal assignment statement was used to define the encoder's outputs. Priority is achieved by the order of the expression tests made after the WHEN clause. The first true result that is encountered will control the value assignment for binout. Since binout is an INTEGER data type with a RANGE of 0 to 7, the compiler will automatically assign three bits to binout. Since the key_press function is to be asserted if any input is LOW (active), it is very easy to define when key_press should be HIGH (inactive). This is done with another conditional signal assignment statement.

Simulation results for either version of this design are shown in Fig. 18-6. Note that a HIGH output for key_press indicates that a switch has been pressed and the correct binary number is produced by the encoder for binout. Higher value input switches have priority.

```
SUBDESIGN  encoder8
(
        sw[7..0]                         :INPUT;
        binout[2..0], key_press          :OUTPUT;
)
BEGIN
        TABLE
                sw[]          =>   binout[], key_press;
                B"11111111"   =>   B"000", 0;   -- no input
                B"11111110"   =>   B"000", 1;
                B"1111110X"   =>   B"001", 1;
                B"111110XX"   =>   B"010", 1;
                B"11110XXX"   =>   B"011", 1;
                B"1110XXXX"   =>   B"100", 1;
                B"110XXXXX"   =>   B"101", 1;
                B"10XXXXXX"   =>   B"110", 1;
                B"0XXXXXXX"   =>   B"111", 1;
                -- listed from lowest to highest priority
        END TABLE;
END;
```

Fig. 18-4 AHDL solution for Example 18-2

```
ENTITY  encoder8  IS
PORT ( sw               :IN BIT_VECTOR (7 DOWNTO 0);
       binout           :OUT INTEGER RANGE 0 TO 7;
       key_press        :OUT BIT );
END encoder8;

ARCHITECTURE vhdl OF encoder8 IS
BEGIN
binout <=    7 WHEN sw(7) = '0' ELSE
             6 WHEN sw(6) = '0' ELSE
             5 WHEN sw(5) = '0' ELSE
             4 WHEN sw(4) = '0' ELSE
             3 WHEN sw(3) = '0' ELSE
             2 WHEN sw(2) = '0' ELSE
             1 WHEN sw(1) = '0' ELSE
             0 ;    -- listed in order of priority

key_press <= '0' WHEN sw = "11111111" ELSE '1';
END vhdl;
```

Fig. 18-5 VHDL solution for Example 18-2

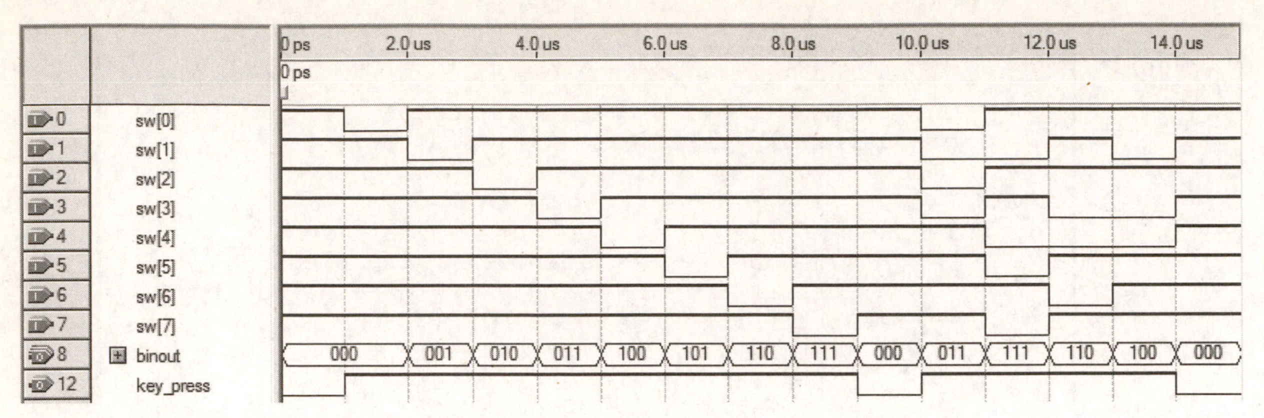

Fig. 18-6 Simulation results for Example 18-2

Scanning Encoder

A scanning encoder can be utilized when there are many switch inputs (more than is practical with cascaded 74148 function blocks) to be encoded in a digital system. Standard functional blocks such as counters, decoders, and encoders can be combined to implement a scanning encoder. The schematic in Fig. 18-7 illustrates a scanning encoder system for a matrix of 64 SPST pushbutton switches. The scanning encoder will output a 6-bit binary number for Z5 Z4 Z3 Z2 Z1 Z0 that is equivalent to a pressed switch's decimal label.

The scan counter (74163) is used as a MOD-8 binary counter that controls the repetitive scanning of the 8 columns of switches. The column scanning decoder (74138) decodes the scan counter's output to determine which switch matrix column is currently being scanned. Only one column at a time is scanned with the active-LOW output produced by the scanning decoder. A closed switch will short one of the rows to one of the columns. Each switch represents a unique combination of a row and column. A closed switch on a particular column is indicated by a LOW input to the 8-input row encoder (74148) when the appropriate column with the pressed switch is scanned with a LOW from the decoder. When the closed switch is detected by a LOW input to the encoder, the GS output (Scan Disable) of the encoder will go LOW and the EO output (DAV – data available) will go HIGH. A LOW on Scan Disable will disable the counter (and stop the scanning of the switch columns) so that the binary code for the switch may be read out as the 6 bits labeled Z5 Z4 Z3 Z2 Z1 Z0. Note that inverters are used to change the inverted binary encoder output into the standard binary value for the corresponding row.

As long as there is no pressed switch in the column that is currently being scanned, all rows will be pulled HIGH with the resistors. With all highs into the encoder, the GS output (Scan Disable) will be HIGH and the counter will be enabled. This allows the counter to step to the next state when it is clocked, and the decoder output will then change to scan the next column. Any pressed switch in a column that is not currently being scanned (with a LOW from the decoder) will have no effect on the encoder or counter because both sides of the switch will be HIGH. The counting and column scanning will proceed until a closed switch is detected in the currently scanned column.

A pressed switch will stop the counter only when that specific column is scanned. Releasing the switch will then allow the count to continue. For example, if switch "21" (located in column 5) is pressed, row 2 will go LOW when the counter reaches state 101 (Z2 Z1 Z0). The LOW on input 2 of the encoder will produce a 010 for Z5 Z4 Z3. This 6-bit number 010101 is the binary equivalent of the decimal number 21.

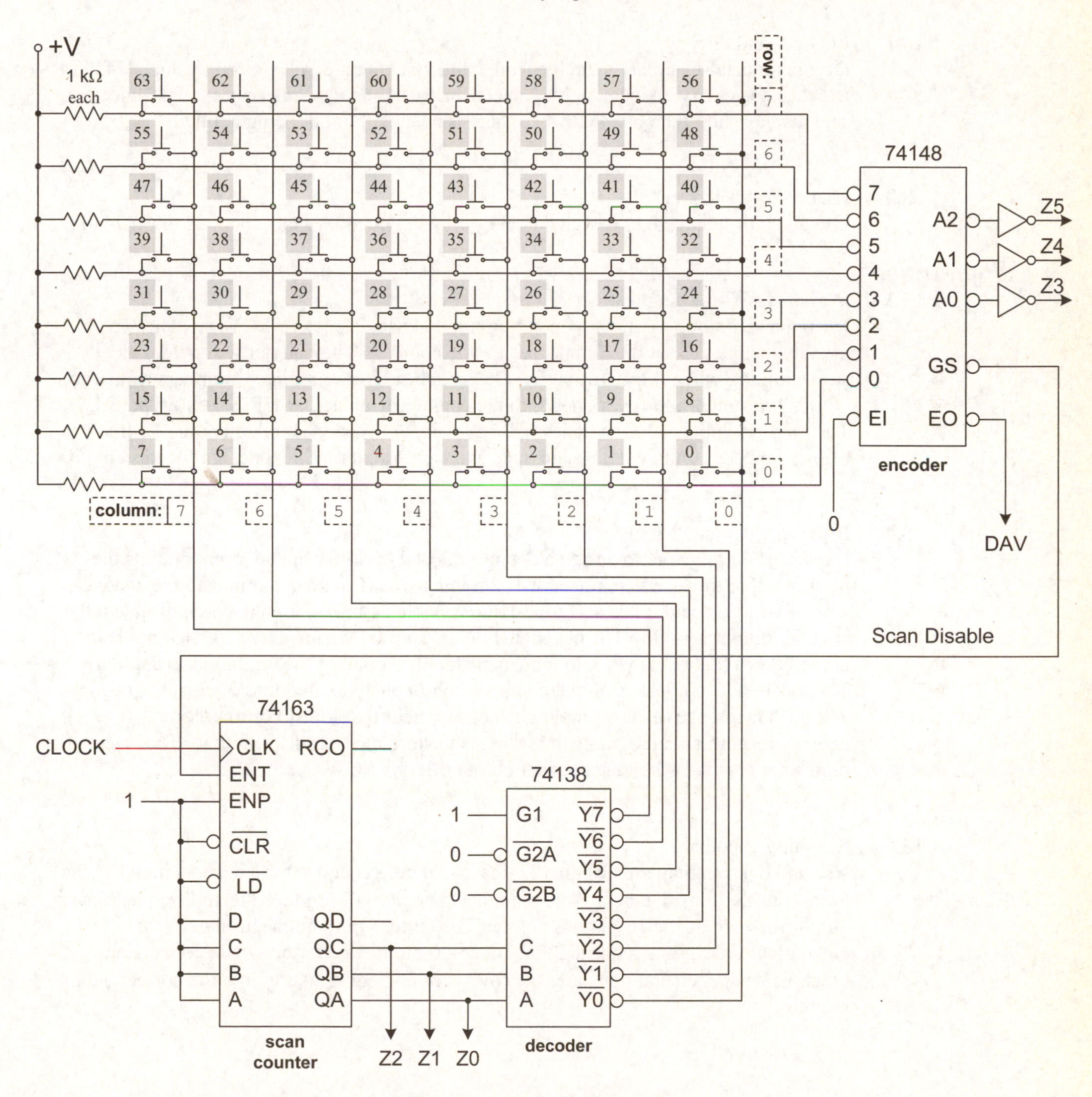

Fig. 18-7 Scanning encoder system

Laboratory Projects

Design encoder circuits for the following applications. Construct and test each circuit design. Functionally simulate circuit designs with Quartus.

18.1 Encoder and display

Design an octal-to-binary switch encoder and 7-segment display circuit using a 74148 maxplus2 priority encoder, a 7447 decoder/driver, and any other necessary devices. The display should be blanked when there is no switch input being applied.

18.2 HDL 74148 encoder

Use an HDL language to design an octal-to-binary encoder that functions like a 74148.

18.3 Variable self-stopping counter

Design a self-stopping down-counter circuit that uses 12 active-LOW switches and a priority encoder to set the initial state for the counter. The counter will automatically stop counting when it reaches 0000. The switches represent binary numbers from 1 to 12 that are synchronously loaded into the counter by an active-HIGH restart signal. The larger number value switches will have higher priority. Implement the 12-line priority encoder with an HDL language and create a hierarchical design for this circuit.

18.4 BCD encoder and 2 registers

Use an HDL language to design a decimal-to-BCD priority encoder that outputs the highest value from switch inputs sw0 through sw9. The switch inputs to the encoder are active-LOW and it has an active-HIGH enable (enable) input. When disabled, the encoder output is 0000. The encoder should also produce an active-LOW output signal named davn (data available) to indicate that at least one of the ten inputs is applied. The 8421 BCD encoder's output can be synchronously loaded into 2 registers (reg0 and reg1). An active-LOW input signal is applied to selectn0 or to selectn1 to choose the corresponding register(s) that is to store the encoder's output. The selected register is enabled when the encoder output davn is LOW.

18.5 Scanning encoder

Use an HDL language to design a 16-input scanning encoder for a hexadecimal keypad (4 × 4 pushbutton switches). The scanning encoder will produce the appropriate binary output for each of the 16 buttons. Note: The buttons may not be arranged in a convenient order, thus requiring that either a custom encoder or a code conversion block be designed (that translates the row × column combination into the correct binary number).

UNIT 19

MULTIPLEXER APPLICATIONS

Objective

- To apply multiplexer circuits in digital system applications.

Multiplexer Circuits

A multiplexer (also called a data selector) is a logic block that steers selected input data to its output. This logic function has multiple data inputs from which to choose, but only one of the inputs will appear on the output. Control signals called select lines are used to determine which data input will be routed to the output. Because of its ability to output different data inputs dependent on a set of control lines, a multiplexer can also easily be used to synthesize a combinational logic function. Multiplexers are available as standard IC chips and also as maxplus2 functions. For example, the 74151 (shown in Fig. 19-1) is a one-of-eight or 8-channel multiplexer. The truth table for this multiplexer is given in Table 19-1. Three select lines are necessary to choose one of the eight possible inputs for the data output. For added flexibility, this multiplexer has both a normal data output and an inverted data output, as well as an active-LOW enable control input. Multiplexers can also be created in Quartus using the LPM_MUX megafunction.

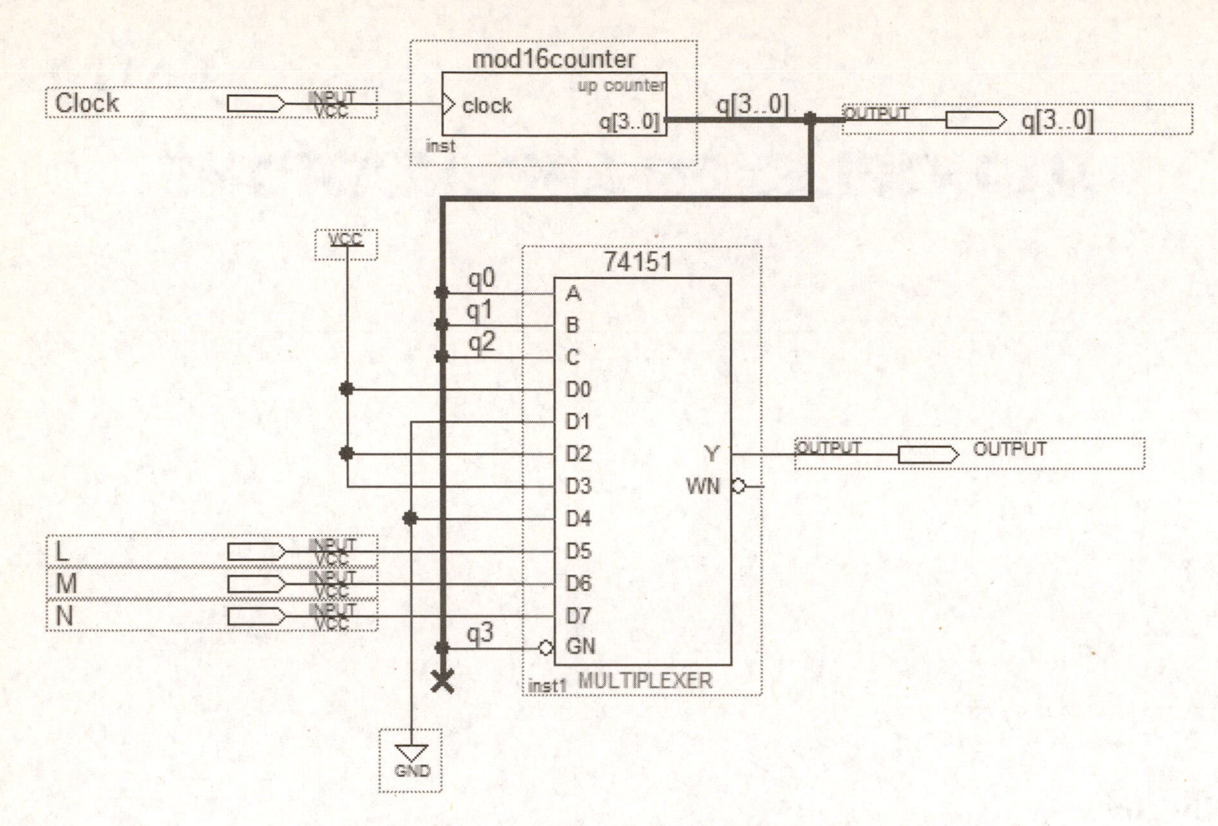

Fig. 19-1 An 8-channel multiplexer circuit

G	C	B	A	Y	W
0	0	0	0	D0	$\overline{D0}$
0	0	0	1	D1	$\overline{D1}$
0	0	1	0	D2	$\overline{D2}$
0	0	1	1	D3	$\overline{D3}$
0	1	0	0	D4	$\overline{D4}$
0	1	0	1	D5	$\overline{D5}$
0	1	1	0	D6	$\overline{D6}$
0	1	1	1	D7	D7
1	X	X	X	0	1

Table 19-1 Truth table for a 74151 8-channel multiplexer

A MOD-16 binary counter is used to automatically sequence the channel selection for the multiplexer in Figure 19-1. The multiplexer will be enabled during the count sequence 0000 through 0111 and it will route the corresponding data channel input to the MUX's output (OUTPUT). Data channel inputs D0 through D4 have fixed inputs while D5 through D7 have variable inputs (LMN, respectively). The multiplexer will be disabled during the count sequence 1000 through 1111, making OUTPUT always LOW. A functional simulation of this multiplexer circuit is shown in Figure 19-2. The output signal is identical to the selected multiplexer data channel input when it is enabled.

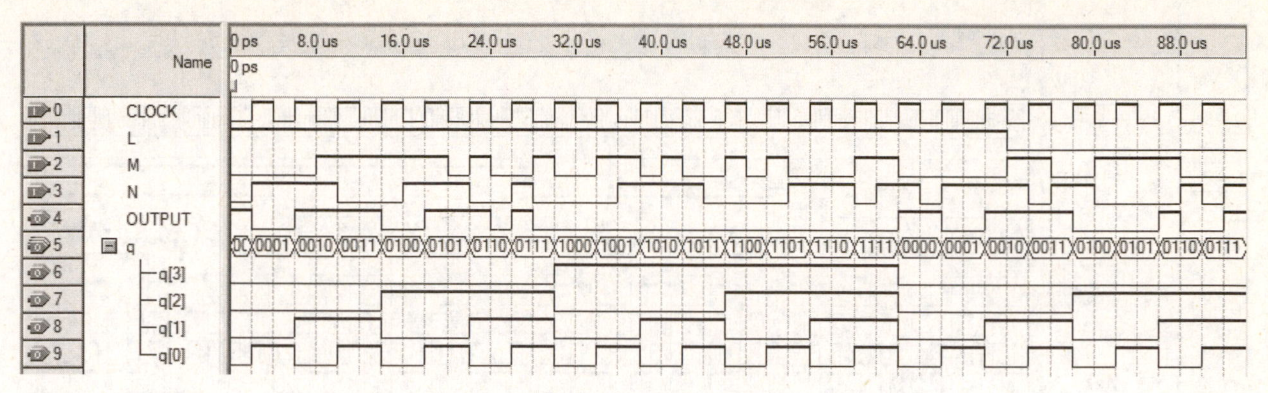

Fig. 19-2 Simulation of multiplexer circuit

Example 19-1

Design a 4-channel multiplexer that will output any selected waveform produced by a 4-bit binary counter.

A schematic solution using megafunctions is shown in Fig. 19-3. The LPM_COUNTER megafunction (mod16) is used to create the 4-bit binary counter. The 4-channel multiplexer function is created by the LPM_MUX megafunction that has been named mux4ch. The counter's four outputs are the data inputs to the multiplexer. Selection of the desired counter signal is controlled by sel[1..0]. Simulation results are shown in Fig. 19-4.

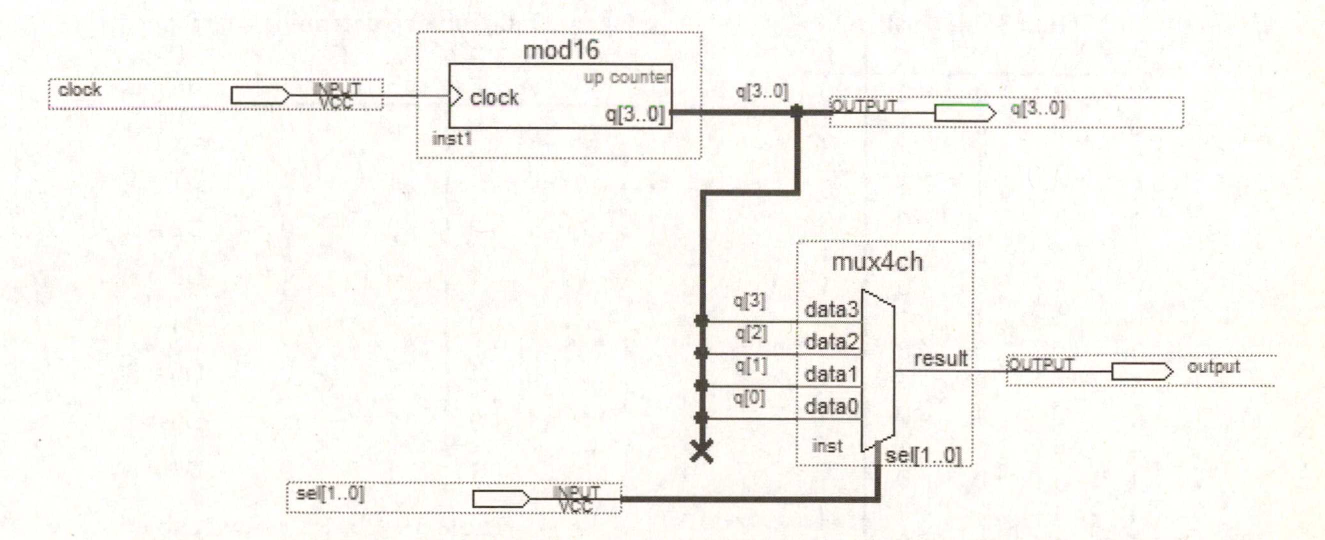

Fig. 19-3 Schematic design for Example 19-1

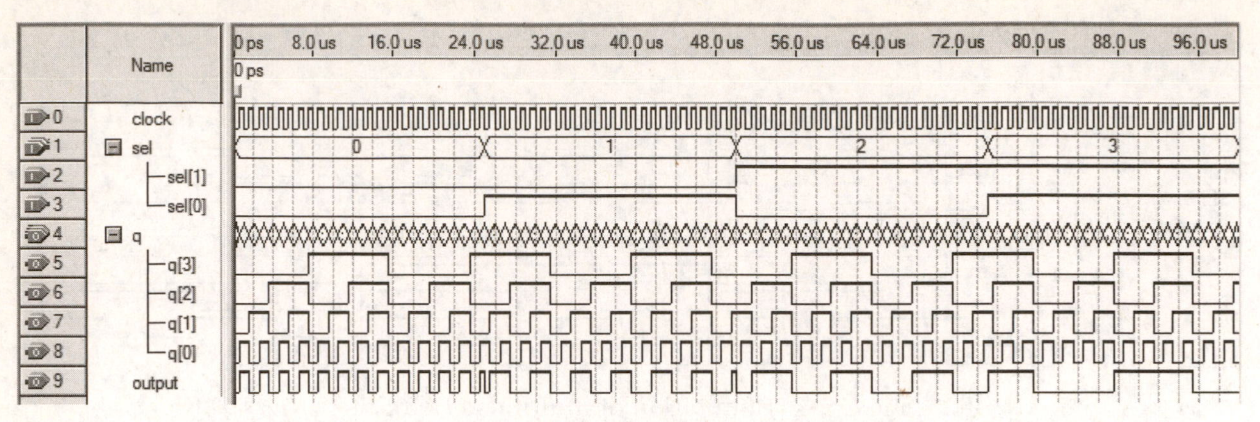

Fig. 19-4 Simulation results for Example 19-1

Example 19-2

Design a circuit that will produce the 4-input variable logic function F given in Table 19-2a using an 8-channel multiplexer.

The key to this design technique is to re-draw the truth table with three of the input variables used as the select inputs to the multiplexer. Then by comparing the two possible outputs for each of the eight combinations, you can determine the necessary data input for each channel. Our solution using a 74151 maxplus2 function is shown in Fig. 19-5 and the circuit's functional simulation is shown in Fig. 19-6.

Truth Table for F

W	X	Y	Z	F
0	0	0	0	0
0	0	0	1	1
0	0	1	0	0
0	0	1	1	0
0	1	0	0	0
0	1	0	1	1
0	1	1	0	1
0	1	1	1	1
1	0	0	0	0
1	0	0	1	1
1	0	1	0	0
1	0	1	1	1
1	1	0	0	1
1	1	0	1	0
1	1	1	0	1
1	1	1	1	0

Table 19-2a Example 19-2 function table

Truth Table to determine data inputs

C	*B*	*A*	F		Data
X	Y	Z	W = 0	W = 1	Inputs
0	0	0	0	0	D0 = 0
0	0	1	1	1	D1 = 1
0	1	0	0	0	D2 = 0
0	1	1	0	1	D3 = W
1	0	0	0	1	D4 = W
1	0	1	1	0	D5 = !W
1	1	0	1	1	D6 = 1
1	1	1	1	0	D7 = !W

Table 19-2b Re-drawn table

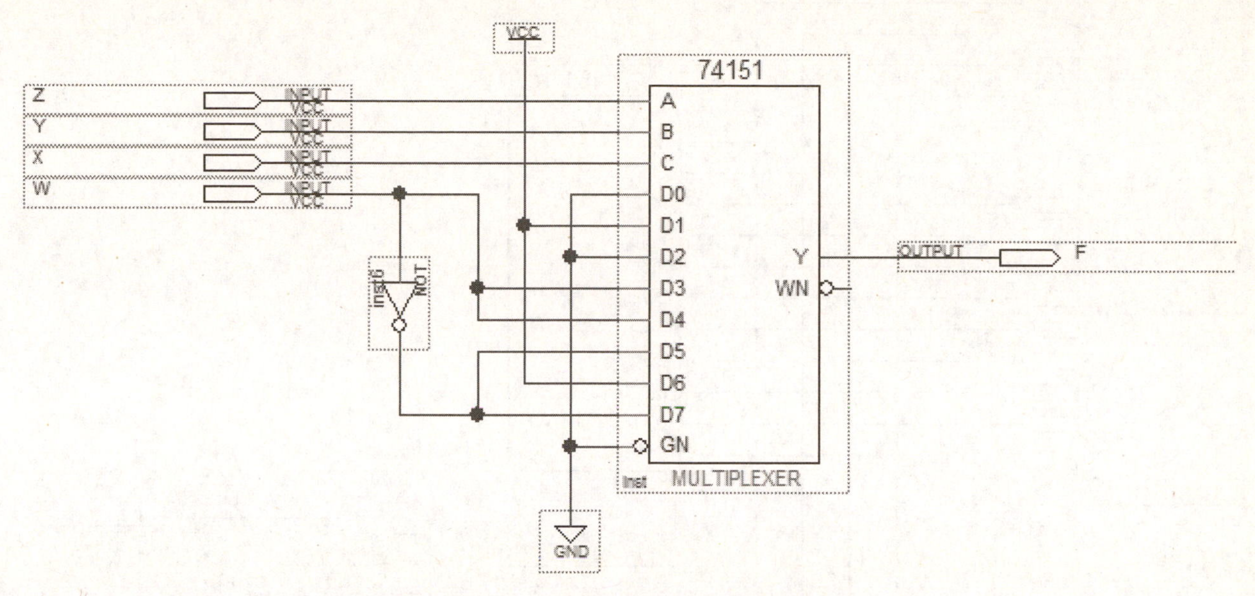

Fig. 19-5 Schematic design for Example 19-2

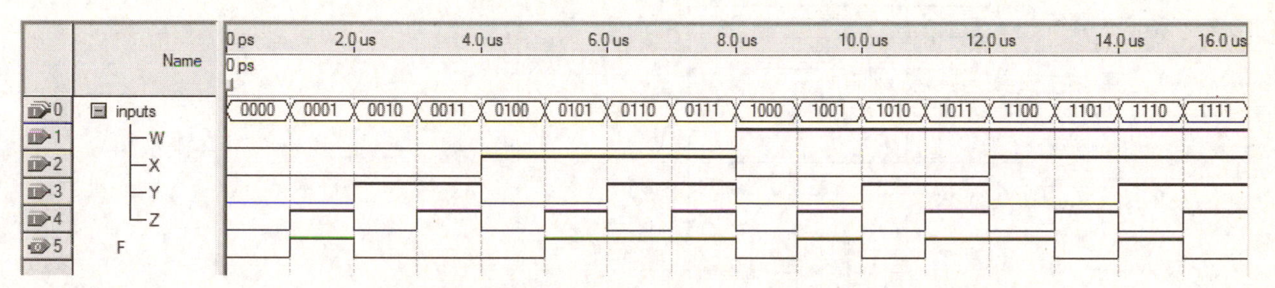

Fig. 19-6 Functional simulation of Example 19-2 circuit design

Example 19-3

Create two designs for a 16-channel multiplexer. Use the 74151 maxplus2 function for one design and an LPM_MUX megafunction for the other design.

The 74151 maxplus2 solution is shown in Fig. 19-7. A 16-channel multiplexer will require four select lines s[3..0]. The most significant select (s[3]) is used to enable one of the two 74151 multiplexers and disable the other. Only one of the multiplexers will be enabled at a time, so when the two outputs are ORed together, the final output will only depend on the one multiplexer that is enabled. The LPM_MUX megafunction solution is shown in Fig. 19-8. Both designs will have identical functions. Functional simulation results are shown in Fig. 19-9.

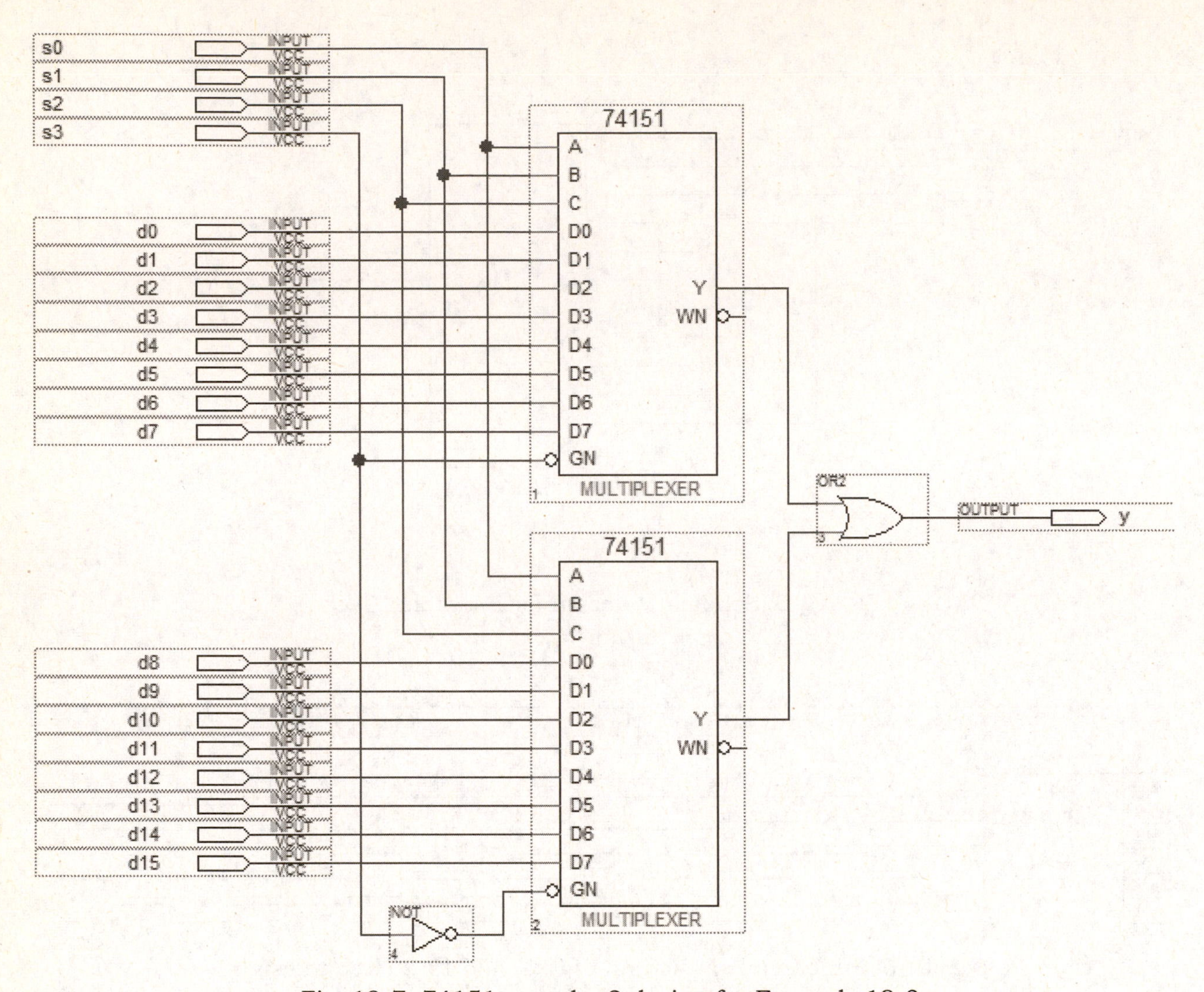

Fig. 19-7 74151 maxplus2 design for Example 19-3

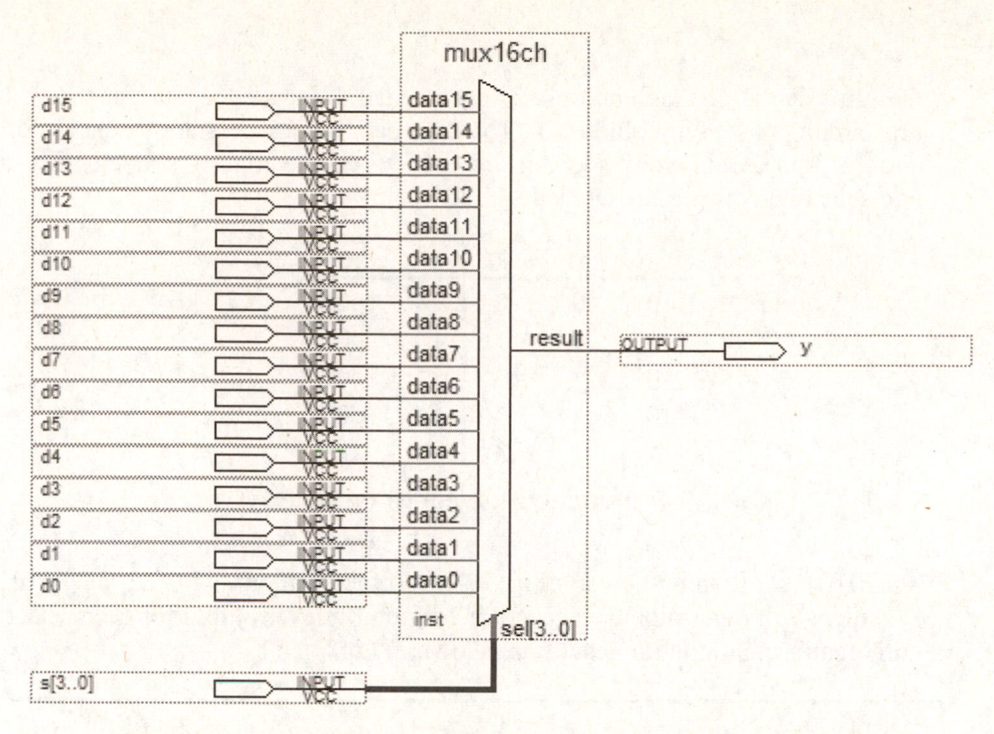

Fig. 19-8 LPM_MUX megafunction design for Example 19-3

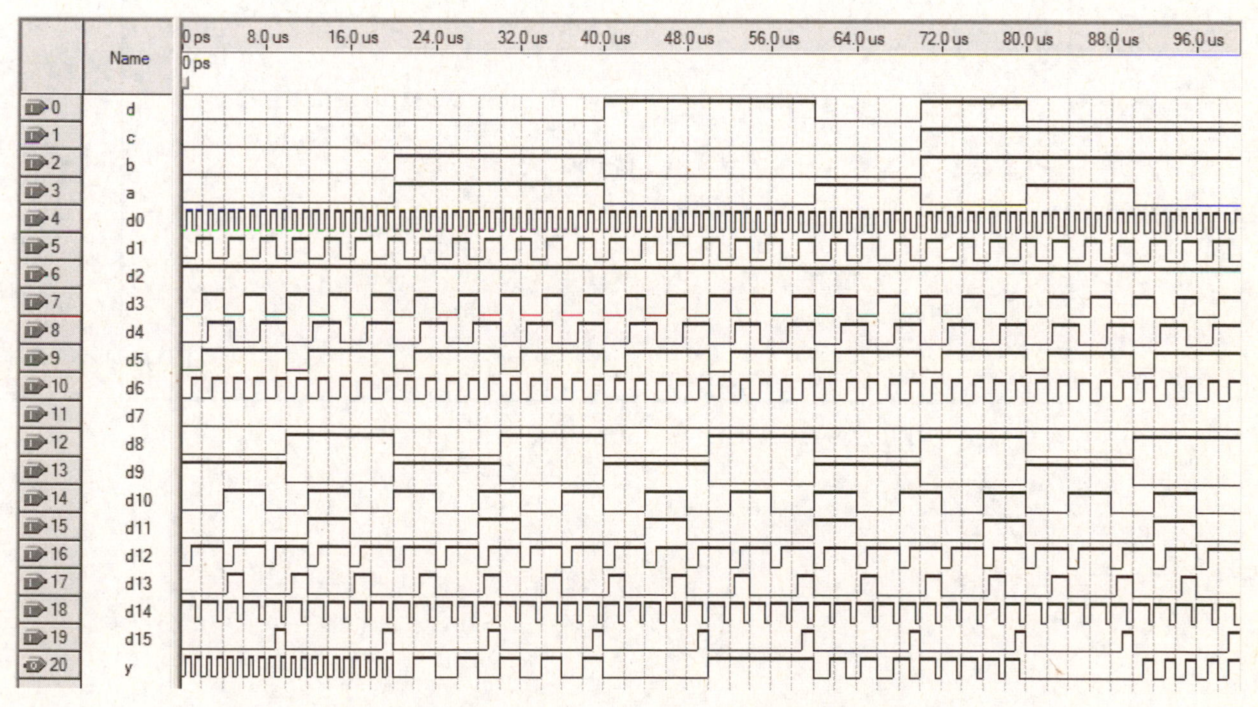

Fig. 19-9 Simulation sample for Example 19-3

Example 19-4

Design a dual 4-channel multiplexer using an HDL language. The function table for this circuit is listed in Table 19-3. The 2-bit data input words are labeled a, b, c, and d and the 2-bit output word is y. An active-LOW enable control is labeled EN, and the two selection controls are labeled s1 and s0.

en	s1	s0	y1	y0
0	0	0	a1	a0
0	0	1	b1	b0
0	1	0	c1	c0
0	1	1	d1	d0
1	X	X	0	0

Table 19-3 Function table for Example 19-4

AHDL An AHDL solution is shown in Fig. 19-10. If the en input is LOW, the CASE statement will determine the necessary 2-bit multiplexer output for each select input combination. Simulation results are shown in Fig. 19-11.

```
SUBDESIGN  dual_mux4
(
      en, s[1..0]                                    :INPUT;
      a[1..0], b[1..0], c[1..0], d[1..0]   :INPUT;
      y[1..0]                                        :OUTPUT;
)
BEGIN
      IF  !en  THEN                  -- active-low enable
            CASE  s[]  IS            -- select input
                  WHEN 0  =>  y[] = a[];
                  WHEN 1  =>  y[] = b[];
                  WHEN 2  =>  y[] = c[];
                  WHEN 3  =>  y[] = d[];
            END CASE;
      ELSE  y[] = 0;                 -- disabled
      END IF;
END;
```

Fig. 19-10 AHDL solution for Example 19-4

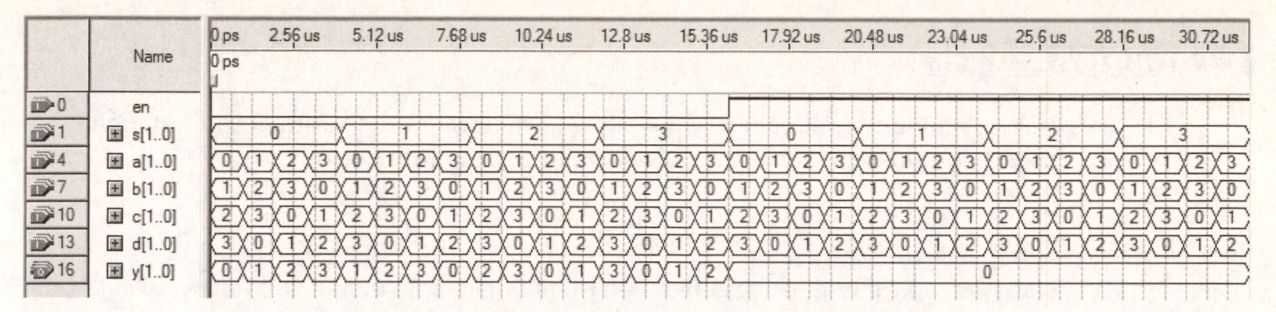

Fig. 19-11 Simulation results for Example 19-4

VHDL

A VHDL solution is shown in Fig. 19-12. If the en input is LOW, the CASE statement will determine the necessary 2-bit multiplexer output for each select input combination. Simulation results are shown in Fig. 19-11.

```
ENTITY dual_mux4 IS
PORT (    en              :IN BIT;
          s               :IN INTEGER RANGE 0 TO 3;
          a, b, c, d      :IN BIT_VECTOR (1 DOWNTO 0);
          y               :OUT BIT_VECTOR (1 DOWNTO 0) );
END dual_mux4;

ARCHITECTURE vhdl OF dual_mux4 IS
BEGIN
      PROCESS (en, s, a, b, c, d)
      BEGIN
            IF en = '0' THEN              -- active-low enable
                  CASE  s  IS             -- select input
                        WHEN 0  =>  y <= a;
                        WHEN 1  =>  y <= b;
                        WHEN 2  =>  y <= c;
                        WHEN 3  =>  y <= d;
                  END CASE;
            ELSE  y <= "00";              -- disabled
            END IF;
      END PROCESS;
END vhdl;
```

Fig. 19-12 VHDL solution for Example 19-4

Laboratory Projects

Design multiplexer circuits for the following applications. Construct and test each circuit design. Functionally simulate circuit designs with Quartus.

19.1 Serial data word generator

Design a logic circuit using a one-of-eight multiplexer to output an 8-bit serial data word. The multiplexer will output the 8-bit data word (whose bit pattern is defined with 8 logic switches) and then the serial output will be blanked (LOW) for 8 more clock cycles. This pattern continues repetitively. View the serial output and clock on an oscilloscope while changing the data input on the logic switches. Trigger the scope with an appropriate signal from the control circuit.

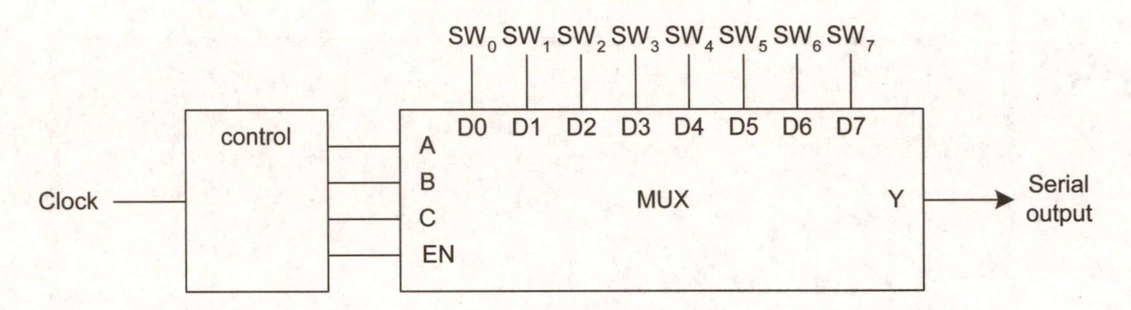

19.2 Prime number detector

Design a 4-bit prime number detector circuit that is implemented using an 8-channel multiplexer. The 4-bit input num[3..0] will allow the binary numbers for 0 through 15 to be applied to the circuit. The output should be HIGH only if prime numbers (1, 2, 3, 5, 7, 11, 13) are being input to the detector circuit.

19.3 Variable frequency divider

Design a circuit that will select (multiplex) a specified frequency to be output (freq_select). A 128-kHz clock signal will be divided into a number of different frequencies by a MOD-256 binary counter. The desired frequency will be selected by a one-of-eight multiplexer, as indicated in the following function table. The frequency selection is determined by the state of S2 S1 S0. Use a frequency counter to measure the signal frequency output by the multiplexer.

S2	S1	S0	Output frequency
0	0	0	500 Hz
0	0	1	1.0 kHz
0	1	0	2.0 kHz
0	1	1	4.0 kHz
1	0	0	8.0 kHz
1	0	1	16.0 kHz
1	1	0	32.0 kHz
1	1	1	64.0 kHz

19.4 Multiplexed BCD display

Design a quad, 2-channel (2-line-to-1-line) multiplexer to route the outputs from two BCD counters (A[3..0] and B[3..0]) to a 4-bit display. Each counter has a separate manual clock signal. Only one of the two 4-bit numbers input to the multiplexer will be displayed, as indicated by the following function table. The gating control G will produce an invalid BCD output (1111) from the MUX when it is disabled.

		MUX Outputs			
G	S	Y3	Y2	Y1	Y0
1	0	A3	A2	A1	A0
1	1	B3	B2	B1	B0
0	X	1	1	1	1

19.5 Quad, 3-channel MUX

Design a multiplexer that can select one of three 4-bit binary words to be output. The inputs are A3 A2 A1 A0, B3 B2 B1 B0, and C3 C2 C1 C0. The multiplexer's outputs are Y3 Y2 Y1 Y0 and the select controls are S1 S0. The function table for this MUX is given below. The output from the MUX will be 0000 when S1 = S0 = 0 (i.e., the MUX is disabled).

S1	S0	Y3	Y2	Y1	Y0
0	0	0	0	0	0
0	1	A3	A2	A1	A0
1	0	B3	B2	B1	B0
1	1	C3	C2	C1	C0

19.6 Alarm system

Design a detector circuit to monitor a set of 8 SPST switches (s[7..0]) in an alarm system. The switches are scanned one at a time by a multiplexer to determine if any switch has been opened (resulting in a logic HIGH for the open switch input). A MOD-8 counter automatically controls the scanning. When an open switch is detected, an alarm lamp will go HIGH and the scanning will stop, with the counter state indicating the location of the open switch.

19.7 16-channel, 1-bit MUX

Design a 16-channel, 1-bit multiplexer using an HDL language. Label the inputs d[15..0] and sel[3..0] and the output y.

UNIT 20

DEMULTIPLEXER APPLICATIONS

Objective

- To apply demultiplexer circuits in digital system applications.

Demultiplexer Circuits

A demultiplexer is also called a data distributor because it has several possible destinations to which the input data may be sent. The single output line that will receive the data is controlled by the specific select code applied to the demultiplexer. The 74138 (see Fig. 20-1 and Table 20-1) is an example of a standard demultiplexer IC chip and maxplus2 function. This device can operate as either a decoder or a demultiplexer. A decoder can be used as a demultiplexer by connecting the DMUX select lines to the data input lines (of the decoder) and applying the DMUX data to the enable input (of the decoder). One bit of input data may be sent to any one of eight possible output destinations with the 74138. Data that is input to one of the G2 enables will not be inverted by the DMUX chip. The G2B input pin has been chosen for the data input pin in the schematic of Fig. 20-1. The other two enable pins may be utilized as enable controls for the DMUX.

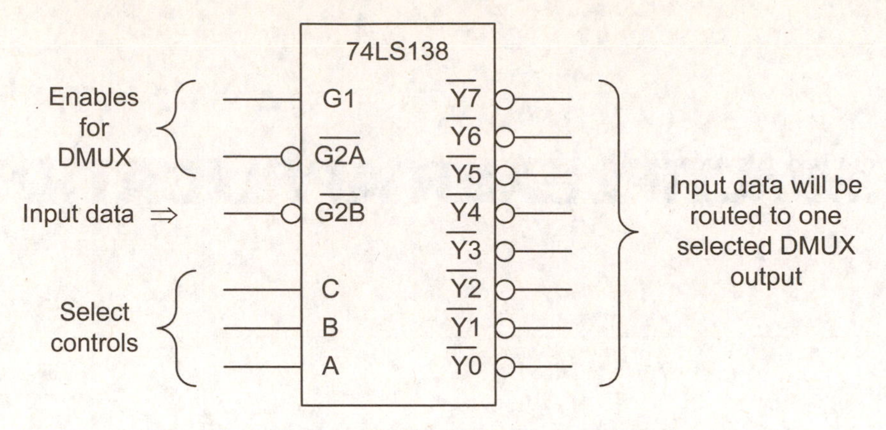

Fig. 20-1 A 74138, 8-channel demultiplexer

G1	G2A	C	B	A	Y0	Y1	Y2	Y3	Y4	Y5	Y6	Y7
1	0	0	0	0	data	1	1	1	1	1	1	1
1	0	0	0	1	1	data	1	1	1	1	1	1
1	0	0	1	0	1	1	data	1	1	1	1	1
1	0	0	1	1	1	1	1	data	1	1	1	1
1	0	1	0	0	1	1	1	1	data	1	1	1
1	0	1	0	1	1	1	1	1	1	data	1	1
1	0	1	1	0	1	1	1	1	1	1	data	1
1	0	1	1	1	1	1	1	1	1	1	1	data
0	X	X	X	X	1	1	1	1	1	1	1	1
X	1	X	X	X	1	1	1	1	1	1	1	1

Table 20-1 Truth table for a 74138 demultiplexer

Example 20-1

Simulate the circuit given in Fig. 20-2 and describe its operation. The block labeled decoder_DMUX is an LPM_DECODE megafunction that is used as a demultiplexer.

Simulation results are shown in Fig. 20-3. One of the four demultiplexer outputs will produce a series of four clock pulses, and then the next demultiplexer output will do the same until all four have done so. The pulsing continues through each output in turn. The two least significant counter bits of the LPM_COUNTER are used to count the four clock pulses for each demultiplexer output. The two most significant counter bits are used to select each of the DMUX outputs in sequence.

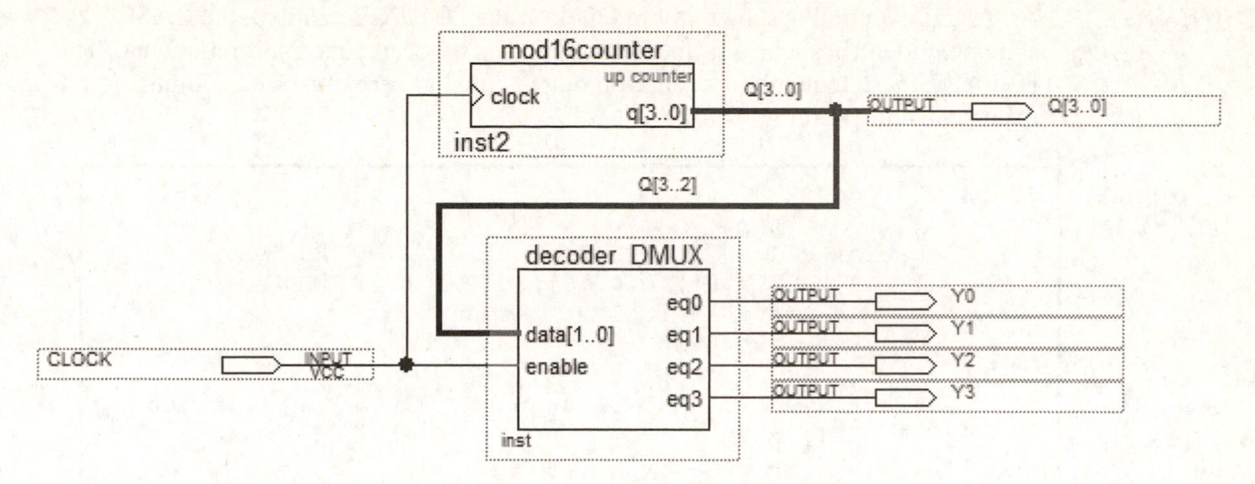

Fig. 20-2 Circuit for Example 20-1

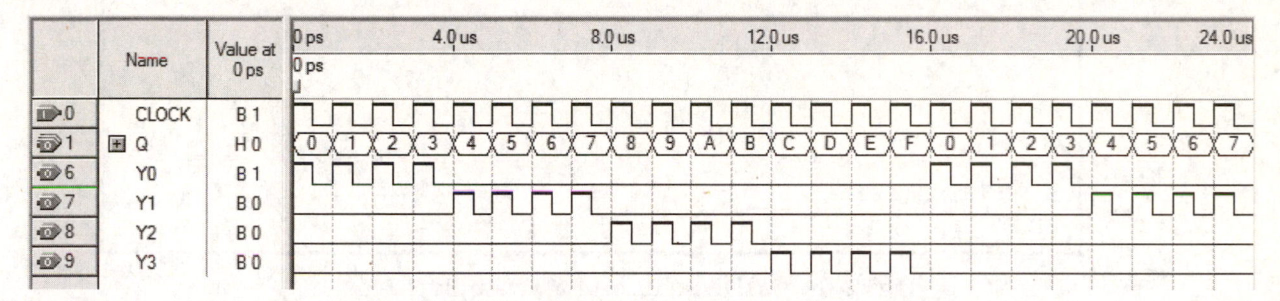

Fig. 20-3 Simulation results for Example 20-1

Example 20-2

Design a 2-bit, 4-channel demultiplexer using an HDL language. The function table for this circuit is listed in Table 20-2. The 2-bit input data i[1..0] can be routed to any of the 4 output channels a[1..0], b[1..0], c[1..0], or d[1..0]. The selection controls are s[1..0], and the DMUX also has an enable control.

enable	s1	s0	a1 a0	b1 b0	c1 c0	d1 d0
0	X	X	0 0	0 0	0 0	0 0
1	0	0	i1 i0	0 0	0 0	0 0
1	0	1	0 0	i1 i0	0 0	0 0
1	1	0	0 0	0 0	i1 i0	0 0
1	1	1	0 0	0 0	0 0	i1 i0

Table 20-2 Function table for Example 20-2

AHDL An AHDL solution is shown in Fig. 20-4. If the DMUX is enabled, the CASE statement identifies which of the 2-bit outputs is to receive the 2-bit input data. The DEFAULTS statement sets each 2-bit output equal to zero otherwise. Simulation results are shown in Fig. 20-5.

```
SUBDESIGN  dmux2_4
(
        enable, s[1..0], i[1..0]                 :INPUT;
        a[1..0], b[1..0], c[1..0], d[1..0]       :OUTPUT;
                -- four 2-bit DMUX outputs
)
BEGIN
        DEFAULTS         -- DMUX output = 00 when deselected
                a[] = 0;
                b[] = 0;
                c[] = 0;
                d[] = 0;
        END DEFAULTS;
        IF  enable  THEN
                CASE  s[]  IS      -- selects DMUX output
                        WHEN  0  =>  a[] = i[];
                        WHEN  1  =>  b[] = i[];
                        WHEN  2  =>  c[] = i[];
                        WHEN  3  =>  d[] = i[];
                END CASE;
        END IF;
END;
```

Fig. 20-4 AHDL solution for Example 20-2

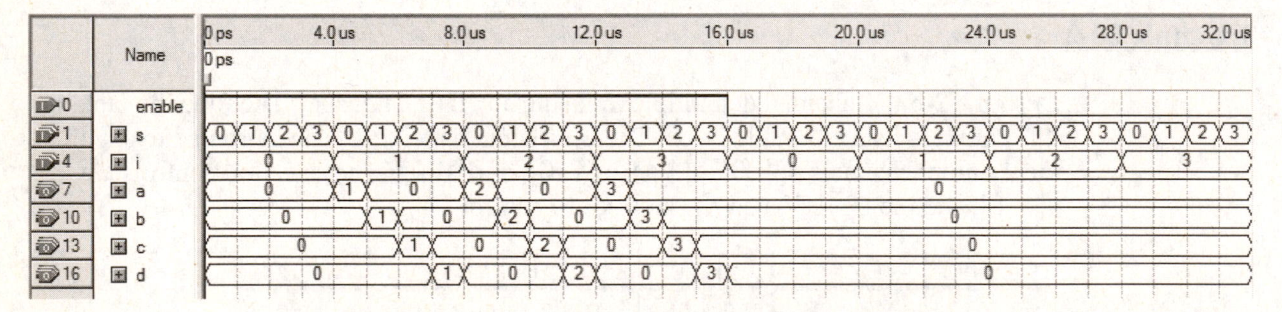

Fig. 20-5 Simulation results for Example 20-2

VHDL A VHDL solution is shown in Fig. 20-6. If the DMUX is enabled, the CASE statement identifies which of the 2-bit outputs is to receive the 2-bit input data while the others output 00. If the DMUX is disabled, all outputs will be 00. Simulation results are shown in Fig. 20-5.

```
ENTITY dmux2_4 IS
PORT (  enable         :IN BIT;
        s, i           :IN BIT_VECTOR (1 DOWNTO 0);
        a, b, c, d     :OUT BIT_VECTOR (1 DOWNTO 0) );
             -- four 2-bit DMUX outputs
END dmux2_4;

ARCHITECTURE vhdl OF dmux2_4 IS
BEGIN
   PROCESS (enable, s, i)
   BEGIN
      IF  enable = '1' THEN
         CASE  s  IS           -- selects DMUX output
            WHEN  "00"  =>
               a <= i;    b <= "00"; c <= "00"; d <= "00";
            WHEN  "01"  =>
               a <= "00"; b <= i;    c <= "00"; d <= "00";
            WHEN  "10"  =>
               a <= "00"; b <= "00"; c <= i;    d <= "00";
            WHEN  "11"  =>
               a <= "00"; b <= "00"; c <= "00"; d <= i;
         END CASE;
      ELSE     a <= "00"; b <= "00"; c <= "00"; d <= "00";
                  -- not enabled
      END IF;
   END PROCESS;
END vhdl;
```

Fig. 20-6 VHDL solution for Example 20-2

Laboratory Projects

Design demultiplexer circuits for the following applications. Construct and test each circuit design. Functionally simulate circuit designs with Quartus.

20.1 Clock DMUX

Design a demultiplexer circuit to route a clock signal to one of three different counters. The counters each produce a different modulus, as shown in the following table. The sel[1..0] controls determine which counter is to receive the clock signal. One of the select combinations (00) does not route the clock to any of the counters.

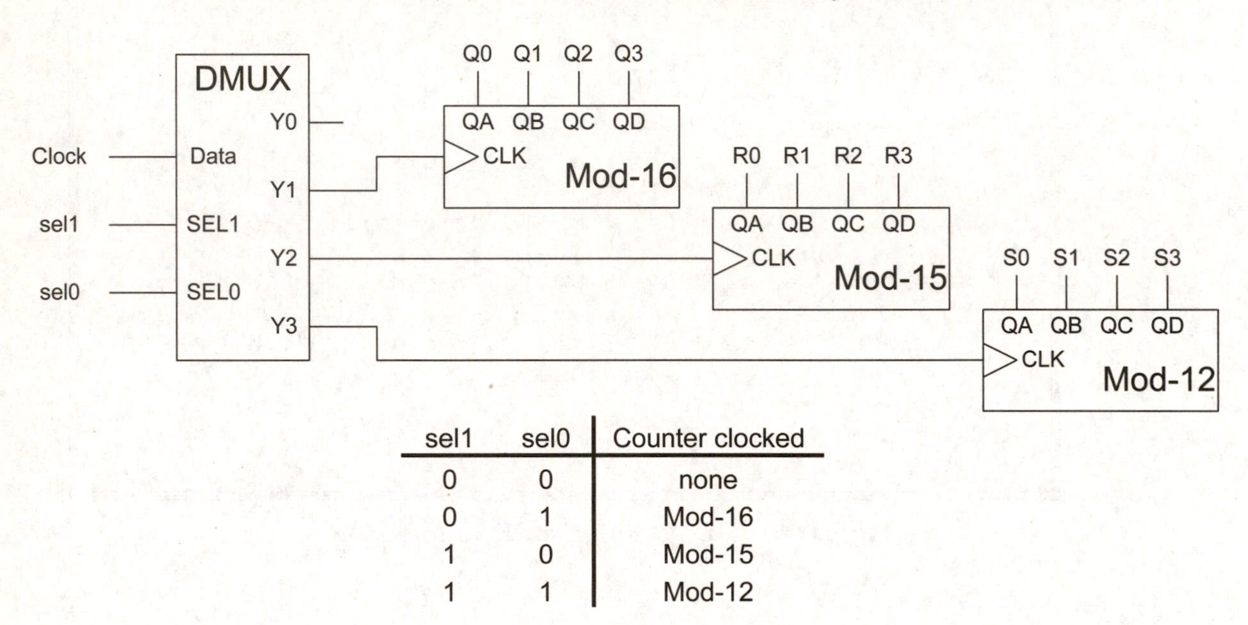

sel1	sel0	Counter clocked
0	0	none
0	1	Mod-16
1	0	Mod-15
1	1	Mod-12

20.2 BCD counter demultiplexer

Design and construct a circuit to route the output from a BCD counter to the selected 7-segment display on the FPGA/CPLD development board. An enable and a select will control a 4-bit, 2-channel demultiplexer for the displays, as shown in the following function table. Blank the display that is not being used. Note: A 7447 will blank the display if the data input is 1111.

Controls		Left display	Right display
G	S		
0	0	Counter	Blank
0	1	Blank	Counter
1	X	Blank	Blank

20.3 Alarm indicator
Implement the alarm circuit given in the block diagram below using standard logic devices for each of the blocks. The switches are to be placed on windows and doors that we wish to monitor. An alarm condition occurs if any of the switches are opened due to the corresponding window or door being opened. The MUX will sequentially check each of the switches and output a signal that is transmitted to the DMUX (that we will assume is some distance away). The DMUX will control a set of lamps to indicate which switch has been opened. All lamps will be off if none of the switches are open. A flashing lamp will indicate an alarm. Use a 200-Hz clock for the counter. Note: If the DMUX is powered with a separate power source (because it is located some distance from the MUX circuit), make sure that you have a common ground.

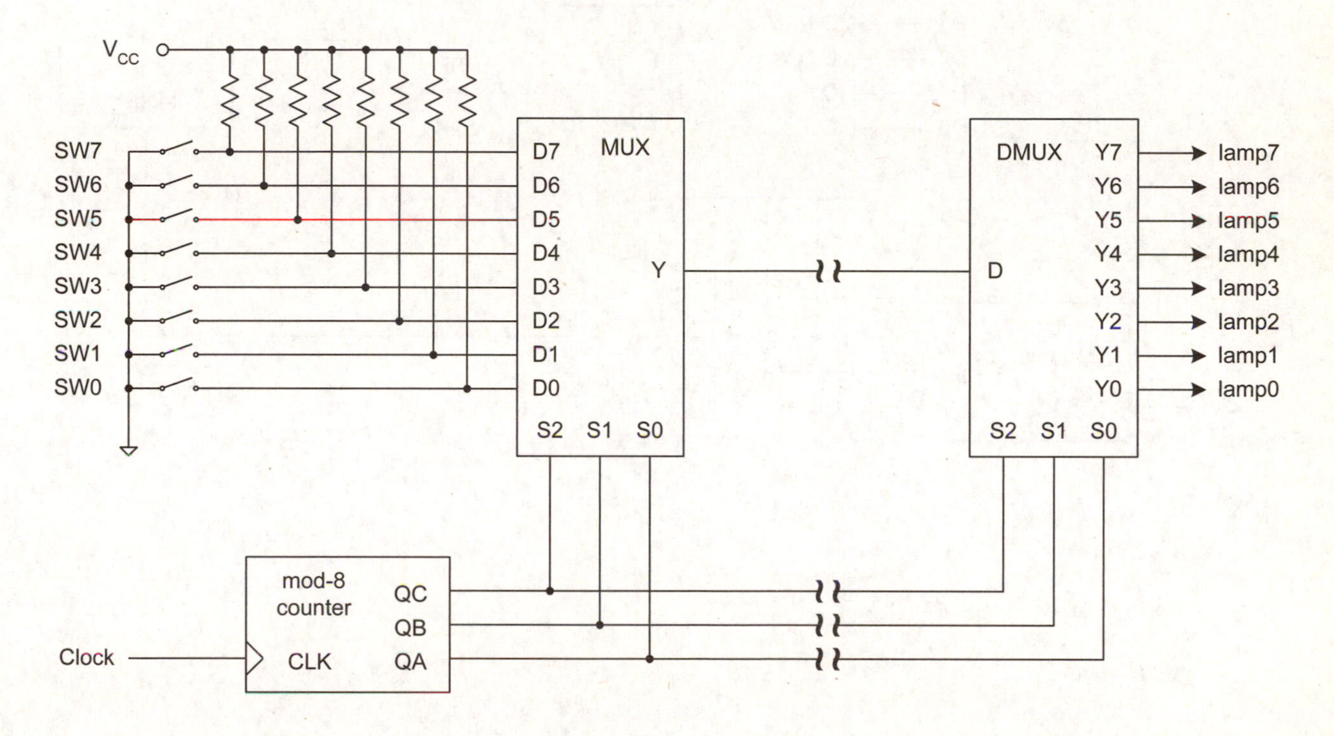

20.4 8-channel, 1-bit DMUX
Design an 8-channel, 1-bit demultiplexer using an HDL language.

20.5 2-channel data transmitter

Design and construct the data transmitter given in the block diagram below using a PLD. The 8-bit data input will be split so that D[7..4] will be transmitted to channel 2 of the oscilloscope and D[3..0] will be transmitted to channel 1. Use the counter MSB to trigger the oscilloscope.

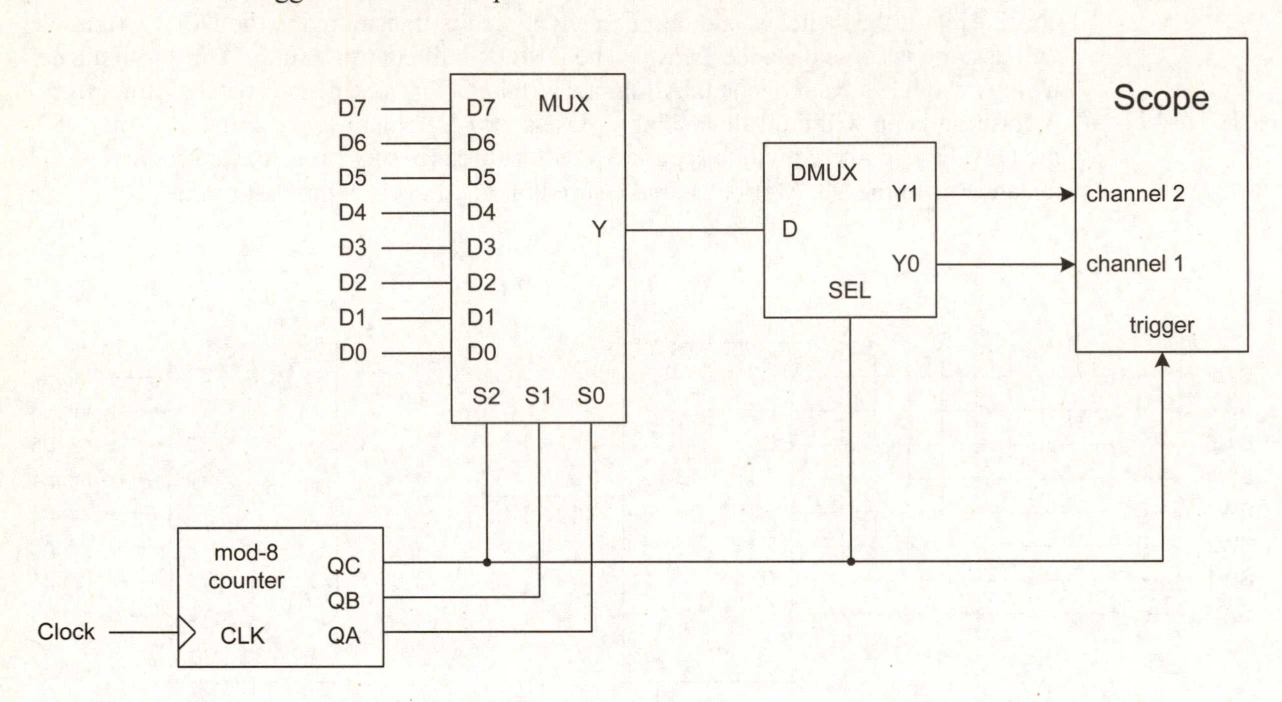

UNIT 21

MAGNITUDE COMPARATOR APPLICATIONS

Objective

- To apply magnitude comparators in digital system applications.

Magnitude Comparators

Magnitude comparators are used to determine the magnitude relationships between two quantities. A typical comparator will indicate whether two input values are equivalent or, if not, which of the values is larger. Logic gates can be used to implement various types of comparator circuits. Standard IC comparators such as the 7485, a 4-bit comparator shown in Fig. 21-1, can be easily used in a variety of magnitude comparator applications. This comparator will determine the magnitude relationship between the two binary values A3 A2 A1 A0 and B3 B2 B1 B0. The outputs produced are A<B, A>B, and A=B. The truth table for this chip is shown in Table 21-1. The three cascade inputs allow multiple 7485 chips to be cascaded together to compare binary quantities that are more than 4 bits long. If the cascade feature is not used, the A=B input will normally be tied HIGH and the other two cascade inputs (A<B and A>B) are tied LOW. The 7485 magnitude comparator is also available as a maxplus2 function.

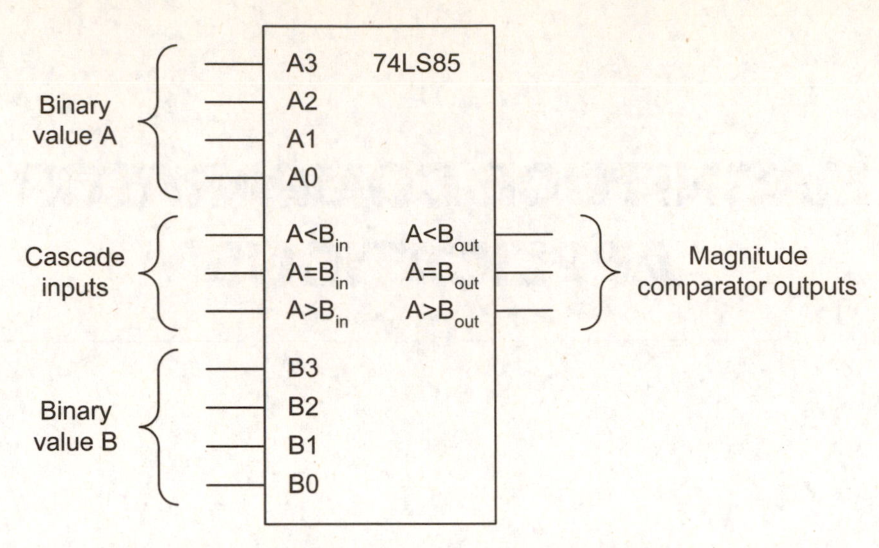

Fig. 21-1 A 7485, 4-bit magnitude comparator chip

Comparing inputs				Cascading inputs			Outputs		
A3, B3	A2, B2	A1, B1	A0, B0	$A<B_{in}$	$A>B_{in}$	$A=B_{in}$	$A<B_{out}$	$A>B_{out}$	$A=B_{out}$
A3<B3	X	X	X	X	X	X	1	0	0
A3>B3	X	X	X	X	X	X	0	1	0
A3=B3	A2<B2	X	X	X	X	X	1	0	0
A3=B3	A2>B2	X	X	X	X	X	0	1	0
A3=B3	A2=B2	A1<B1	X	X	X	X	1	0	0
A3=B3	A2=B2	A1>B1	X	X	X	X	0	1	0
A3=B3	A2=B2	A1=B1	A0<B0	X	X	X	1	0	0
A3=B3	A2=B2	A1=B1	A0>B0	X	X	X	0	1	0
A3=B3	A2=B2	A1=B1	A0=B0	1	0	0	1	0	0
A3=B3	A2=B2	A1=B1	A0=B0	0	1	0	0	1	0
A3=B3	A2=B2	A1=B1	A0=B0	0	0	1	0	0	1

Table 21-1 Function table for a 7485

An 8-bit magnitude comparator circuit constructed with two 7485 functional blocks is shown in Fig. 21-2. The two comparator chips are cascaded together by connecting the outputs from the comparator that is monitoring the least significant 4 bits from each number to the cascade inputs on the comparator that is monitoring the most significant 4 bits. The lower nibbles will affect the 8-bit comparator's output only when the higher nibbles are equivalent. This cascading technique can be expanded to compare the magnitudes of larger numbers (bigger than 8 bits), but the propagation delays can get rather long so other configurations may be desirable. An equivalent 8-bit magnitude comparator using the Quartus megafunction LPM_COMPARE is shown in Fig. 21-3 and the functional simulation results are given in Fig. 21-4.

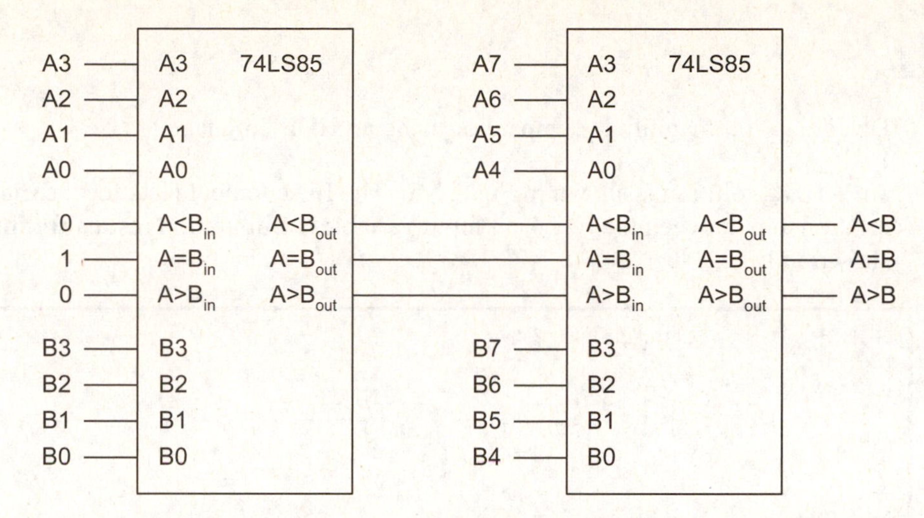

Fig. 21-2 An 8-bit magnitude comparator circuit using two 7485 blocks

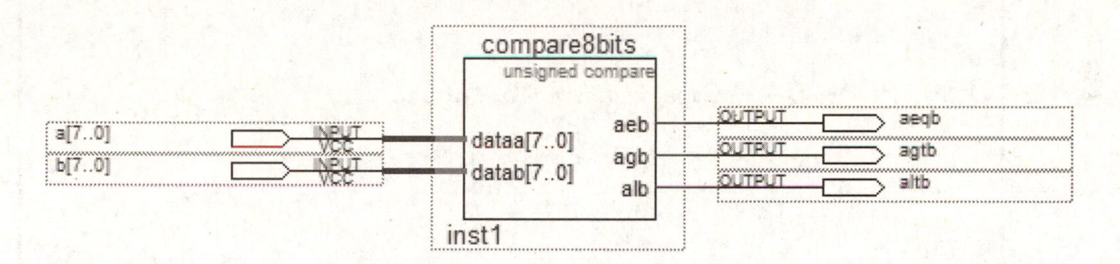

Fig. 21-3 An 8-bit magnitude comparator circuit using LPM_COMPARE

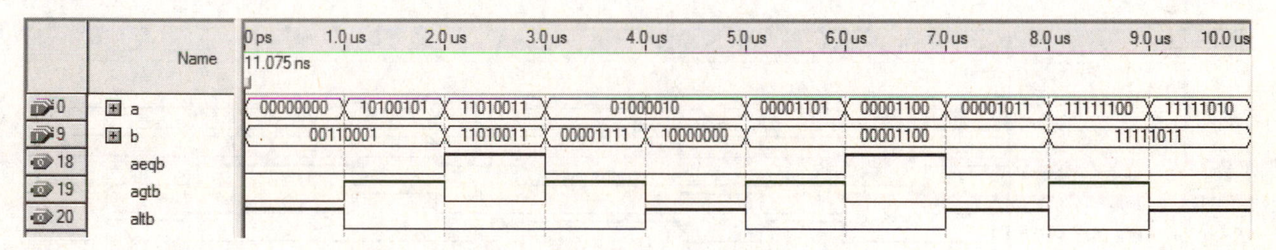

Fig. 21-4 Functional simulation of Fig. 21-3

Example 21-1

Design a 4-bit magnitude comparator using an HDL language.

AHDL An AHDL solution is shown in Fig. 21-5. The IF statement tests for the magnitude relationship between the two 4-bit input variables. Simulation results for this design are shown in Fig. 21-6.

```
SUBDESIGN  compare
(
        a[3..0], b[3..0]            :INPUT;
        agtb, altb, aeqb            :OUTPUT;
)

BEGIN
        DEFAULTS
                agtb = GND;
                altb = GND;
                aeqb = GND;
        END DEFAULTS;

                -- compare a and b inputs
        IF      a[] < b[]   THEN   altb = VCC;
        ELSIF   a[] > b[]   THEN   agtb = VCC;
        ELSE                       aeqb = VCC;
        END IF;
END;
```

Fig. 21-5 AHDL solution for Example 21-1

Fig. 21-6 Simulation results for Example 21-1

VHDL A VHDL solution is shown in Fig. 21-7. The IF statement tests for the magnitude relationship between the two 4-bit input variables. Simulation results for this design are shown in Fig. 21-6.

```
ENTITY compare IS
PORT (
      a, b                :IN INTEGER RANGE 0 TO 15;
      agtb, altb, aeqb    :OUT BIT
);
END compare;

ARCHITECTURE vhdl OF compare IS
BEGIN
      PROCESS (a, b)
      BEGIN        -- compare a and b inputs
         IF    a < b  THEN
               altb <= '1';  agtb <= '0';  aeqb <= '0';
         ELSIF a > b  THEN
               altb <= '0';  agtb <= '1';  aeqb <= '0';
         ELSE
               altb <= '0';  agtb <= '0';  aeqb <= '1';
         END IF;
      END PROCESS;
END vhdl;
```

Fig. 21-7 VHDL solution for Example 21-1

Laboratory Projects

Design magnitude comparator circuits for the following applications. Construct and test each circuit design. Functionally simulate circuit designs with Quartus.

21.1 Count detector
Design a comparator circuit in which the single output (count_detect) will be HIGH if the 4-bit value produced by a binary counter is greater than a lower limit (low_limit) and less than an upper limit (hi_limit). Low_limit and hi_limit are each 4-bit input values set by switches.

21.2 Adjustable range detector
Design a logic circuit using an HDL that will compare a 5-bit input value (value[4..0]) to one of four specified input ranges selected by the control r[1..0]. The comparator should have three outputs to indicate that the input value is either less than (ltr) or greater than (gtr) the range of values in the selected window or within the range (rng) of window values.

R1	R0	Selected range of window values
0	0	16
0	1	15–17
1	0	14–18
1	1	13–19

21.3 Variable modulus counter
Design a recycling counter circuit that can have any selected count modulus from **2 through 16**. The desired count modulus will be determined by a 4-bit input value (mod = n[3..0]). MOD-16 will be selected by n[3..0] = 0 0 0 0.

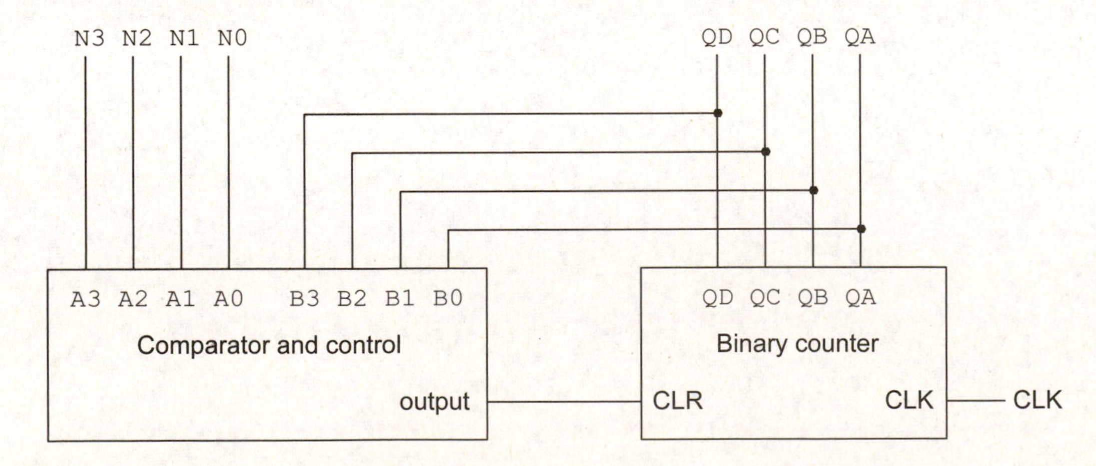

21.4 Variable self-stopping counter

Design a self-stopping counter circuit that can count up to any selected 4-bit binary value and then stop. The desired binary value at which to stop the counter will be determined by a 4-bit input (n[3..0]). The counter should also have a manual, active-LOW restartn control signal that will asynchronously reset the counter to 0 so that the counter can again cycle through its self-stopping count sequence.

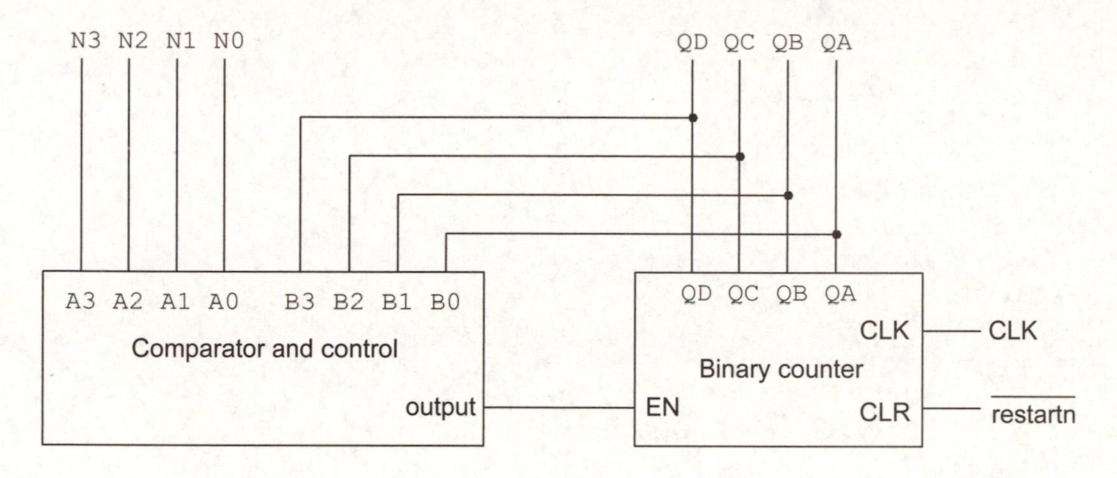

21.5 Number selector

Design a logic circuit that will produce a 4-bit output (y[3..0]) that is equal to either the larger or smaller of two 4-bit inputs (a[3..0] and b[3..0]). The selection of the larger or smaller value is specified by the control signal F, as shown in the following function table. Note that the number selector circuit will output the common value when the a and b inputs are equal. Hint: This circuit is basically a multiplexer in which the output data (y[3..0]) are selected by a function that is dependent upon the control F and the result of a magnitude comparison between a[3..0] and b[3..0] (see the MUX control table below).

Function table

F	Selector output:
0	smaller value of a[ ] or b[ ]
1	larger value of a[ ] or b[ ]

MUX control table

F	a[ ] > b[ ]	y[3..0]	Notes:
0	0	a[3..0]	select a[ ] input
0	1	b[3..0]	select b[ ] input
1	0	b[3..0]	select b[ ] input
1	1	a[3..0]	select a[ ] input

UNIT 22

DIGITAL/ANALOG AND ANALOG/DIGITAL CONVERSION

Objectives

- To construct a digital-to-analog conversion (DAC) circuit using a standard commercial DAC IC chip.
- To construct an analog-to-digital conversion (ADC) circuit using a standard commercial ADC IC chip.

Suggested Standard Parts			
AD557	ADC0804	74LS244	
Resistors: 1.1 kΩ, 10 kΩ, 10 kΩ (10-turn) potentiometer			Capacitor: 150 pF

Digital-to-Analog Conversion

Digital system outputs often must be interfaced to analog devices. One type of interfacing involves converting a digital data word into a representative analog signal (either a current or a voltage). Various digital-to-analog converter chips are available to perform this type of conversion. An example of a complete D/A converter in a single IC chip is the AD557 manufactured by Analog Devices. This 8-bit D/A converter is powered by a single 5-volt power supply and has a full-scale output of 2.55 volts. The resolution or incremental analog step size of the DAC is dependent on the number of digital input bits. The output voltage step size for this device is 10 mV. Fig. 22-1 shows how to use the AD557 to produce an analog output from 0 to 2.55 volts.

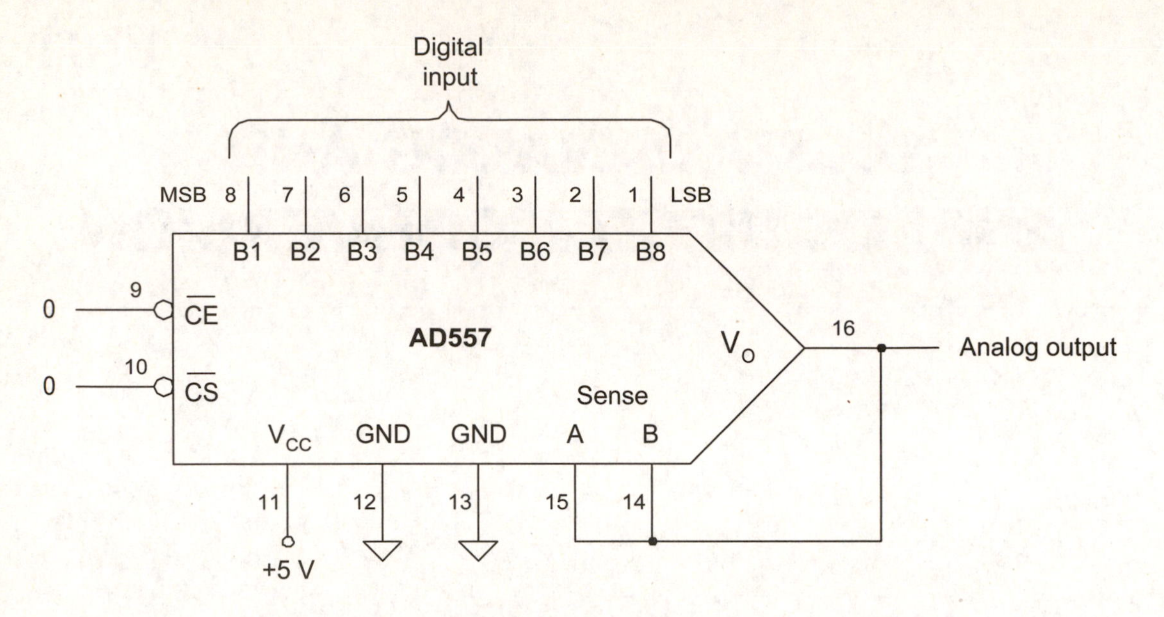

Fig. 22-1 AD557 digital-to-analog conversion IC

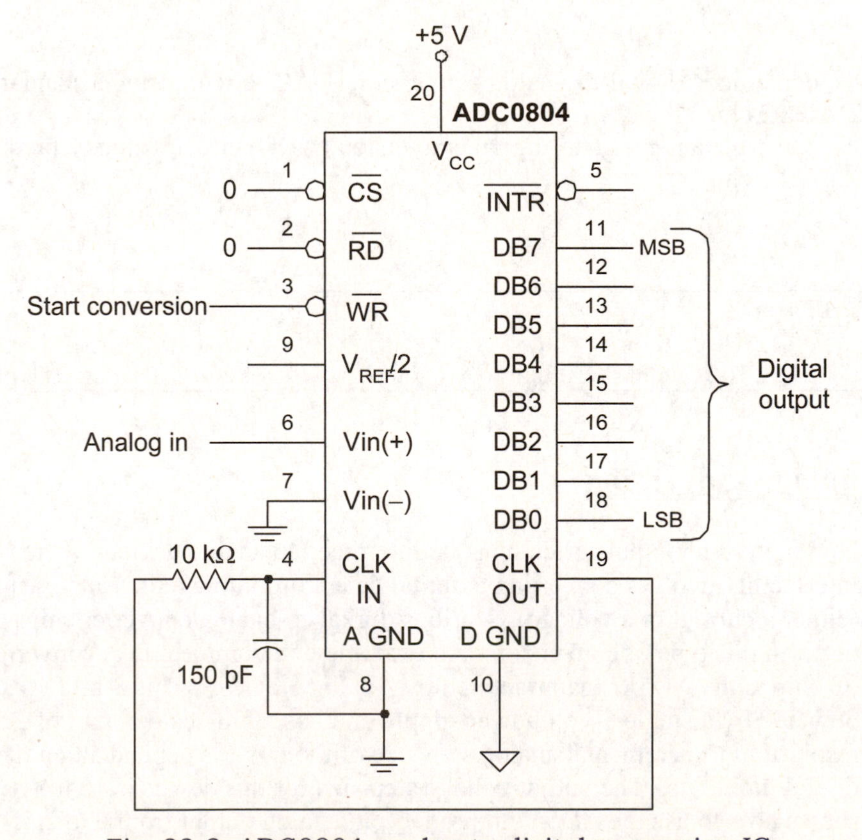

Fig. 22-2 ADC0804 analog-to-digital conversion IC

Analog-to-Digital Conversion

For a digital system to be able to process analog information, the analog signal must first be digitized or converted from analog to digital. The analog input signal is then represented by a digital value in the digital system. There are various analog-to-digital converter chips available to perform this type of conversion. The ADC0804 manufactured by National Semiconductor is an example of an 8-bit successive approximation converter contained in a single IC chip. This device will handle an analog input range from 0 to 5 volts and is powered by a single 5-volt power supply. The quantization error of an ADC is dependent on the number of digital output bits. With a resolution of 8 bits and a reference voltage of 5 volts, the ADC0804 will have a quantization error of $V_{REF}/2^8 \approx 19.5$ mv. Fig. 22-2 shows how to connect the ADC0804 so that it will digitize the analog input when start conversion goes LOW.

The ADC0804 ADC chip can also be operated in free-running mode by connecting the "end-of-conversion" output signal named INTR (interrupt) to the "start-conversion" input signal named WR (write). (See Fig. 22-3.) It may be necessary to momentarily short the WR pin to ground in order to initially start the conversion process, but the ADC should then operate continuously. $V_{in(-)}$ and $V_{REF}/2$ can be changed to modify the ADC's zero and span for the analog input signal's range and the desired step size. The analog differential input voltage ($V_{in(+)} - V_{in(-)}$) is the actual signal that is converted to a representative digital output value. Therefore, the analog input voltage that produces an output of 00000000 (zero) can be adjusted above ground by applying a DC voltage to the $V_{in(-)}$ pin. The input voltage span can also be adjusted by applying a DC voltage to the $V_{REF}/2$ pin. The analog input voltage span (V_{REF}) will be equal to two times the voltage on the $V_{REF}/2$ pin. The new quantization error factor is equal to $V_{REF} \div 2^8$.

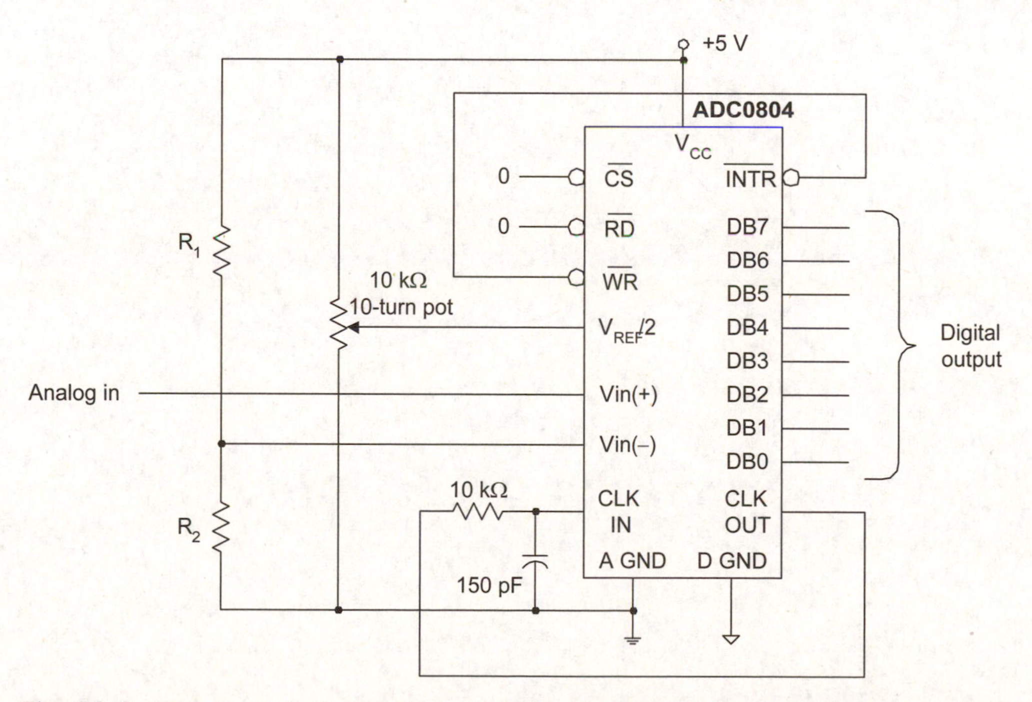

Fig. 22-3 Free-running analog-to-digital conversion circuit with span adjust

Laboratory Projects

22.1 Digital-to-analog converter
Construct a D/A converter using an Analog Devices AD557 IC chip. Determine a procedure to test the operation of the DAC. Enter your test data into a spreadsheet. Use the spreadsheet to calculate the theoretical output and the full-scale error for the test results. Graph the output vs. input test results for the DAC.

22.2 Analog-to-digital converter
Construct an A/D converter using a National ADC0804 IC chip. Determine a procedure to test the operation of the ADC. Enter your test data into a spreadsheet. Use the spreadsheet to calculate the theoretical output and the full-scale error for the test results. Graph the output vs. input test results for the ADC.

22.3 Free-running ADC with span adjust
Design a free-running A/D converter (using a National ADC0804) that produces a digital output of 00000000 for analog input voltages ≤ 0.5 V and has a resolution of 15 mV. Test your design. Enter your test data into a spreadsheet. Use the spreadsheet to calculate the theoretical output and the full-scale error for the test results. Graph the output vs. input test results for the ADC. See Fig. 22-3.

22.4 Digital voltmeter

Construct a simple 0–5 V DC digital voltmeter using the ADC0804 and a PLD. The PLD will convert the binary output from the ADC0804 into BCD, which will then be displayed on two 7-segment displays (using BCD-to-7-segment decoder/drivers). The Binary-to-BCD converter will take the 6-bit binary number and convert it into the equivalent 2-digit BCD value (see Unit 17). The 2 least significant A/D output bits are left unconnected (using only 6 bits of resolution). The tens digit is only 3-bits wide because our meter will only be used for voltages up to 5.0 volts. Calibrate the digital voltmeter by applying 5.0 V to the analog in (measured with a DVM) and adjusting the 10-turn "calibration" potentiometer until the 7-segment display reads 5.0 (the decimal point is assumed to be in the middle). Test the accuracy of the digital voltmeter with several different voltage values for analog in. It may be necessary to readjust the calibration for your digital voltmeter. Note: If your PLD has LVTTL inputs (such as the Cyclone families), use an external 74LS244 chip to interface between the ADC0804 data outputs and the inputs to your PLD chip. The PLD will still only contain the binary-to-BCD converter and BCD-to-7-segment decoder/driver data display functions. The HIGH output voltage for the CMOS ADC0804 will be very close to 5 V, which is a little higher voltage than should be applied to PLDs with LVTTL-compatible inputs. The signal can be easily buffered with a 74LS244 chip which will provide a typical HIGH output voltage of approximately 3.4 V.

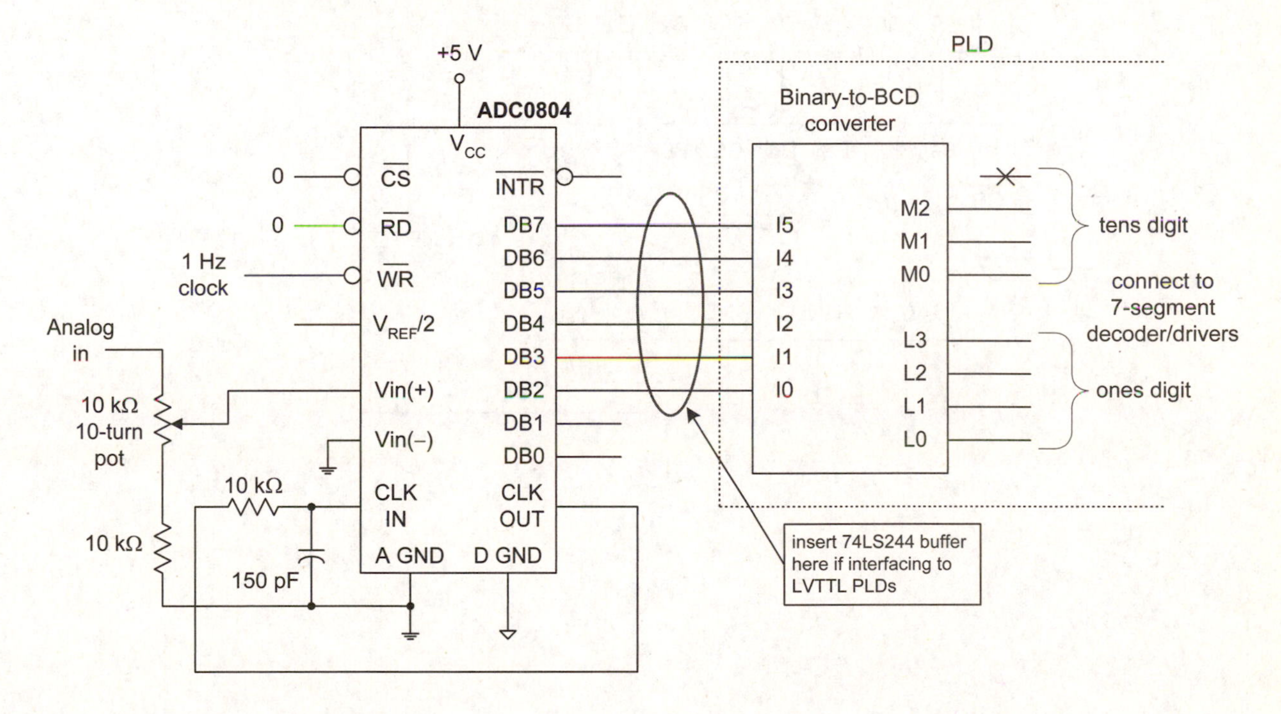

22.5 Digitized sine-wave generator

Construct a digitized sine-wave generator using an AD557 D/A converter, a PLD, and a MOD-36 binary counter. The sine-wave output will be produced by the D/A using data generated by the PLD. A MOD-36 counter will provide 36 counts per cycle of the sine wave. How many degrees are there per count? How can you change the output frequency? Calculate (with a spreadsheet) the D/A digital data needed for each step (count) to generate the desired analog output. Plot the data output from the Sine-Wave Data Generator vs. the angle input (in degrees). The amplitude of the sine wave should be 1.0 volt, and the DC offset should be 1.5 volts. The formula for the instantaneous output voltage is:

$$V_{out} = 1.5 + 1.0 \sin \theta$$

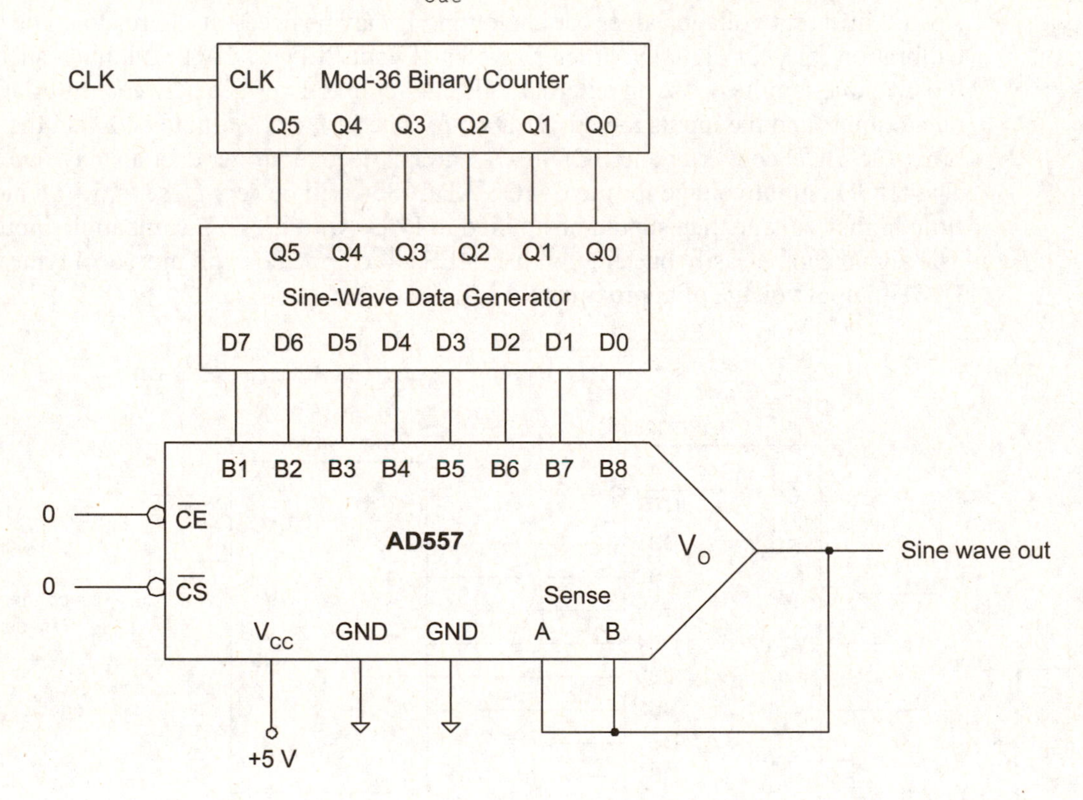

22.6 Reconstructing a digitized signal

Use the A/D converter to digitize a sine-wave input signal and then reconvert the representative digital signal into an analog signal with the DAC. Connect the input of the D/A converter to the output of the A/D converter. Make sure the MSB out is connected to the MSB in. Adjust the input sine wave to 2.5 $V_{p\text{-}p}$ with a DC offset of 1.25 volts (i.e., the sine wave goes from 0 to 2.5 V). Adjust the $V_{REF}/2$ voltage using the potentiometer so that the analog input span on the ADC is 0 to 2.5 V. Use an oscilloscope to compare the analog input and output waveforms. Set the analog input signal frequency to approximately 200 Hz and observe the resultant analog output signal. Sketch the input and output waveforms. Increase the input signal frequency and note the effect on the resultant output signal.

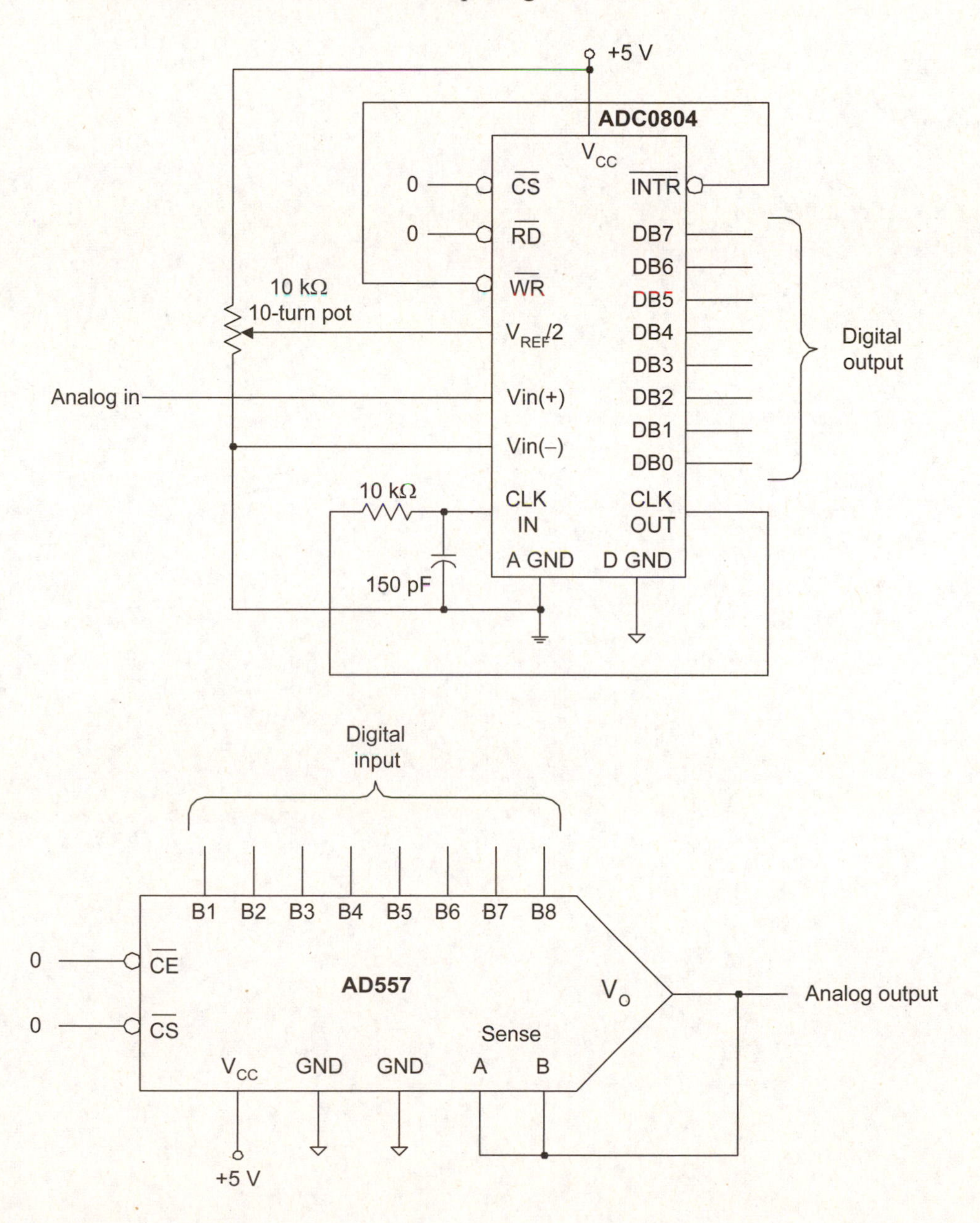

UNIT 23

MEMORY SYSTEMS

Objectives

- To construct a random access memory (RAM) circuit using a semiconductor RAM chip.
- To read and write data in a RAM memory.
- To combine multiple RAM chips to expand word size and/or memory capacity.
- To use embedded memory in a PLD to create a single-port RAM memory block.

Suggested Standard Parts			
7404	74138	74244	2114

Memory Devices and Systems

Memory devices are used in digital systems to store digital information. A specific storage location in memory is identified with a unique binary value called an address. Accessing data stored at a specified address is referred to as a read operation. Storing new data at a specified address is referred to as a write operation.

Random access memory (RAM) is the term used frequently for memory that can be read from or written into with equal ease. Ordinary semiconductor RAM devices are said to be volatile because the information stored in the device is lost if the electrical power for the memory is removed. Semiconductor RAM may be either static, which does not need the stored data to be periodically refreshed, or dynamic, which does require the data to be periodically rewritten into the memory cells. Static RAM memory chips have the following types of pins: address pins that are used to select a specific memory location, data pins that are used to input the data into or output the data from the addressed memory location, a chip select pin to enable a specific chip (or set of chips) in a memory system, and a Read/Write pin to control the chip's read or write function (read = 1 and write = 0). The block diagram for a 2114 static RAM memory chip containing 1024 4-bit words is shown in Fig. 23-1.

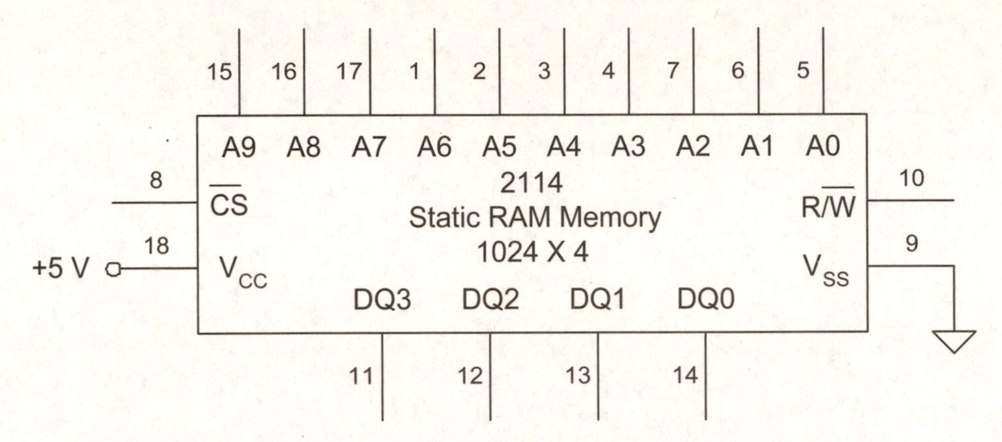

Fig. 23-1 Block diagram for a typical static RAM chip

Read only memory (ROM) often refers to a broad category of memory devices that normally have their data permanently stored in them while they are used in a digital system. There are various subtypes of ROM devices, some of which can have the stored data erased and replaced with new information.

Many different memory devices (with different part numbers) are available that differ in word size and memory capacity. The memory chip's word size is the number of bits that are accessed simultaneously in the chip with a given address. The memory capacity for a chip is the total number of words (2^n, where n = number of address bits) that can be addressed on the chip. Several memory chips can be interconnected to expand the total system memory. This expansion can be in word size, total number of addressable words, or both.

Tristate Bus Drivers

In addition to HIGH and LOW output levels, a tristate output device also has a high impedance output condition. This type of output structure is normally used to connect several possible sources of digital information to a common bus, such as a data bus in a memory system. Only one source of information will be enabled at a time and allowed to place data on the bus. Most memory chips today have built-in tristate output buffers for the data pins that are enabled when the chip is selected and a read operation is performed. The source of data for a write operation must be removed from the data bus (i.e., the output placed in the tristate mode) when a write operation is not being performed. The output for a device in the tristate mode will be a high impedance (acts like an open circuit) when the device is disabled. Separate tristate buffer or bus driver chips are available to accomplish this task. The 74LS244 shown in Fig. 23-2 is an example of a tristate buffer. It contains a total of eight buffers in one chip. The buffers are arranged in two sets of four buffers each. Each set of buffers has a common active-LOW tristate enable.

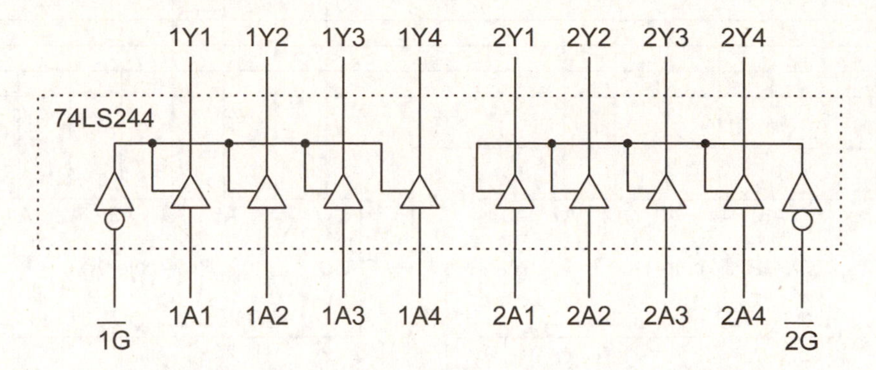

Fig. 23-2 A 74LS244 octal buffer with tristate output

Example 23-1

Design a 256-word RAM memory system in which each word is 8 bits. Use 1024-bit RAM memory chips that are arranged as 256 x 4 memory devices.

It will be necessary to use two of these 256 x 4 memory chips in order to double the word size for the specified memory system. The schematic for this 256 x 8 memory is shown in Fig. 23-3. The chips are permanently enabled with a 0 on the common (active-LOW) chip select line. Both chips receive the same address information to access a 4-bit word in each. The two 4-bit words from each chip are used to double the word size for the memory system. RAM 0 provides the least significant 4 bits, while RAM 1 provides the most significant 4 bits of the 8-bit data word. The octal tristate buffer chip (74244) is necessary so that external data (from some data source such as switches) can be stored in the memory system during a memory write cycle. During a memory write, the data from the switches is placed on the data bus and then is written into the addressed memory location. During a memory read cycle, the tristate buffers are disabled so that the data placed on the data bus comes from the memory chips.

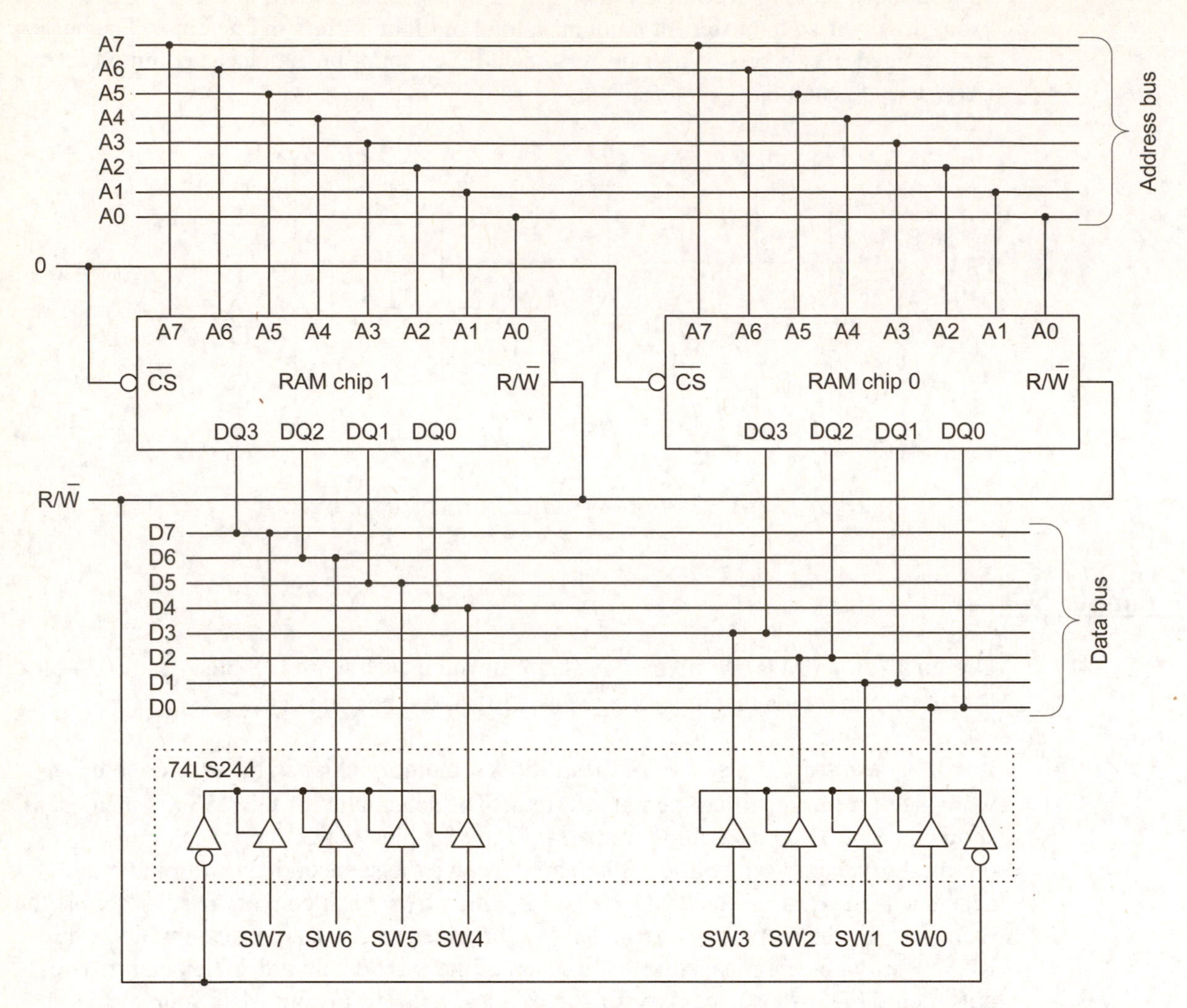

Fig. 23-3 Expanding word size for 256 x 8 RAM memory system in Example 23-1

Example 23-2

Design a 512-word RAM memory system in which each word is 4 bits. Use 1024-bit RAM memory chips that are arranged as 256 x 4 memory devices.

It will be necessary to use two of these 256 x 4 RAM memory chips in order to double the word capacity for the specified memory system. The schematic for this 512 x 4 memory is shown in Fig. 23-4. Another address line (A8) will be necessary to double the word capacity that is provided by only one 256 x 4 chip. The A8 address line is decoded using an inverter so that only one chip is selected at a time. The other eight address lines are connected to each of the chip's address pins. If A8 is LOW, then RAM 0 will be accessed; if A8 is HIGH, then RAM 1 will be accessed. The corresponding data pins on each chip are tied to the four data bus lines. The tristate data output for each of the RAM chips will be enabled when the CS pin is LOW. The decoding gate (inverter) will allow only one of the chips to be enabled at a time. Only one-half (either half) of the octal tristate buffer chip will be needed because the data bus is only 4 bits wide in this memory system.

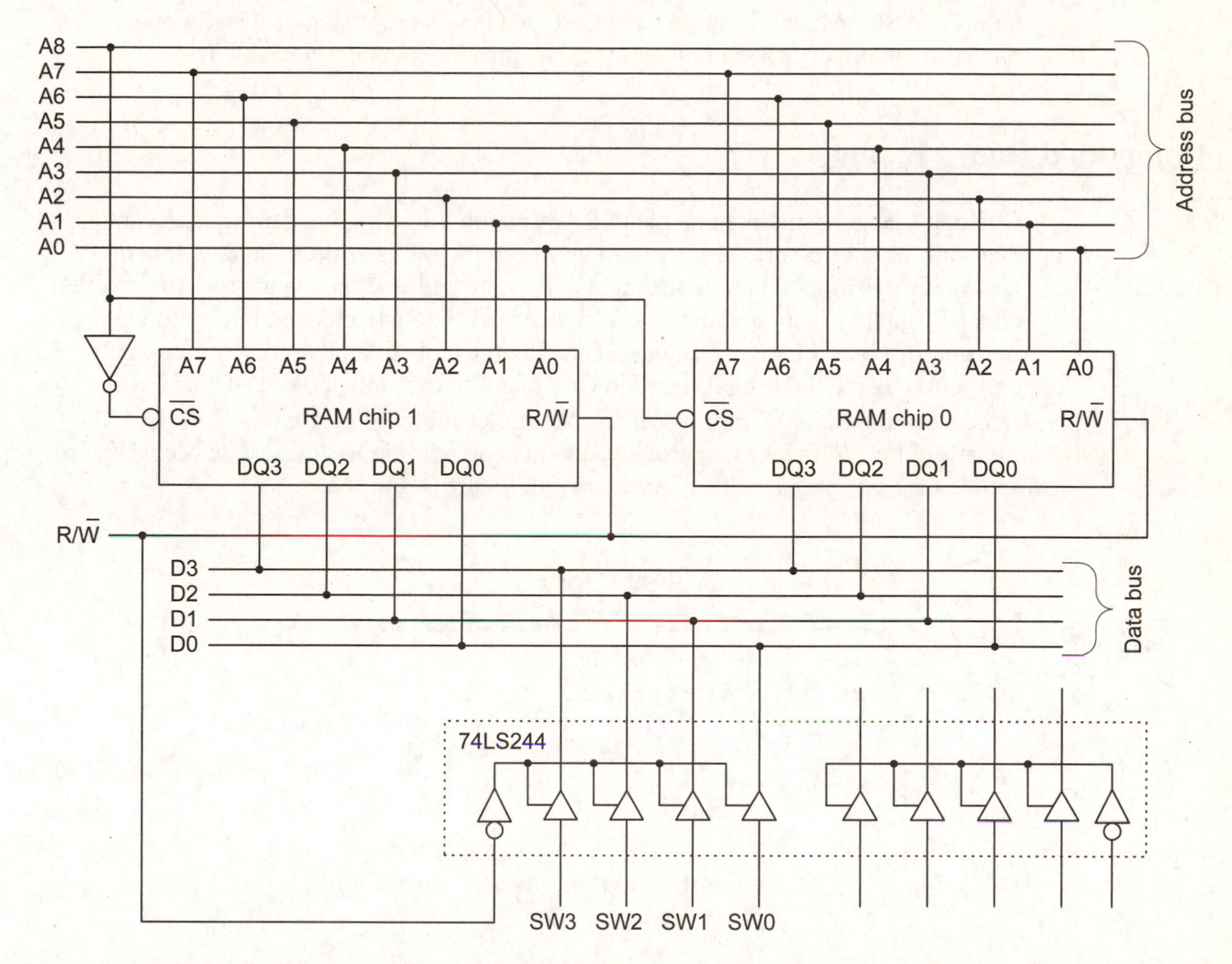

Fig. 23-4 Expanding the memory capacity for a 512 x 4 RAM memory system (Example 23-2)

SRAM Memory Chip

Terasic's DE1 and DE2 development boards each contain a 512K byte SRAM memory chip. The SRAM chip is an IS61LV25616AL that is organized as 256K words by 16 bits with the upper byte and lower byte data pins independently enabled.

Embedded Ram Memory

Cyclone FPGAs contain embedded SRAM memory blocks that can be configured as dual-port or single-port memory with words up to 36 bits wide. The RAM memory blocks always have synchronous inputs that store address, data, and the write enable control signal. The output data can also be synchronously clocked into a storage register or it may be asynchronous. This flexible embedded memory can be easily configured using the MegaWizard in Quartus. For example, RAM:1-PORT was selected in the MegaWizard's Memory Compiler folder to create the 32×4 SRAM memory block with an asynchronous output shown in Figure 23-5. The MegaWizard settings for this memory block are shown in Figure 23-6.

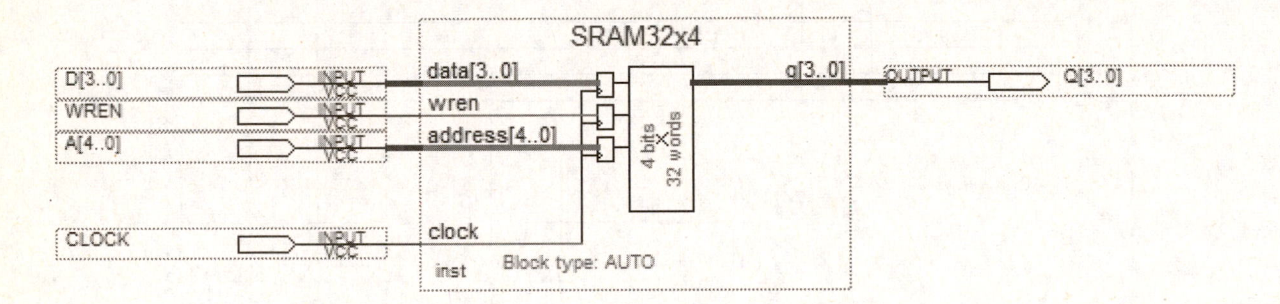

Fig. 23-5 32 x 4 embedded single-port RAM memory

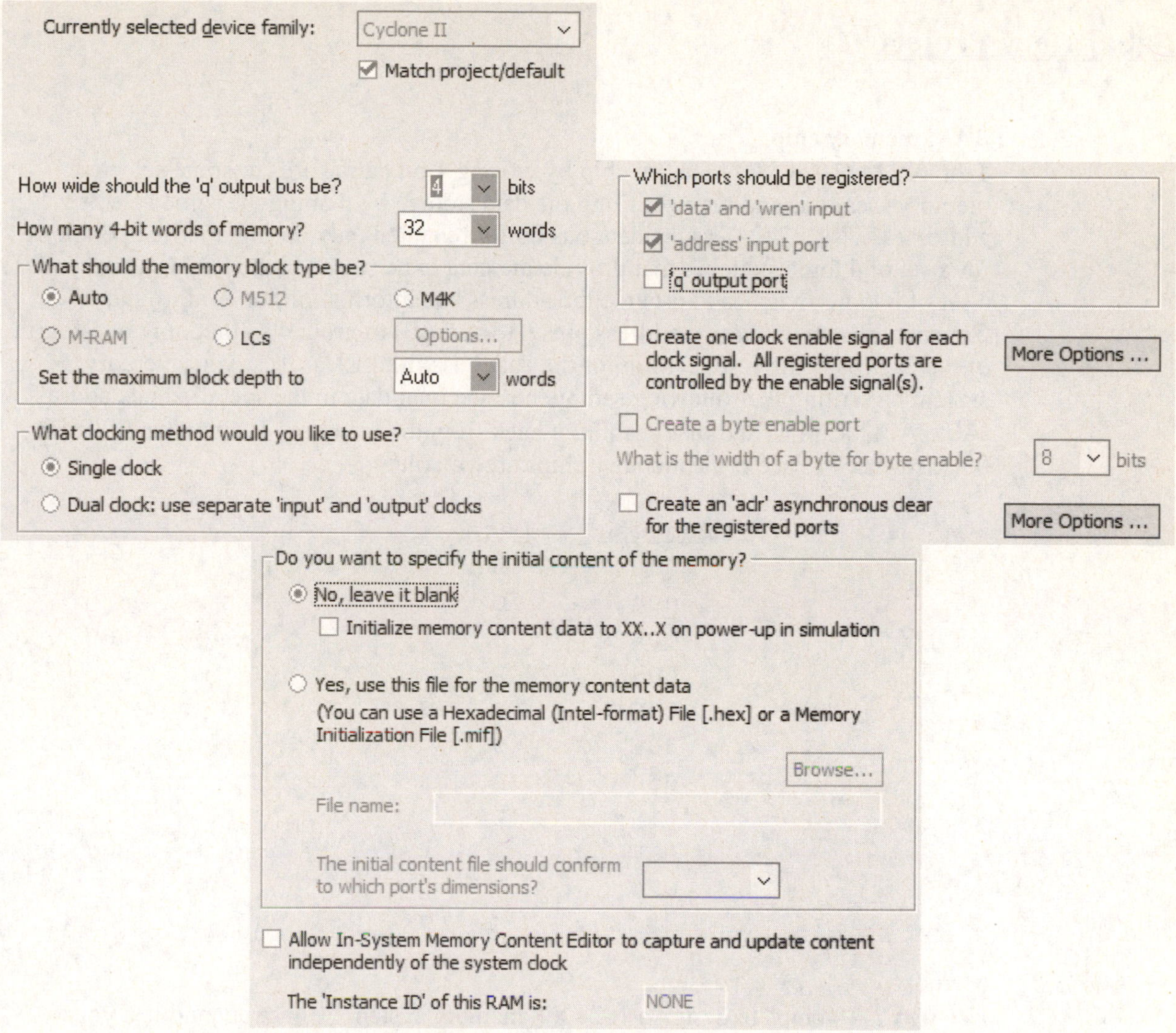

Currently selected device family: Cyclone II

☑ Match project/default

How wide should the 'q' output bus be? 4 bits

How many 4-bit words of memory? 32 words

What should the memory block type be?

◉ Auto ○ M512 ○ M4K

○ M-RAM ○ LCs Options...

Set the maximum block depth to Auto words

What clocking method would you like to use?

◉ Single clock

○ Dual clock: use separate 'input' and 'output' clocks

Which ports should be registered?

☑ 'data' and 'wren' input

☑ 'address' input port

☐ 'q' output port

☐ Create one clock enable signal for each clock signal. All registered ports are controlled by the enable signal(s). More Options ...

☐ Create a byte enable port

What is the width of a byte for byte enable? 8 bits

☐ Create an 'aclr' asynchronous clear for the registered ports More Options ...

Do you want to specify the initial content of the memory?

◉ No, leave it blank

☐ Initialize memory content data to XX..X on power-up in simulation

○ Yes, use this file for the memory content data

(You can use a Hexadecimal (Intel-format) File [.hex] or a Memory Initialization File [.mif])

Browse...

File name:

The initial content file should conform to which port's dimensions?

☐ Allow In-System Memory Content Editor to capture and update content independently of the system clock

The 'Instance ID' of this RAM is: NONE

Fig. 23-6 MegaWizard settings for a 32 x 4 embedded single-port RAM memory

Laboratory Projects

23.1 RAM memory chip

Test a 2114 (1024 x 4) memory chip by writing 4-bit data words to some selected memory locations and then verifying the data storage by reading the same selected addresses. Use a 74244 as the data bus buffer (enabled only during a write operation) for a set of 4 logic switches that provide the data to be stored in the RAM memory. Use a set of 8 logic switches to provide the address bus information for A0 through A7. Connect the remaining two address pins (A8 and A9) to ground so that only one-fourth of the 2114 is being tested. Monitor the data bus with LEDs. Test your memory system by writing the following sample hexadecimal data in the specified hex address (A9…A0) locations and then reading it back from the memory. Remember A9 = A8 = 0. Also check to see if the memory chips are volatile.

Address	Data
000	5
004	C
005	0
008	F
00B	6
00E	9
010	A
015	2
01F	1
020	7

23.2 Memory word-size expansion

Use two 2114 chips to design a 1024 x 8 memory system. Construct and test your design. Test your memory system by writing the following sample hexadecimal data in the specified hex address locations and then reading it back from the memory.

Address	Data
000	23
004	45
005	67
008	89
00B	AB
00E	CD
010	EF
015	01
01F	CC
020	33

23.3 Memory capacity expansion

Use two 2114 chips to design a 2048 X 4 memory system. Construct and test your design. Test your memory system by writing the following sample hexadecimal data in the specified hex address (A10...A0) locations and then reading it back from the memory.

Address	Data
000	1
400	2
007	3
407	4
00A	5
40A	6
010	7
410	8
017	9
417	A
01A	B
41A	C
020	D
420	E

23.4 Memory address decoding

Design a decoding circuit for a 2048 X 4 memory system. One K of the 4-bit words is at hex addresses 1800–1BFF. The other half of the memory is at locations 6800–6BFF. Assume the system has 16 address bits (A15...A0).

23.5 Embedded SRAM memory

Test a 32 × 4 embedded SRAM memory in the Cyclone chip on your FPGA development board. Demonstrate the memory using a data pattern that is one more than the least significant nibble of the address.

23.6 DE1/DE2 SRAM memory chip

Test a 64 × 4 SRAM memory using the IS61LV25616AL on the DE1/DE2 development board with the following *bdf* test circuit. Look up the FPGA pin assignments in the User Manual for your development board. Demonstrate the memory using a test data pattern that is one more than the least significant nibble of the address. Note that the data input/output d[3..0] is a BIDIR (bidirectional) port.

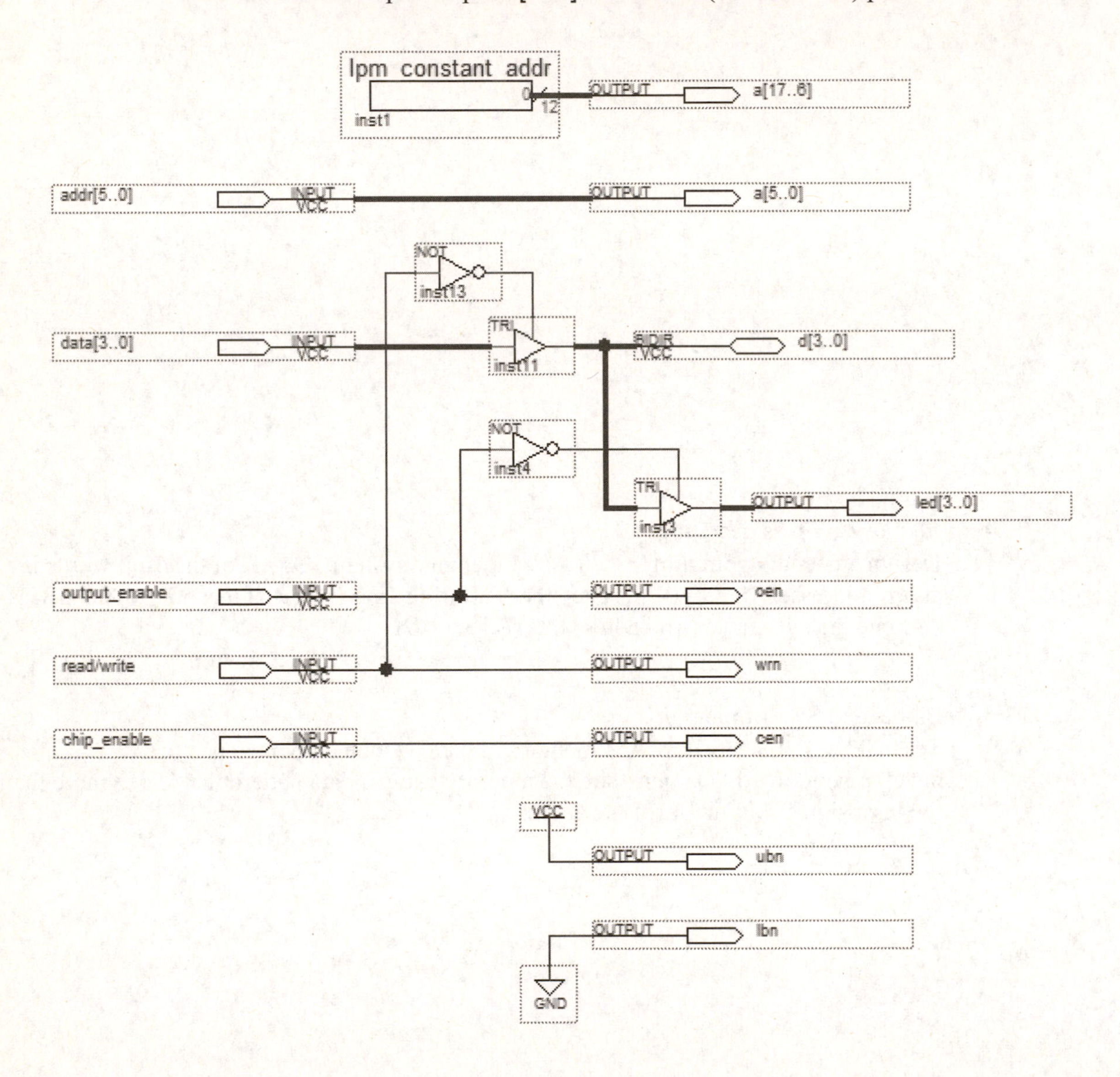

AHDL Language Elements

Names

- Symbolic names specify internal and external nodes and groups, constants, state machine variables, state bits, state names, and instances
- Subdesign names are user-defined names for lower-level design files and must be the same as the TDF filename
- Port names are symbolic names that identify the input or output of a logic block
- Start with an alphabetic character or slash unless quoted ' '
- Can contain numeric characters or underscore or dash (if quoted); cannot contain spaces
- Are not case sensitive
- May be up to 32 characters long
- Cannot contain any reserved keywords unless quoted

Reserved keywords

AND	FUNCTION	OUTPUT
ASSERT	GENERATE	PARAMETERS
BEGIN	GND	REPORT
BIDIR	HELP_ID	RETURNS
BITS	IF	SEGMENTS
BURIED	INCLUDE	SEVERITY
CASE	INPUT	STATES
CLIQUE	IS	SUBDESIGN
CONNECTED_PINS	LOG2	TABLE
CONSTANT	MACHINE	THEN
DEFAULTS	MOD	TITLE
DEFINE	NAND	TO
DESIGN	NODE	TRI_STATE_NODE
DEVICE	NOR	VARIABLE
DIV	NOT	VCC
ELSE	OF	WHEN
ELSIF	OPTIONS	WITH
END	OR	XNOR
FOR	OTHERS	XOR

Reserved identifiers

CARRY	JKFF	SRFF
CASCADE	JKFFE	SRFFE
CEIL	LATCH	TFF
DFF	LCELL	TFFE
DFFE	MCELL	TRI
EXP	MEMORY	USED
FLOOR	OPENDRN	WIRE
GLOBAL	SOFT	X

(*continued*)

AHDL Language Elements

Symbols

% %	enclose comments
--	begin comments (to end of line)
( )	enclose port names in Subdesign section; enclose highest priority operations in Boolean and arithmetic expressions
[ ]	enclose the range of a group name
‘ ’	enclose quoted symbolic names
“ ”	enclose digits in nondecimal numbers
.	separate symbolic names of variables from port names
..	separate MSB from LSB in a range
;	end AHDL statements and sections
,	separate members of sequential groups and lists
:	separate symbolic names from types in declarations
=	assign values in Boolean equations; assign values to state machine states
=>	separate inputs from outputs in truth table statements
+	addition operator
–	subtraction operator
==	numeric or string equality operator
!	NOT operator
!=	not equal to operator
>	greater than comparator
>=	greater than or equal to comparator
<	less than comparator
<=	less than or equal to comparator
&	AND operator
!&	NAND operator
$	XOR operator
!$	XNOR operator
#	OR operator
!#	NOR operator

Numbers

- Default base for numbers is decimal
- Binary values (series of 0’s, 1’s, X’s) are enclosed in double quotes and prefixed with B
- Hexadecimal values (series from 0 to 9, A to F) are enclosed in double quotes and prefixed with H
- Numbers cannot be assigned to single nodes in Boolean equations; use VCC and GND

VHDL Language Elements

Reserved words

ABS
ACCESS
AFTER
ALIAS
ALL
AND
ARCHITECTURE
ARRAY
ASSERT
ATTRIBUTE
BEGIN
BLOCK
BODY
BUFFER
BUS
CASE
COMPONENT
CONFIGURATION
CONSTANT
DISCONNECT
DOWNTO
ELSE
ELSIF
END
ENTITY
EXIT
FILE
FOR
FUNCTION
GENERATE
GENERIC
GUARDED
IF
IN
INOUT
IS
LABEL
LIBRARY
LINKAGE
LOOP
MAP
MOD
NAND
NEW
NEXT
NOR
NOT
NULL
OF
ON
OPEN
OR
OTHERS
OUT
PACKAGE
PORT
PROCEDURE
PROCESS
RANGE
RECORD
REGISTER
REM
REPORT
RETURN
SELECT
SEVERITY
SIGNAL
SUBTYPE
THEN
TO
TRANSPORT
TYPE
UNITS
UNTIL
USE
VARIABLE
WAIT
WHEN
WHILE
WITH
XOR

(continued)

VHDL Language Elements

Identifiers

- An identifier is the name of an object
- Objects are named entities that can be assigned a value and have a specific data type
- Objects include signals, variables, and constants
- Must start with an alphabetic character and end with an alphabetic or a numeric character
- Can contain numeric or underscored characters and cannot contain spaces
- Are not case sensitive
- May be up to 32 characters long
- Cannot contain any reserved words

Symbols

Symbol	Meaning
--	begins comment (to end of line)
( )	encloses port names in entity declaration; encloses highest priority operations in Boolean and arithmetic expressions
‘ ’	encloses scalar values
“ ”	encloses array values
;	ends VHDL statements and declarations
,	separates objects
:	separates object identifier names from mode and data type in declarations
<=	assigns values in signal assignment statements
:=	assigns values in variable assignment statements or to constants
=>	separates signal assignment statements from WHEN clause in CASE statements
+	addition operator
–	subtraction operator
=	equality operator
/=	inequality operator
>	greater than comparator
>=	greater than or equal to comparator
<	less than comparator
<=	less than or equal to comparator
&	concatenation operator

Synthesis Data Types

Type	Description
BIT	object can only have single-bit values of ‘0’ or ‘1’
STD_LOGIC	object with multi-value logic including ‘0’, ‘1’, ‘X’, ‘Z’
INTEGER	objects with whole number (decimal) values, e.g., 54, –21
BIT_VECTOR	objects with arrays of bits such as “10010110”
STD_LOGIC_VECTOR	objects with arrays of multi-value logic, e.g., “01101XX”

DE0 Pin Assignments

Toggle switches

Switch	SW9	SW8	SW7	SW6	SW5	SW4	SW3	SW2	SW1	SW0
Pin#	D2	E4	E3	H7	J7	G5	G4	H6	H5	J6

Pushbuttons

Button#	2	1	0
Pin#	F1	G3	H2

Green LEDs

LED	G9	G8	G7	G6	G5	G4	G3	G2	G1	G0
Pin#	B1	B2	C2	C1	E1	F2	H1	J3	J2	J1

7-Segment displays

Digit	HEX3	HEX2	HEX1	HEX0
a	B18	D15	A13	E11
b	F15	A16	B13	F11
c	A19	B16	C13	H12
d	B19	E15	A14	H13
e	C19	A17	B14	G12
f	D19	B17	E14	F12
g	G15	F14	A15	F13
DP	G16	A18	B15	D13

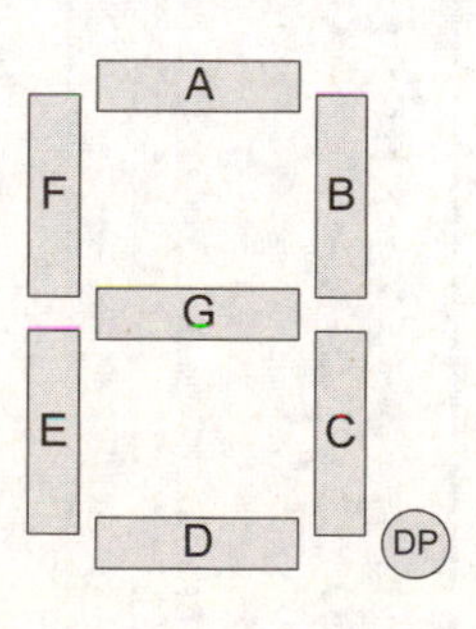

Clock

Freq	50MHz
Pin#	G21/B21

DE0 Pin Assignments

Expansion Headers (board view) [*clock in & clock out pins italicized*]

GPIO 0
J4

FPGA Pin	Header Pin	Header Pin	FPGA Pin
AB12	1	2	AB16
AA12	3	4	AA16
AA15	5	6	AB15
AA14	7	8	AB14
AB13	9	10	AA13
5V	11	12	GND
AB10	13	14	AA10
AB8	15	16	AA8
AB5	17	18	AA5
AB3	19	20	AB4
AA3	21	22	AA4
V14	23	24	U14
Y13	25	26	W13
U13	27	28	V12
3.3V	29	30	GND
R10	31	32	V11
Y10	33	34	W10
T8	35	36	V8
W7	37	38	W6
V5	39	40	U7

GPIO 1
J5

Clock	FPGA Pin	Header Pin	Header Pin	FPGA Pin
CLKIN0	*AB11*	1	2	AA20
CLKIN1	*AA11*	3	4	AB20
	AA19	5	6	AB19
	AB18	7	8	AA18
	AA17	9	10	AB17
	5V	11	12	GND
	Y17	13	14	W17
	U15	15	16	T15
	W15	17	18	V15
CLKOUT0	*R16*	19	20	AB9
CLKOUT1	*T16*	21	22	AA9
	AA7	23	24	AB7
	T14	25	26	R14
	U12	27	28	T12
	3.3V	29	30	GND
	R11	31	32	R12
	U10	33	34	T10
	U9	35	36	T9
	Y7	37	38	U8
	V6	39	40	V7

DE1 Pin Assignments

Toggle switches

Switch	SW9	SW8	SW7	SW6	SW5	SW4	SW3	SW2	SW1	SW0
Pin#	L2	M1	M2	U11	U12	W12	V12	M22	L21	L22

Pushbuttons

Button	Key3	Key2	Key1	Key0
Pin#	T21	T22	R21	R22

Red LEDs

LED	R9	R8	R7	R6	R5	R4	R3	R2	R1	R0
Pin#	R17	R18	U18	Y18	V19	T18	Y19	U19	R19	R20

Green LEDs

LED	G7	G6	G5	G4	G3	G2	G1	G0
Pin#	Y21	Y22	W21	W22	V21	V22	U21	U22

7-Segment displays

Digit	HEX3	HEX2	HEX1	HEX0
a	F4	G5	E1	J2
b	D5	G6	H6	J1
c	D6	C2	H5	H2
d	J4	C1	H4	H1
e	L8	E3	G3	F2
f	F3	E4	D2	F1
g	D4	D3	D1	E2

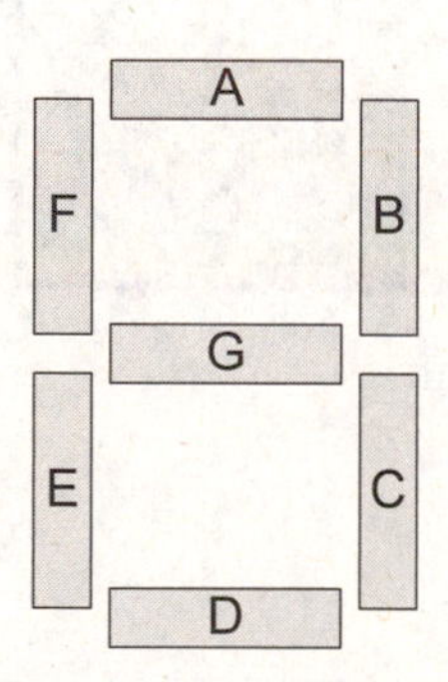

Clocks

Freq	27MHz	50MHz	24MHz(USB)	Ext
Pin#	D12/E12	L1	A12/B12	M21

DE1 Pin Assignments

Expansion Headers (board view)

GPIO 0 JP1

A13	1	2	B13
A14	3	4	B14
A15	5	6	B15
A16	7	8	B16
A17	9	10	B17
VCC5	11	12	GND
A18	13	14	B18
A19	15	16	B19
A20	17	18	B20
C21	19	20	C22
D21	21	22	D22
E21	23	24	E22
F21	25	26	F22
G21	27	28	G22
VCC33	29	30	GND
J21	31	32	J22
K21	33	34	K22
J19	35	36	J20
J18	37	38	K20
L19	39	40	L18

GPIO 1 JP2

H12	1	2	H13
H14	3	4	G15
E14	5	6	E15
F15	7	8	G16
F12	9	10	F13
VCC5	11	12	GND
C14	13	14	D14
D15	15	16	D16
C17	17	18	C18
C19	19	20	C20
D19	21	22	D20
E20	23	24	F20
E19	25	26	E18
G20	27	28	G18
VCC33	29	30	GND
G17	31	32	H17
J15	33	34	H18
N22	35	36	N21
P15	37	38	N15
P17	39	40	P18

DE2 Pin Assignments

Toggle switches

Switch	SW9	SW8	SW7	SW6	SW5	SW4	SW3	SW2	SW1	SW0
Pin#	A13	B13	C13	AC13	AD13	AF14	AE14	P25	N26	N25
Switch			SW17	SW16	SW15	SW14	SW13	SW12	SW11	SW10
Pin#			V2	V1	U4	U3	T7	P2	P1	N1

Pushbuttons

Button	Key3	Key2	Key1	Key0
Pin#	W26	P23	N23	G26

Red LEDs

LED	R9	R8	R7	R6	R5	R4	R3	R2	R1	R0
Pin#	Y13	AA14	AC21	AD21	AD23	AD22	AC22	AB21	AF23	AE23
LED			R17	R16	R15	R14	R13	R12	R11	R10
Pin#			AD12	AE12	AE13	AF13	AE15	AD15	AC14	AA13

Green LEDs

LED	G8		G7	G6	G5	G4	G3	G2	G1	G0
Pin#	Y21		Y18	AA20	U17	U18	V18	W19	AF22	AE22

7-Segment displays

Digit	HEX7	HEX6		HEX5	HEX4		HEX3	HEX2	HEX1	HEX0
a	L3	R2		T2	U9		Y23	AB23	V20	AF10
b	L2	P4		P6	U1		AA25	V22	V21	AB12
c	L9	P3		P7	U2		AA26	AC25	W21	AC12
d	L6	M2		T9	T4		Y26	AC26	Y22	AD11
e	L7	M3		R5	R7		Y25	AB26	AA24	AE11
f	P9	M5		R4	R6		U22	AB25	AA23	V14
g	N9	M4		R3	T3		W24	Y24	AB24	V13

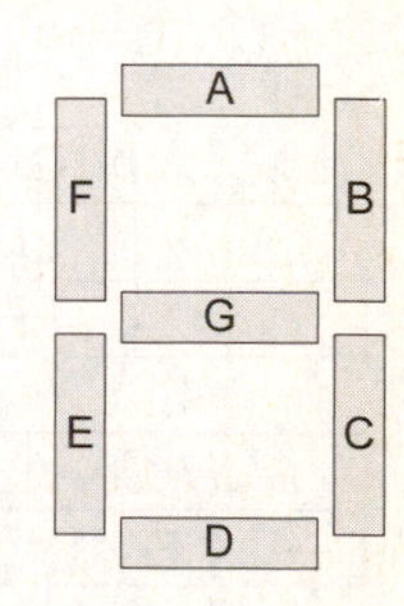

Clocks

Freq	27MHz	50MHz	Ext
Pin#	D13	N2	P26

DE2 Pin Assignments

Expansion Headers (board view)

GPIO 0 JP1

Signal	Pin	Pin	Signal
D25	1	2	J22
E26	3	4	E25
F24	5	6	F23
J21	7	8	J20
F25	9	10	F26
VCC5	11	12	GND
N18	13	14	P18
G23	15	16	G24
K22	17	18	G25
H23	19	20	H24
J23	21	22	J24
H25	23	24	H26
H19	25	26	K18
K19	27	28	K21
VCC33	29	30	GND
K23	31	32	K24
L21	33	34	L20
J25	35	36	J26
L23	37	38	L24
L25	39	40	L19

GPIO 1 JP2

Signal	Pin	Pin	Signal
K25	1	2	K26
M22	3	4	M23
M19	5	6	M20
N20	7	8	M21
M24	9	10	M25
VCC5	11	12	GND
N24	13	14	P24
R25	15	16	R24
R20	17	18	T22
T23	19	20	T24
T25	21	22	T18
T21	23	24	T20
U26	25	26	U25
U23	27	28	U24
VCC33	29	30	GND
R19	31	32	T19
U20	33	34	U21
V26	35	36	V25
V24	37	38	V23
W25	39	40	W23

LCD Module

Data	D7	D6	D5	D4	D3	D2	D1	D0
Pin#	H3	H4	J3	J4	H2	H1	J2	J1

LCD Control	*Pin#*	*Description*
LCD_RW	K4	0 = Write, 1 = Read
LDC_EN	K3	Enable
LCD_RS	K1	0 = Command, 1 = Data
LCD_ON	L4	Power on/off
LCD_BLON	K2	Back light on/off